SPACE:
EXPLORATION AND DISCOVERY

Above: a 16th-century
German woodcut depicting
the astronomer's search
after the truth about the
workings of the universe.

Overleaf: the 15-mile high
Olympus Mons on Mars, bathed in
morning light and wreathed in
clouds to within three miles of its
summit. The mountain is some
375 miles across at the base. Its
cloud-free, multi-ringed volcanic
crater at the summit is over 15
miles across.

SPACE:
EXPLORATION AND DISCOVERY

BY ANTHONY FELDMAN

UPDATE EDITOR: GRAHAM YOST

Facts On File Publications
New York, New York ● Oxford, England

Editorial Coordinator: John Mason
Art Editor: John Fitzmaurice
Editor: Damian Grint
Designer: Adrian Williams
Picture Research: Diane Rich; Frances Vargo
Update Editor: Graham Yost

ISBN 0-8160-1272-5
© 1985 J.G. Ferguson Publishing Company
First published in the United States
by Facts On File, Inc.,
460 Park Avenue South,
New York, N.Y. 10016

Space: Exploration and Discovery is a revised and updated
edition of the book first published in 1980 as *Space*.

Printed and bound in Hong Kong by
Leefung-Asco Printers Ltd.

Special Edition Prepared for Xerox Education Publications by
Facts On File, Inc.

Introduction

Men have always been fascinated by the vastness of the heavens above them. From the earliest astrologers to the Apollo Program of the 1960s, successive generations have gradually built up a picture of the universe and the way it works. In this book, the author explores space in a thorough and exciting way starting with the ancient Egyptians and the Aztecs, and going on to deal with the achievements of giants like Galileo, Copernicus, Kepler, and Newton. One chapter explains the complex nature of our solar system, and includes spectacular photographs recently received of Saturn, Mars, and Jupiter. Another chapter describes the achievements of space exploration, from the first Sputnik in 1957 to the Space Shuttle of today.

In spite of the great deal known about space today, there are still many unsolved mysteries. The author discusses such phenomena as black holes, quasars, pulsars, antimatter, red giants, white dwarfs, and neutrinos, with numerous specially commissioned diagrams to support his explanations. Another chapter deals with the formation of the universe, its present state, and even its possible cataclysmic destruction. There are also artists' impressions of the incredible, but totally feasible, starships and space wars of the future. In the final chapter the author weighs the possibility that life on Earth is not an isolated phenomenon and asks: What are scientists doing to listen – and send radio messages – to other worlds? The complete book not only presents the tremendous amount we now know about the cosmos, but also turns a searching eye toward the unsolved secrets of space.

Contents

1 Birth of Astronomy 9

This chapter describes the development of one of the oldest sciences, from the astronomers of the ancient world, through the discoveries of giants like Copernicus and Galileo, Kepler and Newton, to the amazing findings of modern radiotelescopes and spectroscopy. The result is a complete and thorough survey of the history of astronomy, one of mankind's greatest triumphs over superstition and myth.

2 The Sun and Stars 37

Our own Sun is a star — a seething mass of incandescent gas, fueled by violent thermonuclear reactions. This chapter explores the stars, from their astrological significance right through to more recent aspects such as supernovas, red giants, and white dwarfs. The author shows how the apparently timeless stars, though they may flourish for millions of years, still face death like every living creature.

3 The Solar System 75

On a cosmic scale our solar system is totally insignificant — a cluster of nine planets orbiting a commonplace star. But to us, as the author demonstrates, it is fascinating. With the help of the most recent spectacular color photographs of Saturn, Mars, and Jupiter, the chapter explores the nine planets one by one, their origins and mechanism, and the vast scale of the system, as well as phenomena such as asteroids, meteors, and comets.

4 Galaxies 117

For thousands of years people's vision of the universe was restricted to the stars and planets they could see in the night sky. At the beginning of this century the discovery of countless galaxies, vast island universes beyond the Milky Way, revealed an exciting new perspective on the Earth's place in the immensity of space. This chapter describes the evolution and structure of the galaxies.

5 Origin of the Universe 137

How was the universe created? Was there ever a beginning or has it always existed much as it is today? These are some of the most fundamental questions we can ask about the world about us. Ranging from the big bang to the solid state theories, from the "primeval fireball" to the heat death of the universe, this chapter examines the most recent attempts, first to understand, then to answer, these questions.

6 Man in Space   **159**

It was World War II that provided the sophisticated technology that made possible a real breakthrough in spaceflight. Then the race to land humans on the Moon began in earnest. The hazards faced by the early space explorers, their first close-up sight of the Moon's surface, the unbearable tension of the moments before touchdown on the Moon — all these make the step into space one of the most gripping adventure stories of all time.

7 Mysteries of Space   **197**

Astronomy today is a fast-moving science. Yet the greatest discoveries have often revealed still more fresh mysteries to baffle the astronomers. By tackling and explaining vanishing neutrinos, the antimatter universe, black holes, quasars, and pulsars, this chapter provides a fascinating reminder that the mysteries of space provide a continuing challenge for future generations. Perhaps space will always be a source of mysteries.

8 Man's Future in Space   **227**

Even before the first spaceflights people dreamed about the discoveries that awaited them. Today the dreams are beginning to become feasibilities, if not yet realities. In this chapter, some of the more ambitious schemes are discussed and described — permanent space colonies orbiting the Earth, astronomical observatories high above the Earth's atmosphere, expeditions to the planets, and even the first journeys to the distant stars.

9 Life in the Universe   **273**

Most scientists are convinced that life in some form must have evolved on planets other than Earth. In the immensity of the universe it seems certain that the chemistry that created life here must have been repeated, perhaps thousands of times over. But what do we mean by life? What remarkable chemical process converted nonliving substances into the first living organisms? Will the scientists ever be proved correct?

10 The UFO Enigma   **291**

No one can deny that there are such things as unidentified flying objects. The real problems begin when we attempt to find explanations for them. Only recently have scientists begun to take them seriously. Groups using the latest technological equipment are actively seeking out UFO sightings and phemonema to investigate them. They may even lead to the greatest breakthrough in human history — our first contact with extraterrestrial intelligence. This final chapter weighs up the intriguing possibilities.

Index   **322**

Picture Credits   **333**

Chapter 1

Birth of Astronomy

To the Ancients the skies were the dwelling place of gods. In the brightness of the stars and the wanderings of the planets they thought they could divine omens of the future. For thousands of years mysticism dominated astronomy. But in the late Middle Ages the first real stirrings of modern science brought new ideas that would one day replace the old superstitions. Slowly a rational concept of the universe emerged and mankind at last glimpsed its true place in the scheme of nature.

Today astronomy is an exciting and fast-moving science. Breakthroughs in technology have enabled scientists to probe the most distant regions of the universe and to send remote-controlled spacecraft to Earth's nearer planetary neighbors. Astronomers are now even able to catch a glimpse of how the universe must have been in the instant of its creation.

The birth and development of astronomy has been one of mankind's greatest triumphs over superstition and myth. It remains today perhaps the most awesome and challenging field of scientific enquiry.

Opposite: the Hale Observatory at Mount Palomar, California. It houses a giant 200-inch telescope, which was completed in 1948 and was the largest in the world until the building of a 236-inch telescope in the USSR in the early 1970s. Large reflectors such as this, the 100-inch telescope at Mount Wilson, and that at Zelenchukskaya, USSR, permits astronomers to see farther into the remote universe because they detect faint sources of light.

The earliest astronomy

The origins of astronomy are lost in the remote past. But it is not hard to imagine how man's fascination with the night sky began. Without the glare of modern city lights to dim its magnificence, the great canopy of stars must have seemed an awesome sight to the cavemen of thousands of years ago. It would be natural for them to link the changing patterns in the sky to the seasons on which they depend for survival. The first true astronomers therefore were primitive tribesmen concerned with establishing an accurate calendar with which to predict the passing of the seasons. The importance of this to the cultivation of vital crops made the tracing of celestial events an occupation of religious as well as practical significance. Through the centuries the patterns of bright lights studding the sky, the rising of the Sun by day, and the Moon by night, became inextricably connected with religion and mysticism so that the beginnings of true

Above: a Babylonian record of the positions of Jupiter. Ancient astronomers were concerned with listing the positions of heavenly bodies.

scientific enquiry into the nature of the heavens was regarded as a heresy punishable by death.

From very early times remarkably accurate observations were made and detailed star charts were built up. Over 4000 years ago the Chinese were making fairly accurate forecasts of eclipses – both of the Sun and Moon. As far as they were concerned it was essential to be prepared for such events because they were convinced that eclipses were caused by an invisible cosmic dragon attempting to devour the Sun and Moon. To be sure of frightening the beast away the whole population had to be ready to make as much noise as possible – by shouting, ringing bells, and banging gongs.

Both the ancient Chinese and Egyptian civilizations kept records of remarkable events such as eclipses and comets without understanding the significance of the phenomena they were observing. As far as these ancient peoples were concerned, the signs in the heavens were obviously important heralds of events that affected the course of human affairs on Earth. The early astronomers therefore were priests whose main job was to divine warnings of famine, pestilence, or war in the patterns of the sky. Even the discovery of planets – regarded as "wanderers" because they were seen to move against the apparently fixed background of stars – was the impetus for still more mysticism. Because they were seen to keep within a fixed belt in the sky, the region was eventually called the Zodiac, divided in 12 "houses"

but there are indications that he may have been the first to suggest that instead of being flat the world was in reality a giant sphere. Although the Greeks believed that the Earth was spherical, they were sure that it must be the center of the universe. Even the great 4th-century BC Greek philosopher Aristotle was convinced that the Sun, Moon, and stars turned around the Earth. The only astronomer with the courage to oppose Aristotle was the 3rd-century BC Greek astronomer Aristarchus who was nearly 2000 years ahead of his time by arguing that the Earth revolves around the Sun.

Perhaps the most influential of the Greek astronomers was Claudius Ptolemaeus – better known as Ptolemy. Until his time the idea that the Earth was the center of the universe was a concept founded on religious belief rather than scientific evidence. In about 150 AD Ptolemy attempted to work out a rational framework for the belief and made a careful study of the apparent movements of the planets. He was especially puzzled by the way some planets seemed to move forward at certain times of the year and backward at other times. Dedicated to the idea of an Earth-centered universe, Ptolemy went to great lengths to work out an explanation for his observations. Eventually he came to the conclusion that the planets moved at a steady speed in a complicated system of large and small circles, called deferents and epicycles. He envisaged the Earth at the center of the heavens with the Moon the

Above: to the ancient Greeks the Earth consisted of a circular disk floating on a great ocean above which was the hemispherical bowl of the sky. This concept was popular around Homer's time until the 6th century BC.

Left: the Egyptian cosmos. It shows the goddess Nut arched across the sky with the god Qeb, representing the Earth, lying below. The kneeling god is Shu, and represents the air. The Sun (right) and the Moon (left) are represented as ships sailing over Nut's body as they rise and set.

Right: Aristarchus, the 3rd-century BC Greek astronomer, attempted to work out the distance of the Sun from the Earth by measuring the angles of a right triangle formed by the Sun, Earth, and the first quarter of the Moon. Once the shape of the triangle is known, the ratio of the Sun's distance from the Earth to that of the Moon can be determined. Aristarchus' determinations were incorrect, as it turned out, and he substantially underestimated the distance of the Sun.

and used to predict not only the broad course of human affairs but even the fate of an individual human being.

The first revolution in astronomy was started by the ancient Greeks. Their love of knowledge provided the first real interest in scientific interpretation of celestial events. One of the earliest of the Greek astronomers was Thales, born in about 624 BC. Little of his work survives

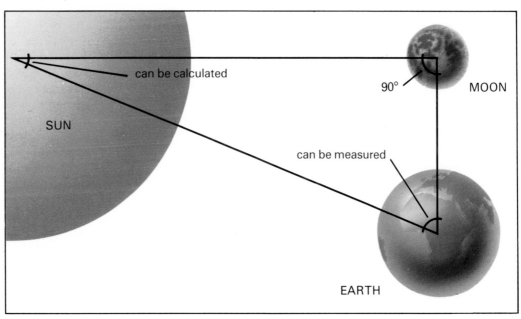

can be calculated

90°   MOON

SUN

can be measured

EARTH

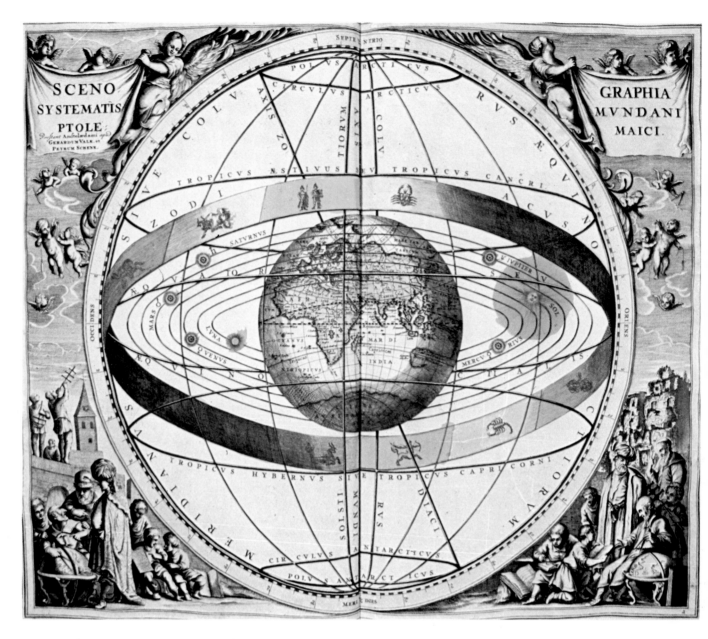

SCENO SYSTEMATIS PTOLE

GRAPHIA MVNDANI MAICI.

nearest object to it. Next came the planets Mercury and Venus followed by the Sun. Beyond the Sun were Mars, Jupiter, and Saturn and beyond these outer planets were the distant stars. Despite the complexity of Ptolemy's proposal it seemed to justify rationally a belief of fundamental religious importance. The Ptolemaic system, as it became known, was not seriously challenged for 1300 years.

Ptolemy's other great work was a vast compilation of knowledge about the heavens called the *System of Mathematics*. Its real importance to astronomy was that through its survival long after Ptolemy's death, the book ensured a remarkable continuity in scientific interest in astronomy. Many of the other Greek writings were lost. But somehow Ptolemy's *Sys-*

Above: the Ptolemaic system of the universe showing the positions of the Sun, Moon, and planets. According to Ptolemaic theory, a planet moved in a small circle, or epicycle, the center of which (the deferent) itself moved around the Earth in a perfect circle. As more and more discrepancies came to light, more and more epicycles were introduced. Finally the theory became hopelessly clumsy and artificial and astronomers abandoned it entirely.

tem of Mathematics reached Baghdad, one of the capitals of the Arab world. Translated into Arabic, Ptolemy's book became one of the cornerstones of Arab astronomy. So highly was the work regarded that the Arabs dubbed it *megistee* meaning the "great work". The word was corrupted into *almagest*, the name by which the book is best known today.

The Arabs were excellent observers and astronomical centers were soon thriving in Damascus and Baghdad. In addition to observing the skies with the unaided eye, Arab astronomers invented a remarkable instrument that helped them fix the position of stars. Called the astrolabe, the first types began appearing in the 9th century AD but different and more sophisticated versions of the original

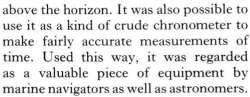

Left: the medieval Italian astronomers in this old Arabian drawing are observing the night sky with the aid of crude mathematical instruments. Although ancient astronomy relied entirely on observations made with the naked eye, many of its measurements were surprisingly accurate.

Right a Persian astrolabe made around the year 1221-2, showing a calendar on the back. Some form of astrolabe was thought to exist in antiquity but the earliest astrolabes to have survived are those of the Persian and Arabian astronomers of the 9th-11th centuries. Such devices, used for observing and timekeeping, remained virtually unchanged until their replacement by more modern instruments in the 18th century.

Below: the greatest of the Maya observatories, the 41-foot high astronomical observatory at Chichen Itza. In the center of the tower is a spiral staircase that winds up to a small observation chamber near the top of the building. From here observations were made, based largely on the movements of Venus, in order to draw up an accurate calendar.

above the horizon. It was also possible to use it as a kind of crude chronometer to make fairly accurate measurements of time. Used this way, it was regarded as a valuable piece of equipment by marine navigators as well as astronomers.

The scientific world's inheritance from Arab astronomy is greater than is realized. Many of the common names of stars are Arabic in origin. For example, Aldebaran in the constellation of Taurus comes from the Arab word for "following," chosen because the constellation seems to follow a bright group of stars known as the Pleiades. Algol, the well-known variable star in the constellation of Perseus, is named for the Arab word meaning "ghoul." Some modern astronomical terms such as "zenith" – the highest point in the sky – are also Arab in origin.

invention were used until the late Middle Ages. It consisted of four sections. First was a circular plate on which was etched a map of the stars surrounded by a scale of hours. Pivoted to the center of this was a cut-away plate with numerous pointers, each bearing the name of a particularly bright and well-known star. Pivoted to the center of this was a metal rule. On the back of the plate, which was marked off in degrees like a protractor, was a sighting arm. Using the astrolabe, it was possible to make extremely accurate measurements of the position of stars

In addition to measuring the position of stars, Arab astronomers were also famed for the detail with which they observed eclipses and planetary movements. One important legacy of these skills was created by the 13th-century Spanish king Alfonso X who called together many of the greatest Arab astronomers and persuaded them to compile a book that came to be known as *The Alphonsine Tables.* Packed with data on eclipses and the movements of the planets, the *Tables* were to become one of the foundations of European astronomy in the Middle Ages and helped to prepare the way for the great revolution in astronomy pioneered by scientists such as Copernicus, Galileo, Kepler and the mathematical genius, Isaac Newton.

The Copernican revolution

Nicolaus Copernicus revolutionized mankind's understanding of the universe. He swept aside the hopelessly complicated arguments that had existed since the time of Ptolemy and replaced them with the simple and dangerously radical proposition that the Sun and not the Earth was at the center of the solar system. With remarkable independence of mind, Copernicus challenged centuries of religious dogma, establishing for all time the cornerstone on which modern astronomical science depends.

Copernicus was born in 1473 in the small Polish town of Toruń. He studied first at Cracow University and then in Italy. His wide researches led him to some little-known writings that showed that not all the Greek astronomers had agreed with Ptolemy that the Earth was at the center of the universe. These discoveries meant little to him at the time and after completing his education he was elected to the cathedral staff, as a canon, of Frauenberg, a large town on the shore of the Baltic Sea. This Church appointment, which did not entail him becoming a priest, was to give him life-long financial independence.

Copernicus was most interested in the theory of astronomy. He was driven by the fascination of how the heavens worked, and quite early in his life became convinced that Ptolemy's version of celestial mechanics could not be right. Ptolemy had worked out a complicated description of planetary movements based on a system of interconnected circles. The main reason why the Ptolemaic description was accepted so unquestioningly was its confirmation of the religiously satisfying idea that the Earth was at the center of the heavens.

Although a devout churchman, Copernicus was deeply unhappy about Ptolemy's work. Instinctively, he felt sure that no description of nature could be as complicated as Ptolemy had made it. He began to sift data on planetary

Top: Nicolaus Copernicus, the Polish astronomer who first produced evidence that the Sun – not the Earth – was at the center of the solar system.

Above: the Geometry Room at Krakow university. The walls still retain the graceful designs that decorated them at the time Copernicus studied there. It was as a young man that he first outlined his own theory of the solar system upon which modern astronomy is based.

positions to see whether he could discover a simpler alternative. The breakthrough in Copernicus' studies came when he remembered the few Greek astronomers who had disagreed with Ptolemy's central assumption that the Earth was the center of the universe. He immediately experimented by placing the Sun at the center and relegating the Earth to a circular orbit around it. At once the need for Ptolemy's complicated explanations was removed. The modern Sun-centered view of the solar system was born.

Copernicus was taking a dangerous step by working on such ideas. Ptolemy's system had been accepted almost as a religious truth by Western Christendom. So Copernicus was discussing more than scientific truth; whether he liked it or not, he was challenging an entire philosophy dedicated to keeping mankind in a central and supreme position in the scheme of the universe. Because people are now so used to thinking of the Earth as just one of the planets orbiting the Sun, it is difficult to realize just how outrageous his idea seemed in the late Middle Ages.

Although by 1533 Copernicus had written his theory in a book called *De Revolutionibus Orbium Caelestium* ("Concerning the Revolutions of the Heavenly Spheres"), he decided against publishing it. Many people knew of its existence however and a few enlightened scholars tried hard to persuade Copernicus to publish it so that the challenge to Church dogma could be made public. In particular, a German mathematician, Georg Rhaeticus, visited Copernicus to discuss the new theory in detail. He left convinced that it must be published and given to the world.

Eventually Copernicus agreed to publish the book but took the wise precaution of adding a dedication to the pope, Paul III. Rhaeticus himself took the manuscript to Nürnberg, but because of opposition from Martin Luther delivered it to the Leipzig theologian and printer Andreas Osiander.

By the time the book was published in 1543, Copernicus was seriously ill. It is said that a copy of his great work reached

his hands only hours before his death. Osiander obviously did not share the courage of the author for included as an unsigned preface was an unauthorized note describing the new theory as no more than a convenient mathematical fiction, a device to help predict planetary positions but having no real basis in fact.

Despite the preface, the book created a storm of controversy. Copernicus had begun a revolution in human thought that he did not live to see triumph. His new vision of man's place in the universe would only slowly replace the old religious prejudices that had survived since the days of Ptolemy. But the publication of his book began a major change in astronomical science that not even the power of the Church could resist.

Below: a chart of the heavens, drawn in the 17th century, and showing the solar system according to the Copernican theory, with the Earth and other planets revolving around the Sun. As can be seen, only six of the nine planets had been discovered at that time.

GALILEVS GALILEI FLORENTINVS
ANNVM AGENS LXXVIII

Galileo

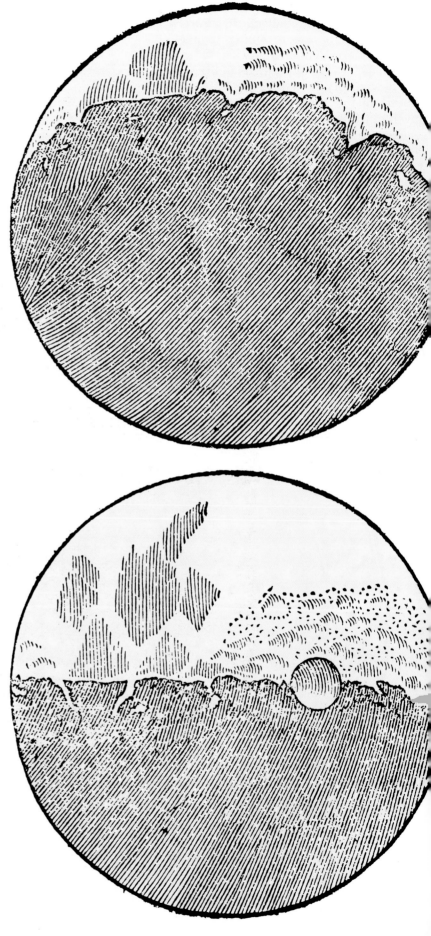

Galileo Galilei was one of the founders of modern scientific method. In an age still preoccupied by mysticism in which occult explanations of natural phenomena were far more popular than rational ideas, Galileo established the fundamental importance of observations and experiments to making true scientific discoveries. His indifference to the dogmas of his time and his unwavering belief in a rational approach to studying the natural world, make him one of the key figures in the history of science.

He was born in 1564 near Pisa in Italy, and even as a child seemed destined for a scientific career. Galileo's parents wanted him to become a doctor and when he was 17 sent him to Pisa University to study medicine. But Galileo was much more interested by natural science and mathematics and soon began attending lectures on these subjects instead of his medical lessons.

While still a student Galileo made the first of his great scientific discoveries. It

Far left: Galileo Galilei; his development and use of the telescope opened up vast new possibilities of observation. He discovered that the Moon shines with reflected light; that its surface is mountainous; and that Jupiter has four large satellites. He also observed sunspots and their regular movement across the Sun's surface.

Right: frontispiece to an early edition of Galileo's *Dialogue on the Chief Systems of the World* in which Galileo supported the Copernican theory. The Church had already reacted violently against Copernicus' ideas. Galileo's uncompromising arguments in its favor led him into serious conflict with religious authorities and almost cost him his life.

Left: two sketches of Galileo's views of the Moon (above and below). Galileo made his observations with one of the first refracting telescopes which, though crude by modern standards, was able to resolve objects in the heavens in greater detail than had been possible before.

Below: a copy of Galileo's design for an escapement, or pendulum clock. Galileo's work on the pendulum was one of the earliest researches into the science of dynamics.

is said that while attending a service in the Cathedral of Pisa he noticed the rhythmic swaying of the huge suspended candelabra, caught in the eddies and drafts near the cathedral roof. Later he built a simple pendulum and by studying the way it swung to and fro was able to work out the basic laws governing its motion. His discovery was later put to practical use in the first pendulum clocks, among the earliest really accurate timepieces ever built.

Galileo's intellectual abilities ensured a rapid success in academic circles. By 1589 he had been appointed Professor of Mathematics at Pisa University. He had acquired a wide reputation for never wanting to take scientific ideas on trust. He insisted on trying to find ways of testing their validity whenever possible. One famous story describes how he set out to test an idea first proposed by the Greek philosopher Aristotle about 2000 years earlier. Aristotle had stated that a heavy body must fall to the ground faster than a light body. But Aristotle had not troubled to test his assertion experimentally. Galileo is said to have climbed the celebrated leaning tower of Pisa and flung cannon balls of different weights from its upper galleries to see which reached the ground first. Whether or not the extraordinary story is true, there is

little doubt that Galileo showed that Aristotle was wrong and that – air resistance apart – all bodies fall at the same rate independent of how heavy they are.

Galileo's reputation for rigorous investigation of the facts made him particularly influential when he turned his mind to astronomy. He had heard about the work of the Polish cleric Nicolaus Copernicus who disputed the long-held belief that the Earth was the center about which the universe turns. Like Copernicus, Galileo was sceptical about Ptolemy's theory – worked out centuries earlier – which tried to fit observed planetary movements to a vastly complex scheme of interconnected circles. Instead of looking at observed facts and working out a theory from them, Ptolemy had tried to fit the facts to a preconceived idea. But despite its complexity, Ptolemy's work had been accepted without question simply because the basic idea of the Earth being the center of all things satisfied the religious dogmas of the time.

Galileo's serious interest in astronomy was probably started by the appearance of a brilliant new star in the skies in 1604. It was almost certainly a supernova – a terrible cataclysm literally tearing a star apart. At the time it was a total mystery. Soon after the appearance of the star, Galileo heard about the invention of the first telescope – built in Holland by Hans Lippershey. Galileo immediately saw the telescope's potential for making detailed observations of the heavens. He began work building his own and in 1609 made his first studies with it. Galileo's use of the telescope was a historic breakthrough in the development of modern astronomy. It was the starting point for a series of fundamental discoveries that would quickly sweep away much of the superstitions surrounding ideas about the heavens.

Judged by today's standards, Galileo's instrument was a weak and inefficient device. But it was vastly superior to using the naked eye and Galileo was able to make observations in greater detail than had been imagined before. The first object he studied was the Moon. He observed the great lunar plains, valleys, mountains, and craters and even tried to measure the sizes of some of the most prominent features.

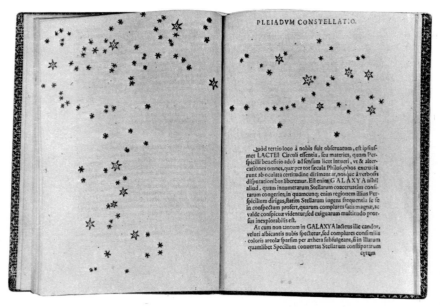

Next he turned his attention to the stars. He soon realized that there were many more than could be seen with the unaided eye. For example, the beautiful cluster known as the Pleiades looks as though it consists of only six stars. Galileo's telescope revealed about 40 — although modern astronomers have revealed the true number to be well over 200. But for Galileo the most stunning sight was the Milky Way, the huge band of light spread across the night sky. Galileo found that it was in fact made up of a vast multitude of individual stars, apparently very closely packed.

More discoveries followed in 1610 when Galileo began to observe the planets. Almost immediately he found that Jupiter was accompanied by four tiny points of light. He realized that what he had observed were four natural satellites orbiting Jupiter in much the same way that the Moon orbits the Earth. The discovery seemed to cast even graver doubts on Ptolemy's work. But Galileo's observations of Venus were the most conclusive evidence as far as he was concerned. He discovered that Venus shows distinct phases – apparent changes in shape rather like the Moon at different times of the month. Since Venus is a planet, visible only by sunlight reflected from its surface, Galileo was able to work out that Venus had to be traveling around the Sun, not the Earth, in order to show its various phases.

In the years that followed, Galileo made other remarkable discoveries. He found that the planet Saturn seemed at

Above: two pages from Galileo's *Sidereus Nuncius*, published in 1610. It shows many "new" stars revealed for the first time with the early telescope. Before 1610 the belt and sword of Orion (left) appeared as a group of only seven stars.

Below: drawings of sunspots from Galileo's *Delle Macchie Solari*, published in 1613. Discovery of sunspots was first announced two years earlier by the German Jesuit astronomer Christoph Scheiner. Galileo was more cautious, wanting to make sure that the spots really were associated with the Sun, and not small bodies that had interposed themselves between the Sun and the Earth.

times to show a triple image in his telescope, a mystery he was never able to resolve. Only in 1659 with the Dutch astronomer Christian Huygen's discovery of Saturn's rings was the puzzle solved.

Galileo's observations were not confined to the stars and planets. He also made the first telescopic studies of the Sun. He was especially interested by sunspots and was the first to state that they were dark areas on the actual surface of the Sun carried across its face by the Sun's overall rotation about its own axis. It was a controversial idea. Astronomers had previously argued that sunspots were simply bodies in space passing between Earth and the Sun becoming visible briefly in silhouette against the bright solar surface. Some religious authorities were angered by Galileo's idea, which seemed to suggest blemishes on the face of one of God's most magnificent creations.

But Galileo's most serious breach with the Church was caused by his growing conviction that Ptolemy had been wrong in placing Earth at the center of the solar system. Galileo was a forthright man who found it hard to suppress views he firmly believed. Soon he began openly teaching that the Sun and not the Earth was the point about which the planets turned. By 1615 the Church was becoming increasingly disturbed by Galileo's work and actually warned him to change his ideas. Guided by more politically aware friends, Galileo agreed to temper his public statements with more discre-

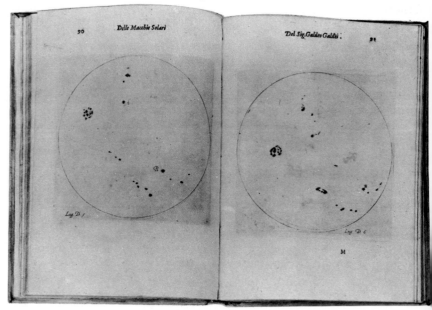

30

Adi 7. di Gennaio 1610 Giove si vedeua co(l) cannone co(n)
3. stelle fisse cosi ✳ ⊕ *oce:* delle quali se l'il cannone.
minina si uedeua. *ori:* ✳ à d. 8. appariva cosi ⊕ ++ era d(un)q
Diretto et nō retrogrado come (s)ogono i calculatori.
Adi 9. si nugolo. à dico si uedeua con ————— ✳ ✳ ⊕ ciò è co(n)
q(ua)ti è la più ocidetale si che à ueultaua f q(ua)ti si può credere.
Adi 11. era in questa guisa ✳✳ ⊕ et la stella più uicina
à Giove era la metà minore dell'altra, et uicinissima all'altra
doue che le altre sere erano le dette stelle apparite tutte tre
di equal grandeza et trà di loro equalm(ente) lontane; dal che
appare intorno à Giove esser .3. altre stelle errāti inuisibili ad
ogn'uno sino à questo tefo.
Adi 12. si uedde in tale costituzione ✳ *sm* ✳ ⊕ ✳ era la stella
ocidentale poco minor della orientale, et giove era i mezo lontane
da l'una et da ll'altra q(ua)to il suo Diametro, è circa: et forse era
una terza hiochia et uicinissa à 4 uerso oriēte; anzi pur ui era
uerome(nte) havedo io co più diligēza oseruato, et uedeo più imbrunita la
notte.
Adi 13. havedo benissimo fermato co strum(en)to si ueddono uicinissa à Giove
4. stelle in questa costituzione ⊕ ✳✳✳ i mezo ui oni ⊕ ✳✳

tion and at about the same time he moved to Florence to work as a resident mathematician to the Grand Duke of Tuscany.

But Galileo could not resist continuing his opposition to Ptolemy's ideas for long. In 1632, and with the pope's permission, he published one of his most famous books. It was probably the most controversial ever to have appeared. Called *A Dialogue Concerning the Two Chief World Systems – Ptolemaic and Copernican*, it took the form of a discussion between two imaginary philosophers. He had been told to discuss the theories noncommittally and to come to conclusions laid down by the Church. But using all the evidence available, Galileo built a convincing case for Copernicus' theory of the solar system. By the time the book was published – and hailed as a literary and philosophical masterpiece – Galileo was an old man with failing sight. Few people believed the Church would punish him despite the controversial nature of his writings.

In 1633, however, Galileo was summoned to Rome for trial at an ecclesiastical court. Scholars all over Europe feared for his life. But confronted by one of the most celebrated scientific figures of the day, the Church was content with merely extracting a statement from Galileo swearing that his former views were false. The story is told that as he rose from his knees after testifying that Ptolemy was right and that the Earth was the unmoving center of the solar system, he was heard to mutter through gritted teeth – "and yet the Earth moves."

Forced into retirement by the Church, Galileo was forbidden to continue his studies of the heavens. But the evidence he had published could not be ignored and astronomers and scientists throughout Europe were rapidly concluding that however heretical the idea might seem, Ptolemy's ideas simply did not stand up to the test of observations. By the time Galileo died in 1642, the Copernican revolution was rapidly overturning centuries of religious dogma. The foundations of modern astronomy had at last been laid.

Kepler's laws

While Galileo was beginning his historic investigation of the solar system, a young German astronomer called Johannes Kepler was working out some basic laws describing the movements of the planets. His remarkable discoveres were conclusive evidence that the Sun and not the Earth was the center of the solar system. Kepler's laws are today still regarded as a fundamental part of celestial mechanics.

Kepler was born in 1571, in Weil der Stadt in Württemberg. Unlike Galileo, he was a theorist rather than an experimenter. Kepler's genius was to be able to work patiently with vast amounts of data and ultimately to be able to resolve it into intelligible scientific principles. The key to his life's work was the great store of astronomical data made available to him almost by chance.

In 1600 Kepler had taken a job in Bohemia as assistant to the great Danish astronomer Tycho Brahe. Tycho was a flamboyant figure with wealthy patrons. He built extravagant observatories and complex instruments with which to study the heavens. Despite his liking for banquets, hunting, and revelry, Tycho was a brilliant observational astronomer. During his lifetime he amassed an unprecedented catalog of data about the

Left: the German astronomer Johannes Kepler, who proved that the planets move around the Sun. He discovered the three laws of planetary motion. Kepler's work as a theorist was helped by the vast store of astronomical data left to him by the Danish astronomer Tycho Brahe (below), who employed Kepler as an assistant between 1600–1601.

Above right: Brahe's variation on the Copernican model of the solar system. He held that the five planets revolved around the Sun, but that the Sun and its planets revolved around a stationary Earth.

Right: the huge quadrant, over six feet in radius, with which Tycho Brahe drew up a very accurate star catalog. Before the telescope was discovered, the quadrant was probably the most important instrument of early astronomy.

positions of the stars and planets. Despite the fact that all his observations were made with the naked eye, his findings were remarkably accurate.

Tycho was a fierce opponent of Copernicus, regarding his views as heresy. But he was equally convinced that the long-established Ptolemaic theory was just as wrong. With typical individualism Tycho worked out his own scheme. He argued that while the planets revolved around the Sun, the Sun and Moon revolved around the Earth.

Tycho saw in his assistant's skills a chance to verify his theory of the heavens. When Tycho died in 1601 he left Kepler his lifetime's accumulation of data hoping that after his death Kepler would use it to establish Tycho's ideas above those of Ptolemy.

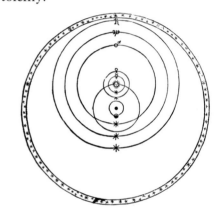

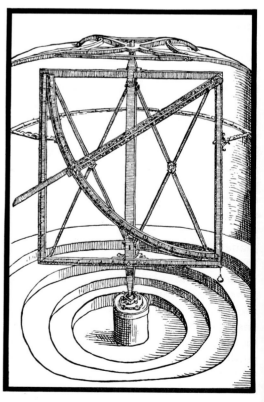

But Kepler was in fact much more inclined to accept the controversial views of Copernicus. Basing his work with Tycho's observations of the apparent wanderings of the planets against the background of stars, Kepler spent years calculating possible orbits. He soon convinced himself that the observed positions of the planets did not fit Ptolemy's idea of an Earth-centered solar system. He concentrated on one planet in particular, Mars, one of Earth's nearest neighbors, determined to establish how it moved in relation to the Sun and Earth. If he could work out how Mars moved, he was sure the key to the paths of the other planets would quickly follow.

The task proved much harder than he expected. Copernicus believed that the planets moved in circular orbits around the Sun but although Kepler could make Tycho's data almost fit this idea, there were always discrepancies. It took all Kepler's patience to continue his painstaking work and then, after thousands of intricate calculations, he hit upon the answer. Mars – and presumably all the planets – certainly moved around the Sun just as Copernicus believed. Its orbit was not a circle, however, but an oval path known as an ellipse.

The idea of the planets describing elliptical orbits around the Sun has become known as Kepler's first law of planetary motion. His second and third laws followed soon afterward. The second law, published in 1609, states that if an imaginary line is drawn between a planet and the Sun, the line describes equal areas in equal times as the planet moves in its orbit around the Sun. This means that a planet moves most quickly when it is closest to the Sun but slows down as it moves farther away. Kepler's third law was not finalized until 1618 and is a complicated-sounding relationship between the average distance between a planet and the Sun and the time the planet takes to complete an orbit around the Sun.

Kepler's work was a major advance in astronomy. For the first time Copernicus' ideas had been verified by a careful analysis of observational data. And the movement of the planets – so long regarded as the preserve of divine powers – was at last shown to be subject to mathematical laws.

Below: the diagram illustrates Kepler's second law. T1, T2, and T3 are equal intervals of time in a planet's elliptical orbit. A, B, and C are the equal areas it sweeps out in those times. The diagram shows how the speed of the planet varies in its orbit. For T2, for example, to the planet covers the largest orbital distance and must be traveling faster than T1 or T3.

Surprisingly Kepler's work attracted relatively little attention from the Church. Perhaps this was because Kepler's scientific ideas were often expressed alongside more occult astrological concepts. Being a strange mixture of rationalist and mystic, he may have been regarded as less of a danger to religious dogma than such uncompromising scientists as Galileo. But in the years to come Kepler's laws of planetary motion would help to establish beyond doubt the supposed heresy of the Copernican theory of the solar system. And before the end of the 17th century, thanks to the remarkable genius of Isaac Newton, the laws would be seen as a demonstration of a fundamental underlying phenomenon of nature – the universal force of gravity.

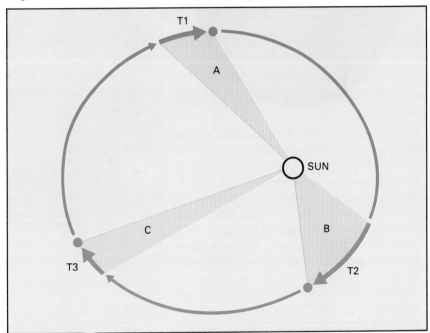

Right: Kepler's determination of the true orbit of Mars. Using a vast accumulation of data about the position of Mars as seen from the Earth, Kepler studied seven so-called oppositions – moments when the Sun and Mars appear opposite each other in the sky. At each opposition the Earth was at a different point in its orbit and by making many painstaking calculations, Kepler was able to use each opposition as a kind of fix of the position of Mars in relation to the Sun. Eventually he was able to show that the Martian orbit was an ellipse.

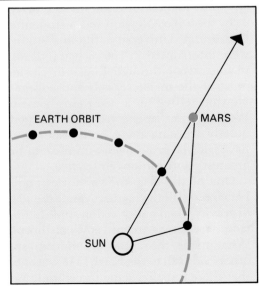

The Newtonian universe

Isaac Newton – widely regarded as one of the greatest scientific thinkers in human history – was born in 1642 at Woolsthorpe, a small village in the heart of the English countryside. During his life he made numerous important contributions to physics, chemistry, and mathematics and even in his own time was internationally acclaimed as a man of unprecedented insight and genius. Perhaps his most famous discoveries concern the laws governing the way bodies move and the existence in nature of a fundamental force known as gravity.

In the 1660s Newton was a student at Cambridge University. But his studies were interrupted in 1665 when a deadly plague swept through England, closing most of the major schools and colleges. Newton retired for a year to his home in Woolsthorpe. In the year that he spent quietly working there he is thought to have laid the groundwork for most of his greatest contributions to science.

One of the best-known stories about Newton tells how he discovered the idea of gravity when he noticed an apple falling from a tree in the orchard at Woolsthorpe. What made the apple or, indeed, any other body fall to the Earth? He reasoned that there must be some force "pulling"

Above left: Isaac Newton, the English scientist and mathematician. His three laws of motion which now form the basis of all mechanical science, and his theory of gravitation are cornerstones of present-day physics.

Above: top: Newton's third law of motion – action and reaction – is demonstrated in this engraving made in 1747 of a "jet engine," driven by a jet of steam. Lower diagram: an experiment from a popular book explaining Newton's *Principia*. It shows an experiment for investigating the motion of bodies on an inclined plane and under the influence of gravity.

the apple toward the ground. But if such a force existed, how far did it extend? Gradually Newton realized the the force that caused the apple to fall could also keep the Moon in its orbit around the Earth or the planets in their orbit around the Sun.

But if the force is the same, why does the Moon not fall toward the Earth like the apple? The reason is because the Moon itself is moving. If the force of gravity between the Moon and Earth could suddenly be switched off, the Moon would hurtle away into space. In other words, the force of gravity is like a chord, binding the Moon to the Earth. The simplest everyday analogy is to imagine a man whirling a ball on a string around his head, with the string taking the part of gravity. If the string breaks, the ball flies off in a straight line. But as long as it remains intact, the ball is pulled into an orbit around the man's head.

Like most analogies, this one is inexact. In reality, the Moon is falling all the time. But because of its speed and distance it actually falls around the Earth with never

Below: Newton showed
that the gravitational pull
of the Moon and Sun cause
the tidal movements of
Earth's seas. When the
Moon and Sun are aligned
as shown in the second
position in this diagram,
their gravitational influence
combine to form
exceptionally high tides on
Earth. Although, for
simplicity, the Moon's orbit
around the Earth is shown
as circular, the Sun's
gravity actually influences
and distorts its shape.

Moon's orbit. He showed that the attraction of the Sun on the Moon varies as the Moon moves around the Earth. Because of this the Moon's orbit is not a perfect ellipse. Newton extended his work to include comets and so enabled his friend Edmund Halley to predict the return of a particular comet last seen in 1682 and named for Halley himself. According to Newton's law of gravity, Halley's comet would be seen again 77 years after its last visit. Its appearance in January 1759 was a triumph for Newtonian mechanics.

Newton's great achievement at last showed that the same laws of nature at work on Earth, can operate in the heavens. But although Newton gave scientists a brilliant insight into the workings of the universe, he was unable to say what causes gravitational force. What is it, for example, that makes the Moon "feel" the presence of the Earth nearly a quarter of a million miles away? Today scientists know that Newton was in fact dealing with a concept far more profound than he could possibly realize. Modern theories tell us that gravity is not merely a force somehow reaching out across space but is in fact a basic quality of the fabric of space itself.

any danger of it plunging into the ground. The same principle applies to the planets as they move around the Sun.

Newton did much more than merely suggest that a force of gravity might exist. He worked out a detailed description of it. In his so-called law of universal gravitation, he states that every particle or body in the universe, from the tiniest grain of dust to the most gigantic star, attracts every other with a force that gets stronger with mass and weaker with increasing distance.

Although Newton did most of the basic work on his theory of gravity in 1665 he did not publish a detailed account of it until 20 years later. Encouraged by friends and admirers he drew together much of his greatest work into a single remarkable book which he called *Philosophiae Naturalis Principia Mathematica* ("Mathematical Principles of Natural Philosophy"), now known simply as the *Principia*. Most experts today regard it as the most important scientific textbook ever written.

A major section of the *Principia* explained Newton's ideas of gravity. Not only did Newton explain how gravitational attraction lay behind all the laws of planetary motion earlier revealed by Johannes Kepler, but showed the way in which the tides in the seas and oceans were caused by the gravitational pull of the Moon, and, to a lesser degree, the Sun. He was even able to use the theory to explain strange variations in the

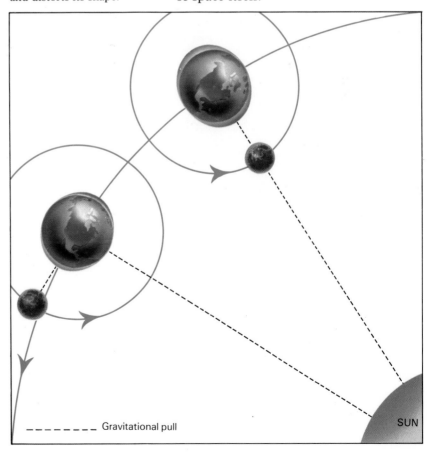

– – – – – – – Gravitational pull

SUN

Birth of modern astronomy

Newton's great work completed a revolution in astronomy. Centuries of superstition and religious dogma had at last been swept away and astronomy had evolved into a true science. The revolution was more than just an intellectual one. At about the same time telescopes came into general use and for the first time astronomers could make really detailed observations of the stars and

planets. Observational astronomy begun by Galileo in 1609 became popular throughout Europe and the beginnings of the modern concept of the universe and mankind's place in it came about very largely to the widespread use of increasingly sophisticated telescopes.

Probably the most famous observational astronomer of all is William Herschel, often called the father of modern astronomy.

Right: one of Herschel's self-built telescopes. He ground varying amounts of copper, tin, and antimony and cast the molten metal into mirrors for his reflecting telescopes. Soon he was constructing telescopes superior even to those of the Greenwich Observatory.

Left: William Herschel, the English astronomer and mathematician. He built his own telescope with which he discovered the planet Uranus in March 1781. He later discovered two of its satellites and found that they, too, rotated in a backward direction. He also discovered the sixth and seventh satellites of Saturn. He is regarded as the virtual founder of stellar astronomy.

Born in 1738, just 11 years after the death of Isaac Newton, Herschel was brought up in Hanover in Germany. He was a gifted musician and worked for some years as a bandsman in the Hanoverian army. In 1757 he left Germany and travelled to England where he was to remain for the rest of his life. He earned

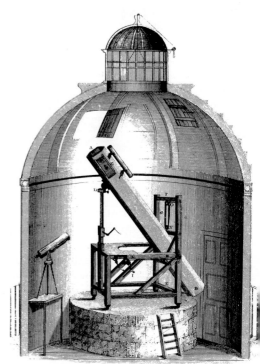

a living as a musician but spent most of his spare time working on his favourite hobby, observational astronomy.

Herschel could not afford to buy the best telescopes and decided to begin making his own. After months of painstaking work he perfected the technique of making the large curved mirrors required by a good quality reflecting telescope. Eventually his telescopes were superior even to those being used at the Greenwich Observatory. Once he could make his own instruments his astronomical work became his consuming interest. Although he continued his musical career to make enough money for food and lodging, his only real passion was the night sky.

Then on March 13, 1781 Hershel made the discovery that was to change his life. While using a self-made 7-inch reflecting telescope he came across an object in the constellation of Gemini that showed a small greenish-colored disk. By watching it for several nights, Herschel realized that it was slowly moving across the sky. He concluded that it must be a comet and reported it as such to the Royal Society in London. But when the Finnish astronomer Anders Lexell worked out the details of the orbit of Herschel's "comet" he was astounded to find that it was not a comet at all but a new planet moving around the Sun far beyond the orbit of Saturn.

It was a momentous discovery. It had not occurred to astronomers that other planets might exist beyond the orbit of Saturn. Herschel became famous overnight and was given a grant by the British king, George III so that he could devote himself to astronomy full-time. In gratitude to his royal patron Herschel named the planet the Georgian star, *Georgium Sidus*, but astronomers all over the world preferred the name Uranus.

In 1784 Herschel began a systematic survey of the heavens. He wanted to work out the way in which the stars were distributed in space and decided that the most direct technique would be to count them. To do this he devised a method called "star-gauging," which involved counting the stars in certain carefully selected regions of the sky. His findings suggested that the stars were grouped in a volume of space shaped like a gigantic lens. It is now known that Herschel's idea of the galaxy's shape was extremely accurate. But his most exciting speculation was made some years later.

Many astronomers had noticed that apart from stars, planets, and occasional comets, strange hazy patches of light could be seen in the sky. Usually called *nebula* – the Latin word meaning "cloud" – the luminous patches had been catalogued by such astronomers as Edmund Halley and Charles Messier. Herschel himself discovered about 1500 of them. With the telescopes then available it was impossible to study them in detail but Herschel made an inspired guess suggesting that some of them at least were separate galaxies, far beyond the bounds of the Milky Way. It was an awesome prospect, one that Herschel did not live to see confirmed. It was not until 1924, just over 100 years after Herschel's death, that the American astronomer Edwin Hubble turned a giant reflecting telescope on the Andromeda nebula and found that it was – as Herschel had suspected – a glittering island universe made up of countless stars.

Above: the Andromeda galaxy. It was Herschel who suggested that this and similar patches of what appeared to be luminous gas might be island universes in deep space. In fact, over 100 years later Herschel's theory was confirmed and the borders of astronomy were extended from interstellar to intergalactic space.

The discovery of galaxies beyond our own added a totally new perspective to man's understanding of the universe. For the first time its true immensity could be glimpsed.

The genius of Herschel and other astronomers like him, laid the foundations of modern astronomical science, preparing the way for a wealth of new discoveries about the universe and mankind's place in it.

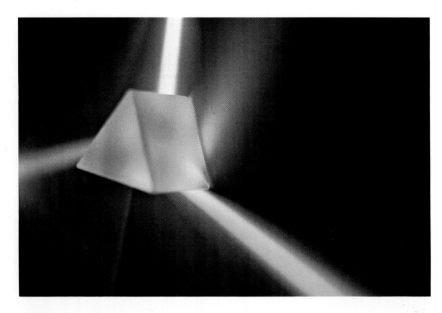

Spectroscopes

The spectroscope is one of the most important instruments in astronomy. By splitting light from distant stars and galaxies into a spectrum of different wavelengths, the spectroscope enables scientists to work out not only the chemical composition of the distant bodies, but also their temperature and their movement relative to the Earth.

The story of modern spectroscopy began in 1665 in the village of Woolsthorpe in England. There in the rural peace of his home, Isaac Newton worked on a wide range of scientific problems and inquiries. Among the most important of the experiments he performed was one designed to investigate the nature of sunlight. He found that by allowing a shaft of light to fall onto the face of a glass prism, the light could be broken into a spectrum of different colors. The apparently white light of the Sun was in reality a mixture of colored light.

Newton could not see the full significance of his discovery and it was not until the 1800s that the modern science of spectroscopy was really established. In 1802 a British chemist, William Wollaston, noticed that if the spectrum from sunlight was examined carefully about seven dark lines could be made out

Above: white light being broken up into its seven colors by refraction through a glass prism.

Below: In 1814 the German scientist Joseph Fraunhofer (left) discovered a large number of dark lines crossing the colored bands of sunlight that emerged from a prism. Although some seven of the lines had been noticed 12 years earlier by the British chemist William Wollaston (center), the superb quality of Fraunhofer's prism enabled him to identify over 500 of them. The true significance of the lines was explained by the German physicist Gustav Kirchhoff (right), who made the first spectroscope in 1860.

crossing the colored bands in various positions. A few years later a German scientist, Joseph von Fraunhofer, took Wollaston's work much further. Fraunhofer allowed a beam of sunlight to fall onto a thin slit. The narrow light ray then passed through a prism and formed a spectrum made up of a large number of colored lines, actually images of the original slit in a tightly packed group. The line spectrum showed the dark lines Wollaston had seen but when Fraunhofer studied the spectrum carefully he spotted not just seven but more than 500. He also realized that whenever he examined the Sun's spectrum the dark lines always appeared in the same positions amongst the other colored lines.

The real significance of the Fraunhofer lines, as they are called, was discovered by Gustav Robert Kirchoff. In 1859, while Professor of Physics at Heidelberg University, Kirchoff conducted a number of experiments to investigate spectra from the light of glowing or incandescent chemicals. His conclusions marked the birth of the science of spectroscopy and placed in the hands of astronomers the most powerful tool since the invention of the telescope.

Kirchoff realized that in a spectrum, many of the lines that appear are actually a kind of "signature" produced by particular chemical elements. If for example, a piece of sodium is heated until it becomes incandescent, the light it produces – passed through a slit and a prism – turns into a bright yellow double-line. When sodium is just one of several elements producing the light, the same double-line will be seen in a characteristic position amongst all the other lines.

The important point about such a spectrum is that each element produces its own unique set of lines in a particular place in the overall spectrum. The double line of sodium cannot possibly be pro-

duced by any other element. Often an element will produce not merely a simple set of lines. Iron for example gives rise to several hundred. But complex though such a spectrum is, the lines always appear in the same position and give positive evidence that iron is present in whatever is producing the light.

The reasons why the chemical signatures in a spectrum are so distinct are complicated. They involve the detailed structure of atoms and the energy processes that take place inside them. Basically when an atom of an element becomes sufficiently hot it gives out energy in the ·form of light. This is a familiar phenomenon. We all know that if an iron bar is placed in a fire it soon glows red hot. But what is less well-known is that a hot element can also absorb light of the same wavelength it usually emits. This means that in a hot cloud of different gases some light will be emitted and some absorbed by the different atoms. Because of this, such a gas cloud will produce two distinct kinds

Above: a spectroscope of 1862, based on the design used by Kirchhoff. It consists of a collimator – a tube with a slit through which the light enters – a viewing telescope, and a tube with a micrometer for measuring spectral lines.

arising from investigations of distant galaxies was that their spectrum is shifted toward the red by varying degrees. This means that all the lines in the spectrum are moved fractionally from their expected positions but by the same amount. We now know that this red-shift indi-

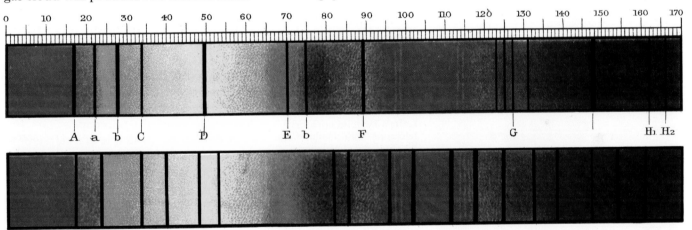

of line spectrum. One, called an emission spectrum, consists of the familiar bright lines. The other, called an absorption spectrum, is made up of dark lines. Both Wollaston and Fraunhofer had noticed absorption lines which, because they occur in characteristic positions in a spectrum, are just as useful in identifying elements as emission lines.

The spectroscope was soon used to analyze the composition of the Sun and so far about 70 of the 92 natural elements have been spotted. Modern high-resolution techniques have also made it possible to examine light from remote stars and even other galaxies far beyond our own. One of the most exciting discoveries

Above: spectra of the Sun (top) and Sirius (bottom). Kirchhoff's discovery that each pure substance has its own characteristic spectrum has been used as a tool by astrophysicists to analyze the composition of the Sun, and later to examine the light from stars to determine the chemical composition of their atmospheres, the physical condition of their surfaces, and the speed at which stars may be traveling away from or toward the Earth.

cates that the galaxies are receding from us. The amount of the shift tells us the speed of recession. This discovery has led directly to our modern concept of an expanding universe.

Because we now understand very accurately what happens inside atoms when they produce a spectrum, extremely detailed spectroscopic studies can even tell us what kind of physical processes are taking place inside distant astronomical objects. Today spectroscopy can look into the heart of some of the most remote phenomena in the universe and provide astronomers with the basic data that may one day unlock their innermost secrets.

Telescopes

The invention of the telescope in 1608 by the Dutch optician Hans Lippershey was a turning point in astronomy. At last astronomers could make detailed observations of the heavens. Quite suddenly a multitude of hitherto unseen stars were revealed and for the first time the true immensity of the universe was glimpsed.

Today telescopes are still among the most important instruments used by astronomers. They are far more powerful than the early models and can be precisely maneuvered to examine any point in the sky. Despite their sophistication however, the principles on which they operate have not changed for over 300 years.

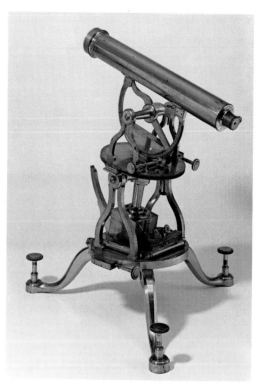

Above: William Parsons, the Earl of Rosse, shown here with the 72-inch reflector telescope he designed and built in the mid-1800s. For many years it was the world's largest telescope and with it he discovered the spiral structure of many nebulae.

Left: a Dollond equatorial telescope. John Dollond and his son Peter developed a non-color-distorting (achromatic) telescope in the mid-1700s. It cured the hazy focus that had been a hindrance to astronomy until then. An equatorial telescope is one that is mounted with its axis parallel to that of the Earth and pointing to the North celestial pole.

There are two kinds of telescope: refractors and reflectors. The earliest instruments were refractors, using a system of glass lenses to bend the light so that a magnified image of a distant object was seen by the astronomer.

The simplest refractor consists of a tube with a large *convex* (outward-curving) lens at the front end. This lens –

called the objective lens because it is closest to the object being viewed – collects the light and bends it (a process scientists call refraction) toward a focus point farther along the telescope tube. At the focus an eyepiece magnifies the image formed there so that the distant object seems much closer to the person looking through the telescope.

Although the principle of a refractor is quite simple, building a really high-quality instrument is extremely difficult. The key problem is making the lenses sufficiently accurate in their curvature and ensuring that the glass from which they are made is free of even the slightest imperfections. In fact, even in the early stages of optical science, extremely good lenses were produced, hand-ground with painstaking care. But a special problem seemed to affect refractors however carefully the lenses were made. Because ordinary light is made of light of different wavelengths, whenever it passes through a lens the wavelengths from which it is made are each bent by slightly different degrees. This means that the various wavelengths are brought to a focus at slightly varying places, producing an image which is always surrounded by a hazy colored edge.

Then in 1729 a British amateur astronomer called Chester Moor Hall dis-

Left: the 36-inch refractor telescope at Lick Observatory on Mount Hamilton, near San Francisco. It was finished in 1888 and was the largest telescope in the world.

Below: William Herschel's 48-inch reflector telescope. King George III granted him £4000 toward the cost of building it. Herschel used it for his study of the Milky Way.

developing in the glass. But astronomers were keen to increase the light-gathering power of their instruments and the demand for ever-larger objective lenses led to breakthroughs in glassmaking culminating in the first giant refractors built toward the end of the 19th century. The most famous of these are the 36-inch telescope at the Lick Observatory and the 40-inch instrument at the Yerkes Observatory, both in the United States. These instruments are still the largest in the world. It seems unlikely that a lens larger than the one used at Yerkes will ever be made. There are two main reasons. First, it is virtually impossible to cast a glass disk of a greater size without imperfections in the glass. Second, a lens must be supported in the telescope by its edges. A large lens is so heavy that it tends to bend slightly under its own weight, distorting any light passing through it. Nonetheless, the 40-inch telescope is an extremely powerful device gathering

covered that the annoying effect could be eliminated by using as the object lens not a single lens but a carefully matched pair known as a doublet. The idea of using a compound lens instead of a single objective enabled Hall in 1733 to build the first refractor to be unaffected by the hazy focus that was so hampering astronomy. But Hall did not publicize his discovery and it was not until 1758 that instrument-maker John Dollond rediscovered Hall's original idea.

Helped by his son Peter, Dollond began manufacturing compound lenses and soon all astronomical telescopes were using them. The double lens system itself is usually called an *achromatic doublet* – the word achromatic implying "no colored edge." The first lens is designed to bend the light strongly toward a focus in the normal fashion. But the second lens works the opposite way, bending the rays very slightly away from the focus. If the two lens are perfectly matched, the second lens brings the different wavelengths bent to varying degrees by the first lens, back to a single sharp focus.

Although Dollond's lenses were extremely successful, they were limited in size to about four inches. At the time it was impossible to cast larger disks from which to grind them without faults

flat secondary mirror inclined at an angle to the main one. The secondary mirror sends the light to the side of the tube bringing it to a focus outside.

The *Cassegrain* system also uses a secondary mirror situated in the main telescope tube. But instead of reflecting light to the side, the Cassegrain secondary bounces light back down the tube. A small hole in the main mirror – too small to affect its overall performance – allows the reflected beam to pass out of the telescope and to be brought to a focus actually behind the main mirror. One of the latest developments in reflectors is the Schmidt telescope. Invented in 1930 by Bernhard Schmidt, an optician at the Bergedorf Observatory, Hamburg, the instrument is primarily a photographic device, rarely used for direct observations by eye. It was designed to provide an exceptionally large field of view. Orthodox reflectors could only offer a fairly restricted field of view to be observed and photographed. Until recently, the largest Schmidt telescope was the 48-inch instrument based at the Mount Wilson Observatory in the United States. Now a 53-inch Schmidt has been built in Jena, East Germany.

40,000 times as much light as the naked eye and making the Moon's surface – more than a quarter of a million miles distant – seem as if it was no more than 80 miles away.

Reflecting telescopes do not use lenses to magnify an image. Instead they employ a system of curved mirrors. In the 1600s when refracting telescopes looked as if they would never be free of the colored edges that always surrounded the images they formed, reflectors seemed an obvious alternative. A major advantage was the relative ease with which large mirrors could be made. Because light does not have to pass through a reflecting medium – it simply bounces off the surface – it was only necessary to ensure that the mirror was curved accurately and that its surface was highly polished. It did not matter if there were a few imperfections in the surface. Another attraction of reflectors was that because it was fairly easy to build quite large mirrors, instruments of much greater power than contemporary refracting telescopes could be built. As long ago as 1789 William Herschel managed to build one with a mirror 48 inches across.

There are three main types of reflecting telescope named for their inventors. Perhaps the best-known is the *Newtonian* system. Light enters the main tube of the telescope and strikes the large, inward-curved mirror at the base. The converging light rays are reflected back along the tube until they strike a small,

Above: a 5½-inch thick, 400-pound mirror blank being examined before its use in the high-altitude telescope system of Project Stratoscope II. The 36-inch mirror is polished to an accuracy of one millionth of an inch.

Right: an astronomer climbs into the prime focus cage of the 200-inch Hale telescope at Mount Palomar, California.

Large reflectors are nowadays fairly common. As long ago as 1845 Lord Rosse, an Irish amateur astronomer, built a 72-inch device and was able to make out spiral structures in many nebulae using it. But probably the most famous of all reflectors is the giant 200-inch instrument at Mount Palomar in California, first used in 1948. Immense difficulties faced the manufacturers of the huge mirror and several prototypes cracked during the delicate cooling process before the huge reflector was suc-

cessfully completed. Galileo – the first astronomer to use a telescope – would be astounded by the giant device. It is so large that the observer has to sit in a cage slung inside the main tube. All the delicate steering movements directing the telescope at particular regions of the sky are controlled by complex electronic mechanisms.

But the Mount Palomar reflector is no longer the largest in the world. Soviet astronomers are now testing a 236-inch telescope near Zelenchukskaya in the Russian Caucasus mountains. Built in Leningrad the huge instrument weighs about 850 tons and has a tube over 80 feet long. To make the mirror, a special furnace was built and the critical cooling stage took place over a period of two years. The massive device is housed on

Right: an artist's impression of the NASA space telescope, which will be put into orbit above the Earth sometime in the early 1980s by the Space Shuttle. The giant telescope will operate outside the Earth's atmosphere, enabling scientists to view distant objects with unprecedented clarity.

Below: an astronomer at the eyepiece of the 48-inch Schmidt telescope of the Royal Observatory, Edinburgh. It has an automatic photoelectric guider that enables the telescope to follow the stars with great precision for many hours at a time – a feature, which with many others, makes it one of the most advanced types of telescope in the world.

an electronically controlled mounting that can swing the telescope through both horizontal and vertical directions at the touch of a button.

The world's major observatories are usually built on the slopes of mountains as far from cities as possible. This makes sure that no glare from city lights interferes with observations and that the atmosphere is as clear as possible. But despite their remoteness, observatories are themselves like small cities, filled with a constant bustle of activity. There are living quarters, libraries, lecture rooms, laboratories, machine shops, and storage facilities. Astronomers are constantly coming and going, spending a few days using the telescope before returning to their homes to spend months or even years analyzing their data.

But however sophisticated modern telescopes have become they are all hampered by having to peer through the dust-laden air of the Earth's atmosphere. One of the most exciting prospects for observational astronomy is the advent of space telescopes, either mounted on orbiting space laboratories such as America's Skylab or orbiting as an individual satellite on their own. With no atmosphere to interfere with the clarity of their observations this new generation of spaceborn telescopes may become an important advance in astronomy, enabling Earthbound scientists to examine the remotest regions of the universe in previously unimagined detail.

Radiotelescopes

When we talk about telescopes we usually mean instruments that "see" by bending and focusing ordinary light. But visible light is only a small fraction of the different kinds of radiation that make up the electromagnetic spectrum. Nowadays a whole generation of telescopes have been devised that use radio waves instead of light to build up images of the universe. These instruments are constructed from huge reflecting aerials that focus radio signals on to extremely sensitive receivers.

The first radio telescope was built almost by chance. In 1931 a young research engineer named Karl Jansky was working in the United States for the Bell Telephone Laboratories. He was investigating "static," an annoying crackling sound that often affects the reception of radio signals. Jansky's experimental equipment was laughingly called the "merry-go-round" by his colleagues. It was a strange contraption looking like the skeleton of an aircraft wing which was rotated on a movable platform.

Jansky picked up plenty of static but he also detected something far more

Above: Karl Jansky with his directional radio aerial system – the precursor of the radio telescope.
Below: Arecibo Observatory, Puerto Rico, the world's largest radio telescope.

interesting. His instrument picked up a steady hiss apparently coming from a particular point in the sky. Eventually Jansky worked out that he was receiving a radio signal coming from the Milky Way, somewhere in the constellation of Sagittarius.

In 1937, using Jansky's findings, an American engineer named Grote Reuber built the first true radio telescope. It used a dish-shaped aerial over 31 feet across to collect radio signals. In addition to confirming Jansky's findings, Reuber located other radio sources in the constellations of Cygnus, Cassiopeia, Canis Major, and Argo.

It soon became clear that by listening to radio emissions from space an entirely new kind of observational astronomy was possible. In Britain one of the most important pioneers of the new type of telescope was Bernard Lovell. He was convinced that radio astronomy could become as important a branch of science as ordinary optical astronomy. But Lovell faced enormous problems. He was determined to build a really large instrument with a dish aerial over 250 feet across. In addition to the design problems involved, Lovell found it hard to raise the money needed to build the instrument. It took 10 years for his dream to be realized. Then in 1957 the giant telescope was completed. Today at its world-famous site at Jodrell Bank near

Manchester, the telescope is to radio astronomy what the 200-inch Mount Palomar reflector has for long been to optical astronomy.

The Jodrell Bank dish focuses radio waves just as an optical telescope collects and focuses light. Its focal point is a 60-foot-long rod of metal called a dipole that extends from the center of the dish. The faint signals from space are focused onto the dipole, which is connected to amplifiers and recorders.

Not all radio telescopes use the familiar dish-shaped aerial. Another type – called a Mills Cross for its Australian inventor, Bernard Mills – consists of two rows of aerials arranged perpendicularly to each other. The largest aerial in the world is a unique kind of dish. At Arecibo in Puerto Rico a large natural hollow in the ground has been bulldozed into a more or less smooth, dish-shaped crater. Its surface, lined with reflecting materials, forms a vast aerial over 1000 feet across.

Radio telescopes have led to many major breakthroughs in astronomy. One of the best known was the discovery of pulsars in 1967. Cambridge radioastronomer Jocelyn Bell detected a quickly varying radio signal made up of regular pulses 1.3 seconds apart. At the time, the discovery caused great excitement. The pulses were so regular that for a short while some people thought they might be signals from an alien civilization. The truth was scarcely less startling. Pulsars are now known to be neutron stars – fantastically dense remnants left over from supernova explosions that have wrecked once normal stars.

Other remarkable discoveries include the mysterious object Cygnus A, one of the first sources of radio waves discovered to lie far outside our own galaxy. The second strongest source of radio waves known, Cygnus A presented its discoverers with a remarkable dilemma. How could a galaxy 550 million light-years away from us produce such a tremendously strong signal? What could be generating such a violent flux of radio energy? The mystery remains unsolved even today although Cygnus A is generally believed to belong to a special class of galaxy suspected of being in the grip of some cataclysmic upheaval.

Even more puzzling than the discovery of Cygnus A is the mystery of quasi-

Above: the 200-foot dish aerial of the Jodrell Bank radio telescope in Britain.

Below: Cygnus A in the constellation Cygnus, 550 million light-years away. Its powerful radio emissions were first detected in 1946.

stellar sources, "quasars" for short. They appear to be among the most distant objects in the visible universe – some are estimated to be thousands of millions of light-years away – and yet are extremely luminous. They were first detected by their radio emissions which, even though they came from the farthest reaches of the universe, were strong and unmistakable. Many have now been linked to visible objects. Radio measurements suggest they are comparatively small bodies, perhaps no more than a light-year across. No one yet understands how compact objects can possibly produce enough energy to be detectable across such vast distances. Quasars, the detection of which remains perhaps radioastronomy's greatest triumph, are one of the most profound, unresolved mysteries in the universe.

Time and distance in astronomy

One of the most awesome features of astronomy is the scale of the universe it reveals. In their everyday lives people become familiar with simple assessments of time and distance using convenient and often traditional measuring rods. For example, the passing of time is marked by using units based on the movement of the Earth around the Sun and the turning of the Earth on its axis. For centuries these units have been important to mankind because they mark the changing seasons and the transition from day into night time. Individuals make subjective judgments about what is a "long" interval of time and what is a "short" one. But broadly speaking a human lifetime, say, is thought of as a long period while the time it takes to eat a meal is, for most people, a short period. Similarly with distance, measuring rods are in readily intelligible units such as inches, feet, and miles. Subjective assessments are made such as "10 miles is a long way to walk, but 1000 miles is a short distance to fly."

All this is good enough for the purposes

Right: the Moon, photographed at 15-minute intervals with a fixed camera. The apparent motion of the Moon over such a short period is due mainly to the rotation of the Earth on its axis, and not to the real motion of the Moon.

Below: the Jantar Mantar Observatory in India. Positions of stars and planets were measured with instruments used with the naked eye. Observers could predict the positions of Sun, Moon, and planets even though the Earth was taken as the central body.

of everyday life. But in astronomical science the units and our grasp of them are almost useless. The time scales and distances that are a routine part of the astronomer's world have little in common with the daily lives of people on Earth. But in order to comprehend the processes that govern the lives of stars and galaxies, astronomers must discipline themselves to have a very special perspective of time and distance. For example, they must be as ready to accept events taking a millionth of a second – such as the break-up of a subatomic particle – as the protracted life processes of stars that stretch over millions of years.

But in working with astronomical distances, astronomers must actually seek new units altogether. On a cosmic scale, the concept of inches, feet, and even miles becomes hopelessly inadequate. Of course any distance can be expressed in miles, but the numbers become so huge that an astronomy textbook would be reduced to long strings of digits.

While astronomy was largely confined to studying only the solar system in any detail it was natural to choose a measuring rod linked to the size of the solar system. The most obvious choice was the average distance of the Earth from the

Sun, a unit astronomers still call one Astronomical Unit or AU, equal to about 93 million miles. Using this measuring rod, the most remote planet in the solar system, Pluto, swings in a highly eccentric orbit which takes it as far as 49.3 AU from the Sun and no closer than 29.6 AU.

Once astronomers look beyond the outer limits of the solar system they encounter gulfs so great that even Astronomical Units become too small to use as measuring rods. New units are needed and the most widely used today takes as its measure of a distance the time it takes for light to travel that distance. Now light travels extremely fast – about 186,000 miles every second – so even a

Below: the Dumb Bell nebula in the northern constellation Vulpecula (Little Fox). It was in 1924 that Edwin Hubble, the famous American astronomer, discovered that the nebula known as Andromeda was not in fact a luminous, gaseous cloud within our own galaxy, but an island universe about 1.5 million light-years away. Astronomers began soon to realize that elsewhere in the universe there were immense star complexes, many greater than our own galaxy.

light-year is roughly 6 million million miles – almost inconceivably large compared to distances on Earth.

But even the stars seem close when we look beyond them into the immensity of intergalactic space. Today astronomers can make fairly detailed observations of galaxies and galaxy-like objects up to 9000 million light-years away. This means that the light now reaching us must have begun its journey some 6000 million years after the universe was formed, an event that occurred about 15,000 million years ago. These staggering distances make it possible to look back into the remote past to a time when the universe was young.

Left: the spiral galaxy NGC4565. It is classed as an Sb type, viewed edge-on, with its spiral arms seen as a disk around the nuclear bulge. There are many types of galaxy, and at the moment scientists still do not have enough data to explain the evolution of these gigantic star systems.

light-minute is a very large distance. But because the stars are so fantastically remote, astronomers need to use light-years to describe the distances between them. For example, light from the nearest star to Earth takes over four years to reach us. In everyday terms a single

Some astronomers believe that in that far off era the physical processes taking place in galaxies were quite different to those occurring around us today. Perhaps future studies of the most distant objects we can detect will someday reveal how the universe was formed in the first place and how it has evolved over the millennia. Ultimately time and distance in astronomy unite to provide a powerful tool with which to probe the most fundamental mysteries of nature.

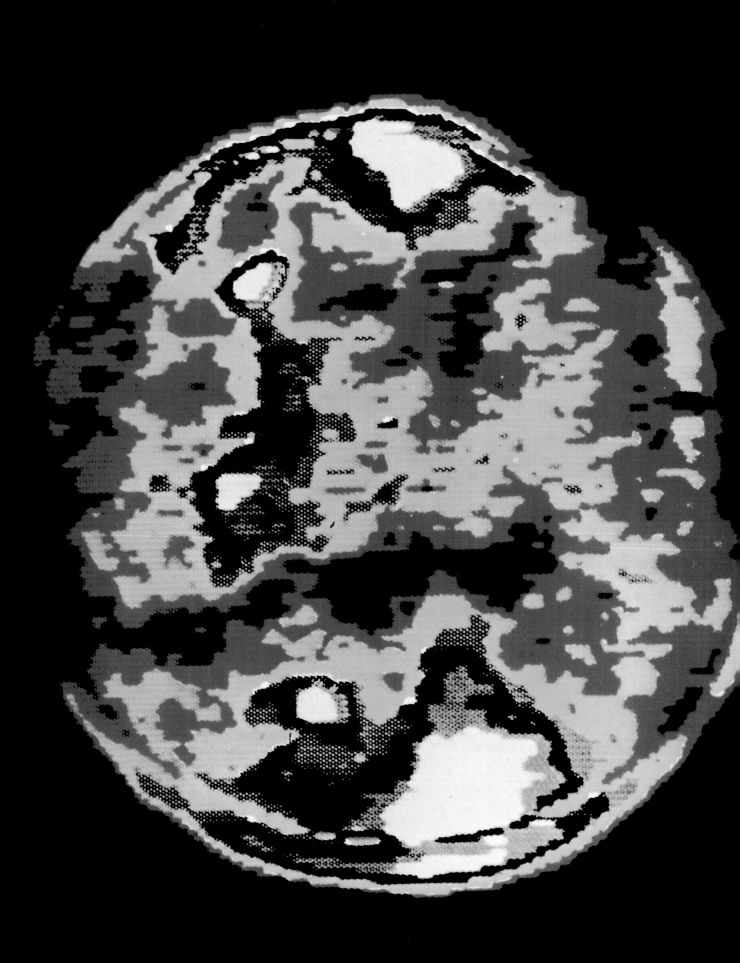

Chapter 2

The Sun and Stars

The origins of modern astronomy lie in people's fascination with the stars and their distinct patterns in the night sky. It is now known that the stars are seething masses of incandescent gas, fueled by violent thermonuclear reactions. Our own sun itself is a star, and it and the other remote and seemingly changeless stars are in reality gigantic powerhouses pouring huge amounts of radiant energy into space.

All stars pass through a life-cycle in which the forces of gravity and radiation pressure contend with each other. To a star a human lifespan is the briefest flicker of time. But although stars flourish for millions of years, they have to face death as surely as any living creature.

The variety of stars – their colors, sizes, and luminosities – makes them a rich source of study for astronomers. But their fascination is heightened by the thought that around some of them planets are orbiting on which intelligent life may have evolved. The stars are distant suns, constant reminders that in the immensity of the universe, mankind is not alone.

Opposite: a color-coded television picture of the Sun. By using specialized instruments such as this to aid their study, solar scientists are able to unravel some of the mystery that still surrounds the workings of the Sun. Because the Sun is a star, the more information that can be gained about it, the more knowledge will be gained about stars throughout the universe.

The Sun

Long before the first rational speculations and the birth of scientific method, the Sun was worshiped as a powerful god – the source of all life on Earth. To the Egyptians the Sun-god was Amen-Ra, to the Greeks, Phoebus-Apollo, and to the Peruvian Incas, Inti, who became incarnate in their king.

The first known scientific speculations about the Sun began around 500 BC, when the Greek philosopher Anaxagoras measured the position of the Sun in the sky at various times of the day and concluded that it was a ball of fire 35 miles in diameter, floating in the air about 4000 miles above the Earth's surface. Anaxagoras' reasoning was sound but he based it on the assumption that the Earth is flat. A more realistic estimate of the Sun's size and position was made in 135 BC by Hipparchos, another Greek with a deep interest in astronomy. Hipparchos believed the Earth was a sphere and his calculations gave much more accurate results. He worked out that the Sun was approximately 10 million miles away with a diameter of just under 50,000

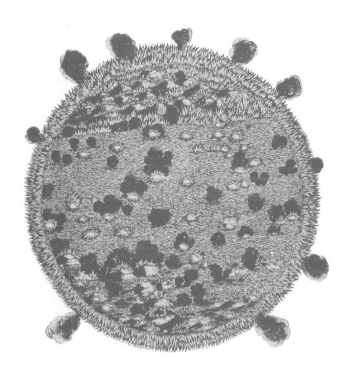

miles. Although far smaller than the true figures, Hipparchos' results remained unchallenged until only about 200 years ago. Then, thanks to the invention of the telescope and the development of precise astronomical instruments, it was possible at last to uncover the Sun's secrets.

It is now known that the Sun is just one of countless millions of stars in the universe and like them is an immense thermonuclear inferno. The Sun is about 93 million miles from Earth. Its diameter is 864,950 miles. Besides being the central body in the solar system, it is also the most massive, containing 99.87 percent of all the matter in it. The planets, their moons, and the asteroids account for the other 0.13 percent.

The one point upon which modern scientists and the Ancients agree is that the Sun is the source of all life on Earth. The vast outpouring of radiant energy constantly bathing the surface of the Earth, has enabled life to evolve and prosper. But the difference between life-giving radiant energy and lethal radiation is quite small. It would take only minor changes in the Sun's output to destroy all life in a fraction of the time it has taken for it to evolve. It is not surprising, therefore, that modern astronomers have made a careful study of the Sun and now know a great deal about its structure. Because it is relatively close, it also provides a unique opportunity to build up

Opposite top left: the prehistoric stone-age circle at Stonehenge, England, which was built between 1800 and 1200 BC may have been used to predict eclipses or measure the times of solstices – which would indicate that the builders had a surprisingly good knowledge of astronomy.

Opposite bottom: an Aztec temple sacrifice to the Sun god Huitzilopochtli. In the 1500s the Aztecs of Mexico had developed an obsessive worship of the Sun that demanded tens of thousands of human sacrifices every year.

Above left: Europeans, having long ago abandoned the Sun as a god, began increasingly to turn a more scientific eye on it. The German Jesuit astronomer Christoph Scheiner studied the Sun by projecting its image onto a screen with a telescope – which is still the best way to study it. In this way he discovered the existence of sunspots, in 1611, independently of Galileo.

Above right: an engraving of the Sun made in 1635 shows it as a burning solid body. It also shows the sunspots, the outer corona, and many solar flares.

an understanding of the behavior of a star, knowledge that provides valuable insights into the nature of the universe as a whole.

Astronomers cannot observe the Sun directly through their telescopes. To do so would invite terrible injuries to their eyes and almost certain blindness. Instead, they arrange for their telescopes to project an enlarged image of the Sun on a screen in a darkened viewing room. Careful studies have revealed that the Sun has no truly solid regions. It appears to be a ball of gas, although in its center gravitational pressures may be so great that the gas becomes a semisolid. It has a luminous yellow surface, known as the photosphere, a 290-mile thick layer of gas – mainly hydrogen and helium – with a temperature range varying from 18,000°F nearest the surface to about 7560°F at the top. In earthly terms, these are extremely high temperatures. But, closer to the core of the Sun, the temperatures make the surface layers seem as cold as an icebox. Calculations show that the mass of all the material making up the Sun is being squeezed inward on its center by the force of gravity to produce immense pressures thereby raising the core temperature to well over 27,000,000°F.

Above the low-temperature photosphere lies a 3000-mile thick shell of pink-tinted gas that ranges in temperature from 7560°F to 1,800,000°F, known as the chromosphere. The outer and most extensive region of the Sun's atmosphere is the corona or crown, made up of streamers, filaments, and rays of pearly white light. Ironically, although it is the Sun's most spectacular and beautiful feature, it is hardly ever visible. The gas that makes up the corona is very hot – about 3,600,000°F – but very thinly spread. Because of this it is far less luminous than the main body of the Sun and pales into invisibility under normal circumstances. However, one quite rare astronomical event gives us the chance to see the corona in all its glory. A total solar eclipse occurs when the Sun, Moon, and Earth are lined up so that the Sun's disk is exactly obscured by the smaller but much closer disk of the Moon. Usually observable only from restricted regions of the Earth, a total solar eclipse is an awesome sight even for experienced astronomers. At about the moment the Sun's disk vanishes, the incandescent corona suddenly appears as a fanlike structure of light extending as far as 10 solar diameters into space.

Because of the very high temperatures of the corona, hot ionized gas, or plasma, is forced from the surface of the Sun and out into space. These coronal particles reach Earth as a "solar wind" at speeds between 200–600 miles a second.

By using modern, satellite-based telescopes to examine the Sun close up, scientists have discovered that its surface has a mottled, grainy appearance. It can be seen as a seething mass of gases in constant turmoil. Frequently, pockets of gas deep within the photosphere become hotter than their surroundings and rise through the nearby layers to break the surface as a brightly glowing region. They then cool, darken, and fall back into the interior.

Sunspots themselves are probably whirling vortices of gas caused by the

complicated flows and circulations of gas within the Sun. They look black only because they are slightly cooler than their surroundings. In fact they are extremely luminous. If the Sun's surface could be obscured except for a good-sized group of spots, they would illuminate the Earth with the brightness of 100 moons.

The detailed structure of sunspots is very complicated. In general they always have a dark central region known as the umbra separated from the rest of the solar surface by a lighter ring known as the penumbra. The boundaries between umbra and penumbra are irregular and fast-changing. Sometimes wing-shaped formations extend almost to the center of the spot. Occasionally, glowing bridges cross the spot from edge to edge.

The size of sunspots varies enormously.

Above: a photograph of solar corona which can be seen in this total eclipse. The form of the corona is related to the mysterious 11-year sunspot cycle. This photograph was taken in 1918, four years after the cycle had reached its 11-year peak.

Below: this graph shows the fluctuations of sunspot activity over the last two centuries. Sharp peaks in activity occur every 11 years with striking regularity.

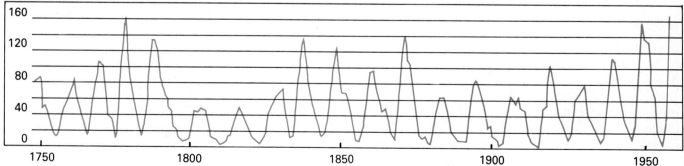

The smallest – known as pores – are only a few hundred miles across. Others have diameters of tens of thousands of miles. Usually, they occur in pairs or small groups, carried slowly across the face of the solar disk by the Sun's rotation about its axis. Sunspots can generally be observed at any time but there seems to be a mysterious time cycle controlling their numbers. Every 11 years their number builds up to a maximum, increasing from perhaps a handful to several hundred. Following the 11-year peak, their numbers begin to decrease. In 1947 a particularly spectacular peak was reached and a single grouping of sunspots appeared covering about 6000 million square miles of the Sun's surface.

Right: a close-up of part of the photosphere. Dark areas are boiling gases nearly 700 miles across and 200 miles in depth.

Below: the spots seen here on the face of the Sun are sunspots – huge whirlpools of gas brought about by the whirling and tunneling action deep inside the Sun. When these whirlpools reach the photosphere, they burst out in the form of spots that look black because they are cooler than their surroundings.

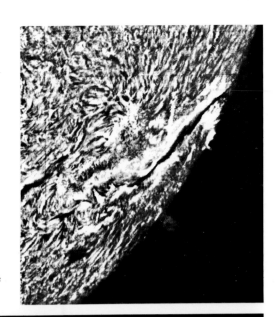

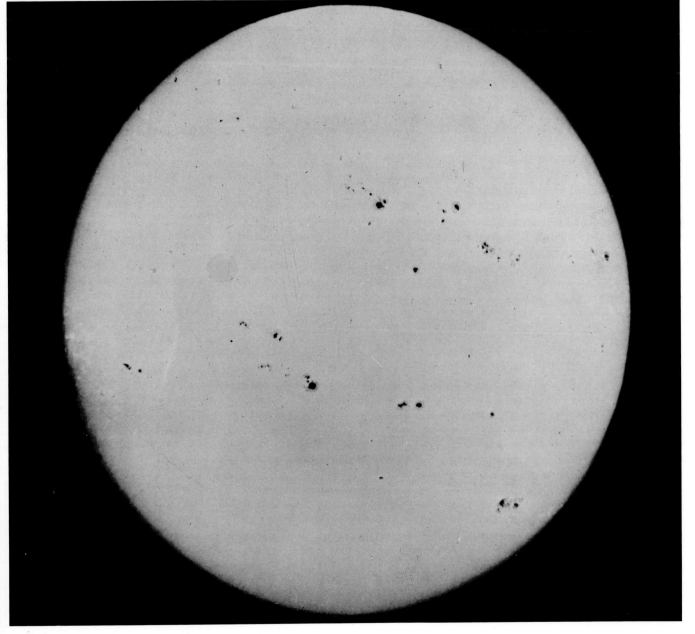

No one knows why this 11-year cycle occurs but it seems to be closely linked to other major upheavals on the solar surface that build up to a maximum at about the same time. Most spectacular of these are *solar prominences* and *flares*.

Prominences are huge flame-shaped masses of hot gas that erupt from the Sun rising over 100,000 miles above the solar surface. Flares are gigantic superheated areas that occur near sunspots. They often cover several million square miles of the surface. One current theory accounting for the flares likens their formation to the bursting of a kind of energy dam. The idea is that the high sunspot activity leads to localized formations of solar energy. These "bubbles" of exceptionally high energy are contained for a time by magnetic fields in the Sun's gaseous atmosphere. But at a certain point, the energy breaks free of its restraints. During the five to ten minutes a flare is at its maximum brilliance, it sends out a tremendous surge of radiation together with a burst of charged particles. If the flare is in an appropriate position, lined up with the Earth, the radiation and particle emission strikes the atmosphere producing spectacular effects. In the polar regions, aurora – bright discharges in the upper atmosphere – become especially magnificent. More important, the air layers we rely on to reflect radio waves become saturated with charged particles. This causes static interference in radio transmissions and sometimes a total blackout. Also, especially violent electrical storms are known to increase a day or two after a solar flare.

Sunspots – and the solar wind – affect more than the upper reaches of the Earth's atmosphere, however. Many remarkable correlations have been discovered that suggest these phenomena have a profound effect on our environment. Examination of the rings formed annually in the trunks of trees shows that trees grow faster during years of high sunspot activity. This also seems to be true of wheat, because the smaller the number of sunspots in any year, the higher wheat prices become. It is even possible to link sunspot cycles to political upheavals although a fairly inconsistent picture emerges. For example, the French Revolution of 1789 and the Russian Revolution of 1917 both occurred at times

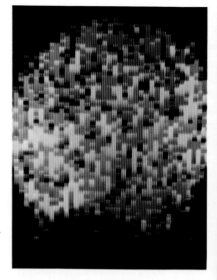

Above: a record of the Sun's x-rays taken by Skylab's Apollo telescope mount, which was fitted especially to observe the Sun. Such pictures show scientists that the corona of the Sun is far more complex than has been thought until now.

Right: a spectacular solar eruption. By color-coding the television pictures obtained from the Skylab, scientists are able to study the make-up of the Sun and gather much new data. The eruption here sent an arch of intensely hot rarefied gas some 500,000 miles into the solar corona.

of maximum solar activity. The American Revolution of 1776, however, coincided with a minimum.

Modern scientists tend to take the increase in growth of plants and crops more seriously than a possible link with social crises. Some evidence exists indicating that sunspots affect the Earth's weather which, in turn, affects growing patterns. Usually the solar wind is deflected by the Earth's magnetic field. But

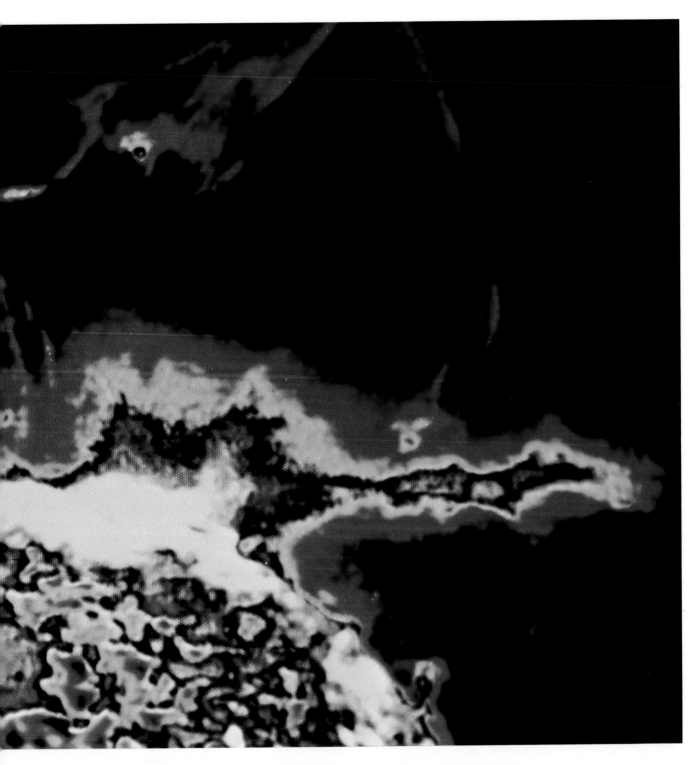

there is evidence suggesting that on rare occasions the entire field suffers a total reversal of direction. When this happens the protective magnetic shield around the Earth weakens. Perhaps for a period of several centuries, the tiny electrically charged particles of the solar wind are free to pepper the atmosphere. One possible side-effect of this is a depletion of the vital layer of ozone in the upper air, the gas that normally absorbs ultra-violet radiation in the Sun's rays. With the strength of this important filter temporarily reduced, the Earth's surface would be drenched with a high concentration of ultraviolet radiation. There is evidence that whole species of animals may have become extinct during these periods.

Perhaps the most fundamental question about the Sun is: how does it sustain its vast output of energy? More important still for mankind is the question: how much longer can the Sun survive?

Geologists know that the Sun must have been shining with its present intensity for at least 500 million years. This is the time separating us from the earliest eras to leave recognizable fossils of living organisms. Nowadays, we can be fairly sure that the Sun has remained more or less unchanged since the solar system formed, over 4700 million years ago. The Sun is radiating energy at a rate equivalent to burning about 10,000 million million tons of coal every second. In itself this is a staggering amount of energy but the idea of such an output

Below: a cross-section of the Sun. *A* denotes the corona, the Sun's extensive outer atmosphere. *B* represents the Sun's chromosphere. The upper edge of the chromosphere is uneven and is made up of jets of high-temperature gases called spicules. In the interior, the heat is convected through the outer layers (*C*). Below these layers, heat probably travels as pure radiation (*D*). It is in the Sun's core (*E*) that the nuclear transformations take place giving rise to temperatures in the region of 27,000,000°F.

being sustained over millions of years is almost unbelievable. Clearly the source of such colossal stores of energy cannot be a chemical fuel like coal.

When we look at the Sun all we see is the activity on its surface. The turmoil we observe is only a pale reflection of the much more fundamental cataclysms taking place closer to the Sun's core. Since the early 1900s when Albert Einstein showed that matter can be converted directly into energy, scientists suspected that the extremely high temperatures of the solar core might be causing this to happen. Einstein's work showed that a fairly small quantity of mass could transform into a huge amount of energy. Using this principle, it is easy to calculate that the Sun is losing about four million tons every second, barely noticeable in such a gigantic object with a total mass of about one million, million, million, million, million, million tons!

The exact details of what is going on deep inside the Sun was not fully understood until after World War II when the principle of nuclear fusion was discovered. We now know that the chain of nuclear reactions depends on the direct conversion of hydrogen to the next-heavier element in nature, helium. This "hydrogen burning" is quite different to ordinary burning in which atoms stick together and release energy. In solar nuclear processes, hydrogen nuclei are fused directly with each other making entirely new atoms of helium. In the course of this fusion some mass is converted directly into energy creating the inferno on which we all depend for survival. The life process of the Sun is the same one that supplies the energy of the hydrogen bomb. The Sun is therefore equivalent to the controlled explosion of 10 million million large hydrogen bombs every second, running continuously for countless centuries.

Can such a massive powerhouse keep running indefinitely? It must fall low on fuel at some stage. The answer, of course, is that everything, including the universe itself, has a limited lifetime. However, the Sun is likely to maintain its present energy output for several thousand million years more before its hydrogen fuel is seriously depleted. Even when this happens it is possible that temperatures may rise high enough to allow the secon-

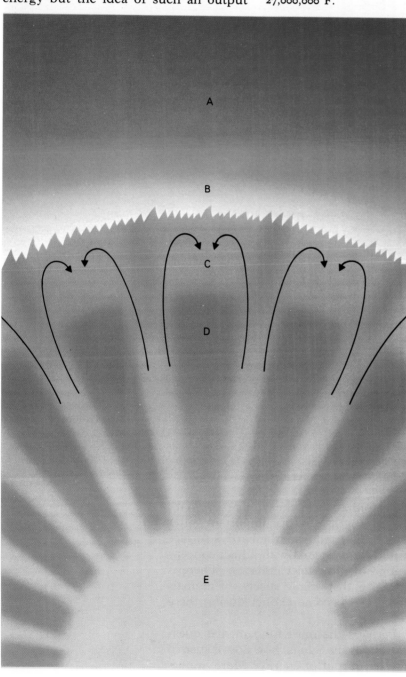

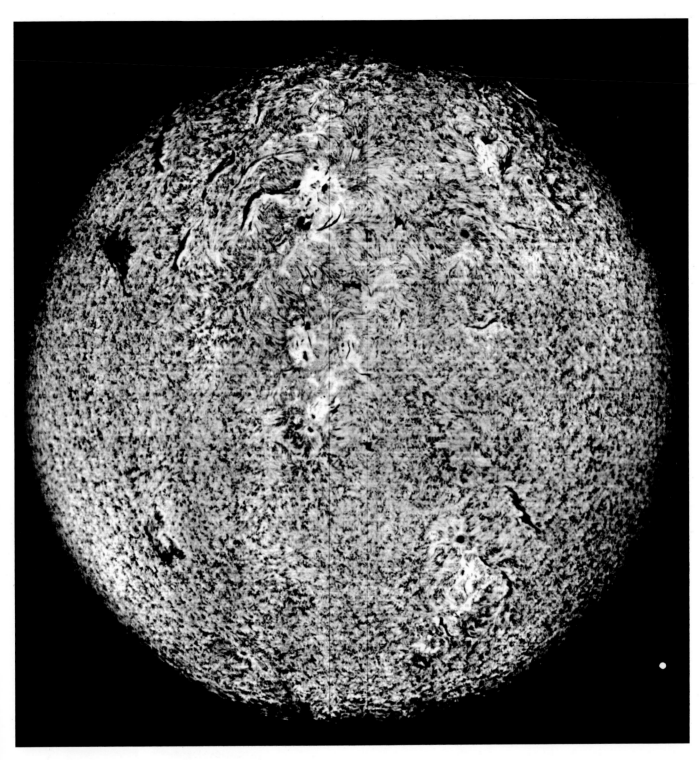

Above: the Sun photographed in hydrogen light and showing the distribution of hydrogen on its surface. The photograph suggests the unceasing violence of the solar surface and the even greater violence deep within the Sun's core.

dary burning of helium to begin, so extending the Sun's existence still further. But the life cycles of stars are now fairly well understood. Astronomers know that stars not only form and evolve. They also die. The extinction of the Sun is therefore an inevitable catastrophe that will signal the end of life in our solar system. All being well, however, there are many thousands of millions of years still ahead of the human race and since we are now only at the dawn of our existence, we may well have spread human civilization throughout the galaxy by the time the Sun dies. If this is so, the Sun's passing may be only a minor loss to us. By that time, mankind may even have forgotten its remote origins. The destruction of the third planet in the course of the Sun's death would probably pass without comment in the galactic empire that may by then be flourishing.

Observing the stars

The stars visible in the night sky are other suns, their apparent dimness due to their extreme remoteness. The nearest star, in the constellation Centaurus, is so far away that its light takes over four years to reach Earth. But despite the vast distances that separate the stars from the Earth and from each other, careful observations of them have provided a wealth of knowledge about how they were formed, how hot they are, and the elements from which they are made.

Anyone who looks at the night sky will notice that stars differ in brightness. As long ago as 130 BC, the Greek astronomer Hipparchos cataloged the stars by simply assessing their relative brightnesses. He divided them into six magnitudes, the brightest he could see being of the first magnitude and those barely visible, of the sixth. Modern astronomers still use a version of Hipparchos' system, although refined so that the scale of magnitudes can be extended in either direction. The dimmest stars known today are of magnitudes 22 and 23 and can be detected only on photographs taken by the most powerful telescopes. The brightest star apart from the Sun is Sirius with a magnitude of −1.4. If the Sun were measured against the same scale, its magnitude would be −27.

The scale is purely comparative. It is arranged so that there is a difference of 2.5 times between each magnitude. This means that a star of the first magnitude is 2.5 times brighter than a second magnitude star. A difference of five magnitudes therefore amounts to a difference in brightness of 100 times. But the system tells us only about the apparent brightness of the star and gives us no idea of how luminous a star actually is. Sirius, for example, may be an extremely luminous star seen from a great distance or merely an average one quite close by. If the distance of stars can be estimated, their true brightness can easily be worked out by correcting their apparent bright-

Right: the diagram depicts the types of radiation that penetrate the Earth's atmosphere through the optical and radio "windows." The more harmful radiation, such as ultraviolet, x-ray, and gamma rays are screened out. Nevertheless, all the radiation that reaches across space from the Sun and other stars has to be studied, which is why probes, rockets, and satellites are essential to research.

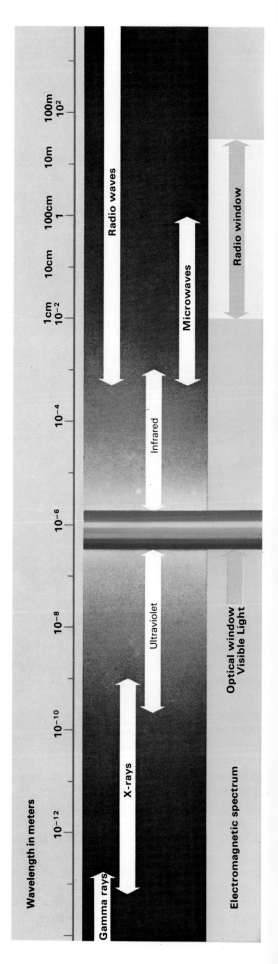

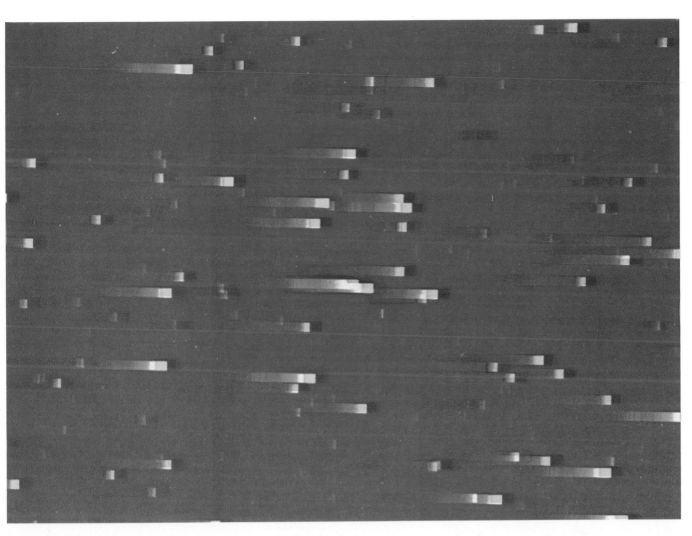

ness with a simple mathematical factor.

One of the most important techniques for measuring stellar distances makes use of an everyday phenomenon called *parallax*. To see how parallax works, hold your thumb about six inches from your face and sight along it to some object across the room using your right eye only. Without moving your thumb, sight along it with your left eye. Your thumb will have appeared to have shifted to the right of the object with which it was first lined up. The shift is called parallax. By measuring the amount of the shift you can work out the distance the object is away by relating it to the distance between your eyes.

In order to observe parallax in anything as distant as a star, sightings have to be taken at two widely separated points. On Earth the greatest separation that can be achieved is to make observations from opposite ends of the Earth's orbit around the Sun. This creates a baseline of about 186 million miles. As

Above: starlight from the Hyades open cluster in the constellation Taurus, as viewed through a spectroscope. It is one of the most conspicuous star clusters in the sky and is about 135 light-years away. Spectroscopy is valuable tool for astronomers, enabling them to obtain information about objects that are in most cases billions of miles away. They can tell, for example, a star's temperature, the elements it contains, what kind of magnetic field it possesses, and its speed of rotation.

seen from each end of this, a comparatively nearby star will appear slightly displaced against the background of apparently fixed and therefore more distant stars.

The nearest star, Proxima Centuri, a magnitude 11 star seen in the Southern Hemisphere, has a parallax of under 1/4800 of one degree, less than a second of arc. To understand how small this is, remember that a circle is divided into 360 degrees, each degree into 60 minutes and each minute, 60 seconds of arc. A complete circuit of the sky is well over a million seconds of an arc. A second is therefore a tiny displacement, equal to about one two thousandth of the Moon's diameter. Even Jupiter, which looks like a mere point of light, measures over 30 seconds of arc.

Despite the difficulties of observing such minute parallax, astronomers have been able to make good estimates of stellar distances from the technique. The scale of distance they encountered, however, was so vast they had to devise special units. In much of the English-speaking world the most common measuring rods are one yard long. These small units are built into larger units such as miles for measuring geographical distances. It is obviously more convenient to measure the Earth's diameter as about 8000 miles rather than 14 million yards. In interstellar space, all terrestrial units become meaningless. Distances are so vast that the measuring rod has to be the distance traveled by light in a given amount of time. The most common unit, of course, is the light-year, the distance light travels in a year. An alternative unit is the distance for which the parallax seen from Earth is one second of arc. This unit, called the parsec, turns out to be equal to 3.26 light-years.

Using this system of measurement, Proxima Centuri is 4.3 light-years distant, nearly 7000 times as far from the Sun as the planet Pluto. The second closest star, Barnard's Star, is about 6 light-years away. Sirius, the brightest star visible, is 8.6 light-years away. Once the actual distance of a star is known, astronomers can easily calculate its true luminosity or, conversely, determine how bright it would appear at any distance.

Below: one method of determining whether objects in the universe are approaching, receding, or remaining stationary in relation to the Earth's position is to observe light from celestial objects through a prism. Light from a moving source shows itself in a shortening of wavelength in an approaching object, a lengthening of wavelength in a receding one. Light displaced toward the red end of the spectrum means that the source is receding; toward the blue end, that the source is approaching. In the diagram below *A* illustrates a light source from an object traveling at the same speed as Earth. The spectral lines are in a normal position. *B* illustrates a light source approaching Earth; the spectral lines shift to the blue end of the spectrum. *C* is a light source receding from Earth; the spectral lines shift to the red end of the spectrum.

Astronomers compare the luminosities of stars by imagining that they are all 10 parsecs away. The brightness of a star at this standard distance is called its absolute magnitude. Using this system of comparison our Sun has an absolute magnitude of 4.9 while Sirius, more than 23 times as luminous, has an absolute magnitude of 1.5. The star Deneb with an absolute magnitude of 1.26 is about 47,000 times more luminous than the Sun.

More still can be learned from starlight by splitting it into a spectrum of different wavelengths. The existence of certain characteristic spectral lines can tell the observer of the presence of well-known chemical elements. The spectroscope, the commonest device for recording these spectra, has become a basic tool of the astronomer enabling him to work out stellar compositions even from distances of hundreds of light-years.

This kind of spectroscopic measurement is similar to a color analysis since the color of light is due to the wavelengths it contains. It is possible to link the overall color of starlight with the surface temperatures of stars. For example the hottest stars are bluish, indicating temperatures of up to 72,000° Fahrenheit.

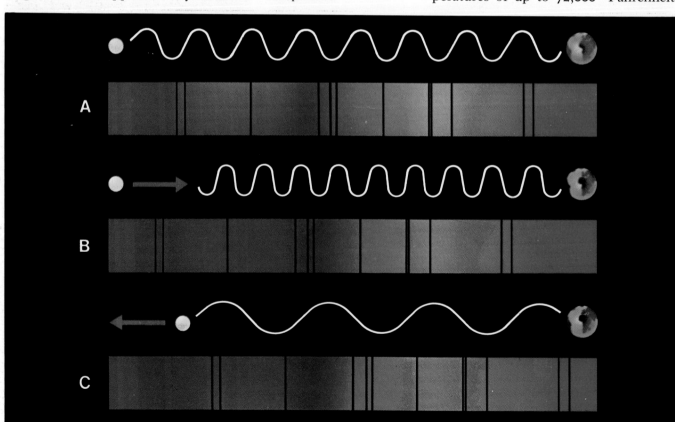

Reddish stars are the coolest, their surfaces usually being little more than 5400°F. The Sun is a predominantly yellow star with a surface temperature of around 10,800°F.

Astronomers often use a series of letters known as the *spectral sequence* to classify stars by their distinct types of spectra. These different spectra are linked with temperature and are identified by letters of the alphabet. In order of decreasing temperature the letters are: O, B, A, F, G, K, and M. Within each main class of stars, there are subdivisions indicated by numbers ranging from 0 to 9. Using this system, the Sun is a class G2 star. Other G-type stars will therefore have a similar surface temperature to the Sun, but be different in size and composition, depending on their numerical classification.

The spectra of stars can also reveal which way the star is moving in relation to the Sun. As long ago as 1867, the British astronomer William Huggins, while making a spectroscopic study of Sirius, discovered that the spectral lines were all slightly out of place. They appeared to have all been shifted uniformly toward the red end of the spectrum.

This so-called *red shift* is much like the change in pitch heard as a fast-moving source of sound passes a listener. As a through train approaches a station whistling to warn people on the platform that it is not stopping, the note of its whistle rises as it approaches, and falls as it passes and recedes. Similarly, a star receding from us at a high velocity has its light wavelengths "stretched" toward the red end of the spectrum. If it approaches us, the light is shifted toward the blue.

Calculations show that Sirius is moving away from the Sun at about five miles each second. Or it may be that the Sun is moving away from Sirius at that speed – there is no way of telling. The spectral shift requires only a relative motion of some kind.

Thousands of stars have so far been studied. All appear to be moving – some approaching, some receding. Velocities vary from five to 25 miles each second. These relative motions should be observable if we keep a careful watch on those stars that are not merely directly approaching or receding but moving

Right: because they are so remote, most stars have very slight individual or proper motions. The star with the largest proper motion is Barnard's Star, in the constellation Ophiuchus. It takes the star 180 years to cross a portion of the sky equal to the apparent diameter of the full Moon. These photos taken in 1894 (top) and 1916 (opposite), clearly show the motion. Barnard's Star is arrowed.

across our line of sight. The amount of movement each year is termed a star's proper motion. The star with the largest proper motion is Barnard's Star, named for the American astronomer Edward Barnard who discovered it in 1916. It moves 10.3 seconds of arc each year. Although only a very tiny displacement, most stars have proper motions of less than one second each year.

Proper motion accounts for two important phenomena. First, it explains why the constellations we see today are not those seen by the earliest astronomers. The seemingly fixed canopy of stars in the night sky is very slowly changing as the centuries pass. Second, the measurement of proper motion tells us that the Sun is itself moving and calculations show that its motion is just a part of an overall rotation of our entire galaxy. Our best estimate is that we are about 27,000 light-years from the galactic center and revolving around it at about 140 miles per second.

Spectra, Temperatures, and Brightnesses of Main-Sequence Stars

spectral type and typical spectrum

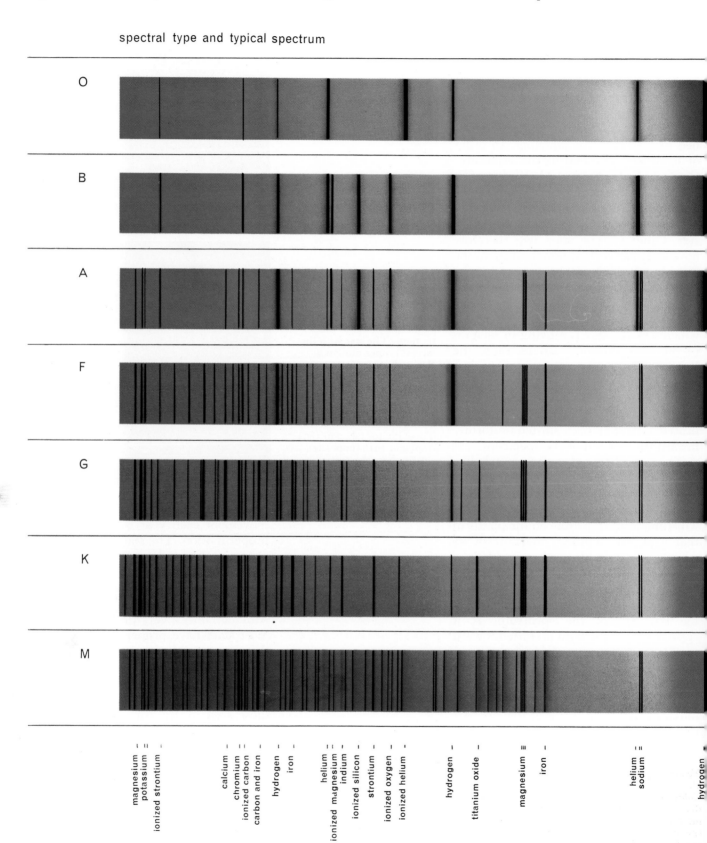

atoms producing main lines of spectrum	colour of main radiation	temperatures at surfaces of stars	brightness (absolute magnitude) of specimen stars	
ionized helium, neutral hydrogen and helium		35,000°–40,000°C	λ Cephei	−6·5
neutral helium; ionized silicon, magnesium, oxygen and nitrogen; neutral hydrogen		11,000°–35,000°C	η Orionis Alkaid	−5 −1·5
metals (especially calcium) giving weak lines; hydrogen giving very strong lines		7500°–11,000°C	Sirius Vega	−1·3 −0·5
metals (especially calcium) giving strong lines, hydrogen giving fairly weak lines		6000°–7500°C	Procyon	3
potassium giving a strong line; neutral metals (fairly strong lines) ; hydrogen (weak lines)		5100°–6000°C	Sun Capella	−5 −0·5
neutral metals giving strong lines; hydrogen (very weak lines)		3600°–5100°C	Arcturus 61 Cygni	0 8
molecules of titanium oxide (strong bands)		2000°–3600°C	Betelgeuse Antares many others	−5·6 −5·1 12 or 13

Above: visible light from stars tells the astronomer a great deal about their make-up. Color indicates roughly how hot the surface of that star is. In very bright stars, the most intense radiation, as observed with a spectrograph, is in the bluest part of the spectrum – which means that very bright stars must also be the hottest. Astronomers use letters (extreme left in diagram above) to classify stars, with the brightest and hottest stars falling into the O group and the least bright and hot into the M group (our Sun, incidentally, is a G group star). These seven groups are known as main sequence stars. Next to the spectral type is the typical spectrum. The lines that interrupt the band of color indicate the presence of chemical elements. They always fall in the same place for each element, which enables scientists to tell just what element a star contains. The intensity of the lines shows the predominance of the various elements.

The lifespan of stars

In the early 1900s two astronomers, a Dane, Ejnar Hertzsprung and an American, Henry Russell, invented perhaps the most useful way of charting the characteristics of different kinds of stars. They drew a graph that plotted the absolute brightness of stars against their colors. The bottom axis of the graph was made up of the spectral classes of stars in order of descending temperature, starting with spectral class O on the left and ending with M on the right. The upright axis plotted the brightness or absolute magnitude. Called a Hertzsprung-Russell diagram – or H-R diagram, for short – it provides a remarkable picture of the various types of stars in the galaxy.

In general, the hotter the star, the brighter it is. On the H-R diagram therefore, the farther left a star is in spectral class – and therefore the higher in temperature – the greater its absolute magnitude. Because of this most of the stars plotted fall into a roughly diagonal line stretching from the upper left to the lower right. This huge group, probably accounting for about 99 percent of all stars, is called the *main sequence.*

The fact that most stars are located on the main sequence suggests that they are all made roughly the same way – that is, the same sort of events led to their birth and account for their life cycles, which makes the H-R diagram an important key to our understanding of the evolution of stars.

The birth of stars is still cloaked in mystery. They appear to form in the depths of huge gas clouds or nebulae many of which can be easily observed from Earth. One of the best known is the Great Nebula in the constellation of Orion, a glowing, fan-shaped cloud of gas, 17 light-years across and 1500 light-years away. Nearly all the young, bright stars forming in the galaxy are situated on the spiral arms that extend from the dense galactic center. When they form out of the giant gas clouds, the new stars

tend to appear in well-defined clusters, sometimes numbering many thousands. They are often loose groupings like the beautiful cluster of several hundred stars making up the Pleiades, 490 light-years away in the constellation Taurus.

The process by which the stars form is probably a kind of condensation. The principal force behind the phenomenon is gravity, operating between the countless molecules of gas and dust motes in the nebulae. As an accumulation of matter builds up in a particular region of the nebula, the beginnings of a localized gravitational field emerge and the process accelerates, more material being drawn into the region by the gravitational pull of the increasing build-up of dust and gas.

After a few million years a rotating cloud of gas and dust will have formed. The most common ingredient of the embryo stars is hydrogen, the element that will provide the basic fuel enabling the star to burst into life. The spinning gas cloud is held together by its own gravitational field, pulling its surface layers toward the central core. The core itself is tightly compressed by gravitational pressure and as the pressure rises, its temperature begins to climb. When it reaches about 970,000°F nuclear fusion reactions begin and the star begins to glow.

During this phase, the nuclei of hydrogen atoms are being fused together forming nuclei of the next heavier element, helium. In the course of the fusion, a quantity of matter is completely annihilated, releasing huge amounts of energy. As the star's nuclear furnace begins to generate its vast output of heat and light, the outward force of all the radiated energy reaches a balance with the inward pull of gravity. In this condition the star is very stable and appears for the first time on the main sequence of the H-R diagram. The exact point at which the star enters is determined by its mass. An extremely massive star will join the sequence at the hot upper end because its large gravitational field will create much greater internal compression than a lighter star and therefore a much higher temperature. This also means that its fuel will be consumed more quickly. Because of this some stars – perhaps a hundred times more massive than the

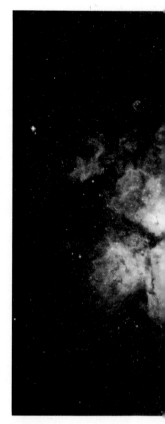

Above: the Keyhole nebula around the star Eta Carinae. This nova-like object exploded in 1843 and rivaled the brightest stars in the sky until it faded from naked-eye view some years later. It is now surrounded by glowing gas, which shows red in the photograph, and dusty regions that show blue in the reflected light of nearby stars.

Below: two early 20th-century astronomers, the American Henry Russell, and the Dane Ejnar Hertzsprung. Their famous "H-R diagram," which links a star's luminosity with its spectral type, is now one of the most important tools of astrophysical theory.

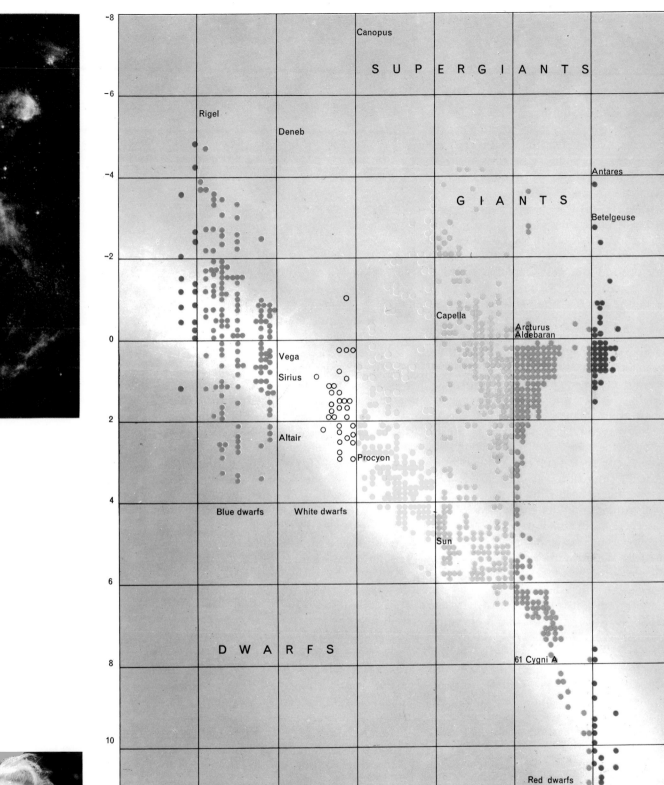

Above: the Hertzsprung-Russell diagram showing the positions of some well-known stars – 61 Cygni (a red dwarf), the Sun (a yellow dwarf), Procyon, Vega, Sirius, and Altair (main sequence), and some very luminous giants and supergiants (Arcturus, Aldebaran, Betelgeuse, Antares, Canopus, Rigel, Deneb). Capella is a typical yellow giant, with a spectrum like that of the Sun, but with 200 times the Sun's luminosity. A typical star probably evolved from main sequence to giant, then collapses into a small superdense white dwarf.

Sun – might have lifetimes as brief as 100,000 years.

Average stars like the Sun spend virtually all their lives around the center of the main sequence, steadily converting their hydrogen into helium. The Sun will remain in this stable condition for at least another 4000–5000 million years until the hydrogen in its core runs low. When this happens, the first critical stage marking a star's decline is reached and major changes begin to take place around its central region. The exact nature of the changes is extremely complicated but the key to what occurs is a build-up of helium in a kind of secondary core around the now depleted original core of hydrogen. No nuclear processes are taking place in the helium to oppose the inward pull of gravity and as the hydrogen fires begin to fail, the now massive core of helium begins to contract under its own weight. As the helium core shrinks, its temperature rises. But the effect of this inner change is to cause the outer gas layers of the star to cool and gradually expand.

As the surface swells and cools, its color turns to a dull red. The overall diameter increases spectacularly. The Sun in this phase of its life will become so distended that its outer layers may actually engulf the Earth. It will now have become a *red giant*. But before this occurs, a medium mass star like the Sun will also become increasingly luminous. As it cools it will radiate growing amounts of energy until it is hundreds of times as bright as it is today. Long before its physical expansion destroys the Earth, the increased levels of solar radiation will have melted the ice caps, and made life extremely difficult. Over millions of years the slow roasting of the Earth will continue until the oceans are vaporized and the entire planet becomes an arid, lifeless desert.

As the comparatively slow transition into a red giant is completed, pressure in the new helium core of the star reaches a critical level. When it is about 1000 times as dense as water, the atoms of which it is made undergo a sudden "stiffening" that prevents any further gravitational contraction. This curious stiffening, known as degeneracy pressure, stops the core shrinking and allows the gravitational energy to be converted directly into heat.

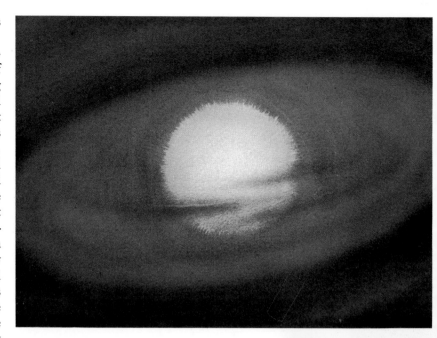

The core temperature soars. When it reaches a staggering 176 million degrees a new nuclear reaction starts. In the same way as hydrogen nuclei were once fused to form helium, helium is now itself fused together to produce nuclei of a still heavier element – carbon. But as soon as the helium burning begins, a spasm shakes the heart of the star. The core, stiffened by degeneracy pressure, cannot expand quickly enough to cope with the surge of new energy being produced and within a few minutes the temperature rises to such an extreme level that the entire core ignites. In an instant the star's energy output increases 100,000 million times in a kind of stellar flash fire.

Despite the violence of the event, all the energy produced is trapped by the outer layers of the gas surrounding the stellar core. Almost as swiftly as the spasm developed, it is brought under control. The vast energy output is safely dispersed and the core itself swells and a phase of stable helium burning begins.

In a star like the Sun this new stage in its life cycle carries it up the main sequence of the H-R diagram. Its luminosity increases a hundredfold and as its surface temperature rises its color changes from red to blue. The star becomes a *blue giant*.

Most stars with moderate mass will now pass through a number of complex stages. Gradually, inside the star, numerous cores of increasingly heavy ele-

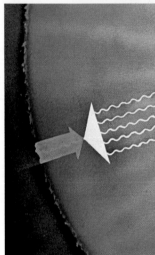

Right: When the star's reserves of hydrogen are exhausted, the star contracts. Eventually, the helium, too, is burned and turned into carbon. The carbon heats, pressure overcomes gravity, and the star expands. When it has used nearly all its reserves of nuclear fuel, it becomes unstable and puffs off its outer layers. Radiation pressure reduces, and gravitational forces crush it to a tiny white dwarf. It will continue burning until its nuclear reactions die out and it becomes a cold black dwarf, and completely invisible.

All artwork on these two pages by David Hardy, FRAS, from the filmstrip *Life and Death of a Star*, published by Visual Publications.

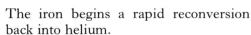

Left: a ball of dust and hydrogen gas has developed from the clouds present in a galaxy. Mutual gravitational attraction of atoms causes the gas to become denser and the temperature to rise, to turn the ball first into a glowing protostar (as here), then into a bright star. The hydrogen is gradually turned into helium, which in turn heats up and increases the radiation pressure. This pressure overcomes the inward pull of gravity and the star swells to 50 times its normal size (below). It has now become a red giant.

ments are built up as different types of nuclear burning start up, flourish, and fade. At each stage, an increasingly higher ignition temperature is required to keep the star alive. As each new fuel runs out, the core contracts under its own weight until its temperature is high enough for a new kind of nuclear fusion to begin.

By the time an element such as oxygen has begun to burn the core is close to 1000 million degrees. But at this fantastic temperature in addition to oxygen fusion a new kind of nuclear reaction begins that produces a vast flux of fast-moving particles known as neutrinos. These pour out of the star draining energy from its burning core.

To compensate for the sudden huge losses, the nuclear fusion reactions accelerate. The temperature rises still higher and the core continues to shrink. A rapid and bewilderingly complex sequence of nuclear reactions occur and heavy elements such as iron begin to be synthesized. Iron is usually regarded as the final stage of the chain of nuclear fusion reactions. Its heavy nuclei cannot be built up into still heavier atoms and under other circumstances, the energy-producing processes in the star would begin to falter. But at this stage, the conditions in the star are rapidly approaching a major crisis. The temperature reaches 8000 million degrees and not even iron nuclei, normally stable and unreactive, can withstand the inferno.

The iron begins a rapid reconversion back into helium.

Suddenly a vast amount of matter has vanished from the core. The heavy elements have been transformed into the comparatively light nuclei of helium. In the seconds that follow, catastrophe overtakes the star. The core collapses, contracting in an instant to a vastly dense ball of nuclear material. In the cataclysm that follows the implosion of the core, the star increases in brightness 1,000,000 million times. Perhaps for a few days, it outshines an entire galaxy. At the same time, its outer layers are blown into space by a violent explosion forming an expanding cloud of luminous gas.

This stellar eruption, known to astronomers as a *supernova*, is one of the most dramatic events in nature. But despite its violence it does not entirely wreck the star. It leaves behind a tiny superdense remnant known as a neutron star. This remaining body is tremendously compressed. No individual nuclei can be identified, all having been crushed into a solid ball of neutrons. The material of which it is made would be so dense, a thimbleful would weigh thousands of tons.

The *neutron star* is the final stage in the life cycle of a normal star. After thousands of millions of years and a complicated sequence of thermonuclear processes, the star dies leaving nothing more than a barely glowing ball of neutrons, perhaps a few miles in diameter.

Although this general pattern of evolution is followed by most stars, the exact life cycle they follow is determined largely by their initial mass. In some cases the surface layers of the star may not explode violently enough to destroy the star. Often these minor upheavals simply expel a circular or spherical shell of luminous gas, which astronomers call a planetary nebula. If the remaining mass of the star is less than about 1.5 times the Sun's present mass, the star will contract to an equilibrium state and become a dense, extremely hot *white dwarf* star. The fate of the dwarf is a gradual decline into darkness. It continues glowing until its energy supplies run out. Then after perhaps 100,000 million years it simply fades out becoming a *black dwarf*, the darkened remains of a totally dead star.

Red giants, white dwarfs and neutron stars

Red giants are the largest stars in the universe. They are also the coolest, their outer layers reaching only a few thousand degrees, no hotter than an industrial furnace.

Spectacular though they are, their immense size is far from being an indication of vigor. A red giant is a star approaching death. The stores of hydrogen, once providing the fuel that sustained the nuclear fires around its core, are virtually gone. A secondary core of helium, produced as the end-product of the fusion of hydrogen nuclei, has begun collapsing, its density and temperature steadily rising

as it does so. While these inner changes take place, the star's outer layers cool to a dull red and the great envelope of gases slowly expand to form a giant star.

The ultimate size of a red giant depends on its initial mass. A star like Aldebaran in the constellation of Taurus, with a mass less than that of the Sun, has a diameter 45 times larger. Although Aldebaran is 52 light-years away it still appears extremely bright. In fact, despite its comparatively low surface temperature it is about 200 times more luminous than the Sun. This is a characteristic of all red giants. Although the coolest stars known, their sheer physical size provides a vast surface area from which energy can radiate, ensuring an extremely high luminosity.

It seems reasonable that where there are giants there are also dwarfs. But because dwarf stars are so small – some may be no larger than a good-sized planet – they are difficult to observe directly. The first one discovered was predicted from observations of the brightest star in the sky – Sirius, the Dog Star.

In 1834 the Prussian astronomer Friedrich Bessel studying Sirius' motion in the sky noticed that it performed an unusual spiral and calculated that it took 50 years to complete one turn. This peculiar behavior led Bessel to conclude that Sirius must be accompanied on its journey around the galaxy by a companion star, too dim for him to see. In 1862 the American astronomer Alvan Clark spotted the faintly glowing companion star and labeled it Sirius B. It is sometimes referred to simply as the Pup.

Although Sirius B is dim it appears to be a fairly hot star. The only explanation is that it is extremely small – about .02 of the size of the Sun. But despite its size, it is sufficiently massive to affect the motion of the Dog Star itself. This can only mean that Sirius B is remarkably dense, an enormous amount of matter squeezed into a small volume. Calculations show that a matchboxful of the material from which Sirius B is made would weigh several tons on Earth.

Scientists today name this class of star a white dwarf. They are fairly common and Sirius B is by no means the smallest known. Some appear to be about the size of the Moon and are so dense that a handful of their substance would weigh thou-

Below: the orbit of Sirius (red curve) and its binary companion (blue curve) plotted over a period of 50 years. If Sirius had been a single star it would have made a straight track, indicated by the dotted line. In fact it makes the snaking track of the red curve. This led astronomers to suspect the presence of a companion, which was first observed in 1862.

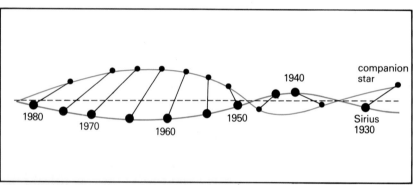

Left: Sirius, the brightest star in the heavens, is a white star with a luminosity of about 26 times that of the Sun. The tiny dot to the lower right of the star is the companion star to Sirius known as Sirius B and is an example of a white dwarf. White dwarf stars are the incredibly dense, small luminous objects that are often the final phase of a star's life. At this stage nuclear fires are on the wane, having almost ceased to function, and the once-high surface temperature will continue to fall until light is no longer emitted.

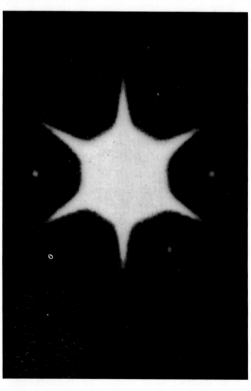

Below: sizes of some giant stars compared with the Earth's orbit around the Sun. The Sun's diameter is 864,950 miles, and the radius of the Earth's orbit is roughly 93 million miles. The red giant stars Antares and Betelgeuse are so large that they could contain the Earth's orbit comfortably. They are also very luminous, though their surface temperatures are much lower than that of the Sun. Hot white stars, such as Sirius and Vega, are not nearly so large as the red giants.

sands of tons.

White dwarfs are the remains of stars of 1.4 solar masses, or less that collapsed when their nuclear fires faltered. This ceiling of 1.4 solar masses is known as the *Chandrasekhar limit,* after Indian scientist Subrahmanyan Chandrasekhar. Stars above the 1.4 limit, but no greater than 2.5 solar masses, collapse farther than white dwarfs and become neutron stars – densely packed spheres, no bigger than the Earth, with gravity so intense that an astronaut landing on one would be flattened in a second to a pancake a millimeter thick. But what happens to stars above the 2.5 solar mass

limit when they collapse? Theorectically they will be crushed beyond the stages of white dwarf and neutron star to a point where their gravitational field is so strong that nothing can escape – not even light. However, there is still no proof that such *black holes,* as they are called, exist.

Red giants, white dwarfs and neutron stars are all stars in the final phase of their lives. But while red giants often die in cataclysmic supernova explosions, white dwarfs and neutron stars just fade away as their nuclear fires dwindle to the point where they become cold black rocks, drifting in space.

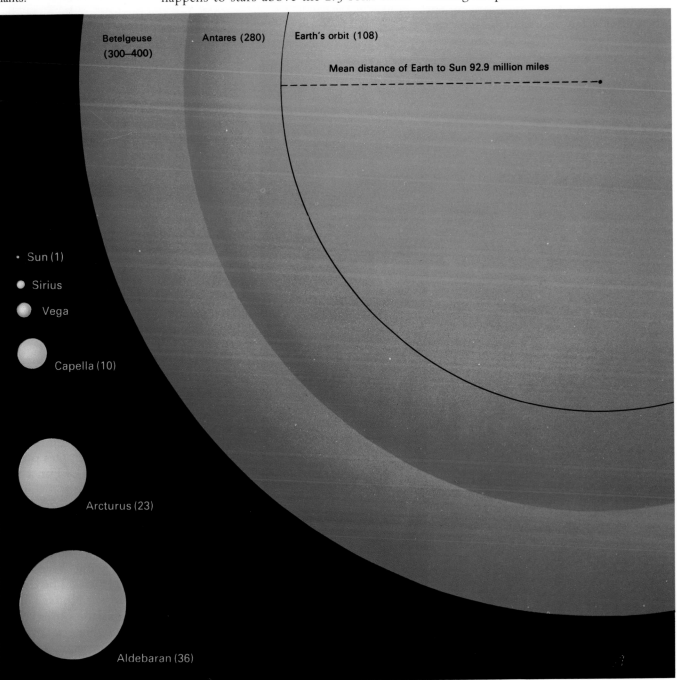

Betelgeuse (300–400)

Antares (280)

Earth's orbit (108)

Mean distance of Earth to Sun 92.9 million miles

Sun (1)

Sirius

Vega

Capella (10)

Arcturus (23)

Aldebaran (36)

Supernova

In 1054 AD Chinese astronomers watched a remarkable celestial event. A new star seemed to have been born, so bright it was visible even by daytime. They recorded its position in the constellation we call Taurus. But the excitement of the astronomers changed to astonishment when after a few months, the new star waned and eventually disappeared.

Below: the Crab Nebula in the constellation Taurus. This vast and still expanding cloud of luminous gas is the remains of a supernova explosion observed by Chinese astronomers in 1054.

Today we know that what the Chinese astronomers saw was not the birth but rather the death of a star. They had witnessed one of the rare cataclysms that can virtually destroy a star, the devastating explosion we call a supernova. We can still see the debris of the catastrophe, the Crab Nebula, a ragged cloud of gas hurtling outward at several miles a second even after nine centuries.

Studies of historical records show that only two other supernova have been observed in our own galaxy – one in 1572 and another in 1604. There is some evidence that there may have been one more, in 1006, but the records are by no means clear. Of these the brightest and closest was the supernova in 1054. In addition, it left an obvious remnant in the form of a huge gaseous nebula. The supernova of 1572 observed by the Danish astronomer Tycho Brahe and the one in 1604 spotted by Brahe's German assistant Johannes Kepler, have left no apparent trace.

Perhaps the most exciting supernova of all times was one seen in another galaxy altogether. In 1885 an individual star suddenly became visible in the hazy blur of light that we now know is the neighboring galaxy M31 in Andromeda. It was visible briefly to the naked eye as a star of about magnitude 6. But it quickly faded and soon vanished. The real significance of the event was not appreciated until early this century when the remoteness of the Andromeda galaxy was realized. S Andromedae – as the star was called – had become visible at a distance of about 1,800,000 light-years. This means that at its peak, S Andromedae outshone an entire galaxy – a single star brighter than the combined brilliance of thousands of millions of ordinary stars. Modern estimates put its absolute magnitude at −19. For a few days at least, S Andromedae was shining with the light of 10,000 Suns.

The scale of the explosion creating such luminosity can scarcely be imagined. What is really surprising, however, is that a supernova does not necessarily destroy a star altogether. The latest evidence suggests that supernovae are a violent release of the gaseous outer layers of the star. When the blast is over, a small remnant is left – the original core of the exploded star.

In 1967 the first of these extremely

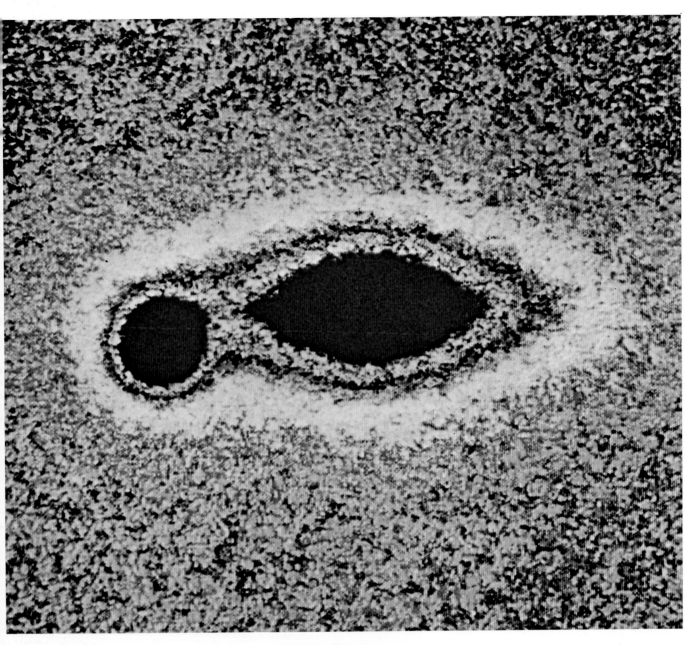

dense remnants was detected by the regular pulses of radio waves it emitted. Called *pulsars*, we now know that they consist of material so tightly packed that their atoms are crushed together into a solid ball of neutrons. These so-called neutron stars rotate, leaking energy in the form of radio waves as they do so. The confirmation that pulsars were the remains of a supernova came in 1968 when one was discovered in the heart of the Crab Nebula.

We now know that pulsars – or spinning neutron stars – are the last phase of the life cycle of a star massive enough in the first place to produce a supernova. Once its rotational energy has been released as pulses of radiation, the

Above: this brightness contour map was especially created at the Kitt Peak Observatory in Arizona to compare the brightness of luminous bodies. The round red area on the left is a supernova discovered in 1975, the elongated red area on the right is a galaxy in the constellation of Ursa Major. The color code on the map is that used at Kitt Peak to indicate brightness – red for the brightest areas, blue for the dimmest. As can be seen, the supernova is as bright as the whole parent galaxy.

neutron star fades away and can no longer be detected.

Within the last few years evidence has come to light suggesting that on rare occasions an even stranger remnant may be left after a supernova. Its density would be much greater than even a neutron star – so dense that the gravitational field surrounding it would not even allow light itself to break free of its grip. This means that no information about the space inside the gravitational field could ever escape and reach us. The supernova will have left in its wake a small region cut off from the rest of the universe. Scientists predicting this bizarre phenomenon call it a *black hole* in space.

Evolution of a Supernova

Right: the evolution of a supernova depends on a star's mass.

A represents a star only a few times more massive than the Sun.

1: hydrogen in its core is fused into helium. After several billion years the hydrogen is used up. 2: the star's core contracts, its exterior expands and the star becomes a red giant. 3: the outer layers of the red giant are eventually expelled and become a planetary nebula. 4: all that remains of the star is a white dwarf.

B: some supernovas are members of double-star systems. Stages 1 and 2: these are the same as their solitary counterparts, except that as one star reaches the red giant stage it begins to lose matter to its companion. 3: the companion expands to a red giant, the other star becomes a white dwarf. 4: matter is suddenly transferred from the red giant and added to the white dwarf, which increases its mass beyond the critical limit of 1.44 solar masses. 5: the core of the white dwarf collapses, releasing energy as a supernova. 6: a binary system remains in the form of an ordinary giant star and an x-ray source (wavy blue arrows).

C: If a Star is much more massive than the Sun, the evolution to supernova is different after stages 1 and 2. At stage 3, hydrogen in the red giant continues to be burned in a shell (red) around the core, and the core contracts and heats up until helium (yellow) fuses into carbon. At the next stage one of two catastrophes may overtake the star. When the helium is exhausted the core begins to burn the carbon. 4: if the ignition of the carbon (black) induces instabilities, the star explodes cataclysmically as a supernova (5) leaving behind only an expanding gaseous remnant (6). If, after stage 3, the carbon is safely ignited, extraordinarily high temperatures build up in the core. 7 and 8: the star throws off radiation particles (blue arrows) at an ever-increasing rate, sapping its energy so that its core plunges to total collapse. 9: a final burst of radiation might cause the red giant to throw off the outer envelope of the star in a gigantic explosion. It could leave behind a nebulous collection of gas, at the center of which would be either a pulsar, or (10) a black hole.

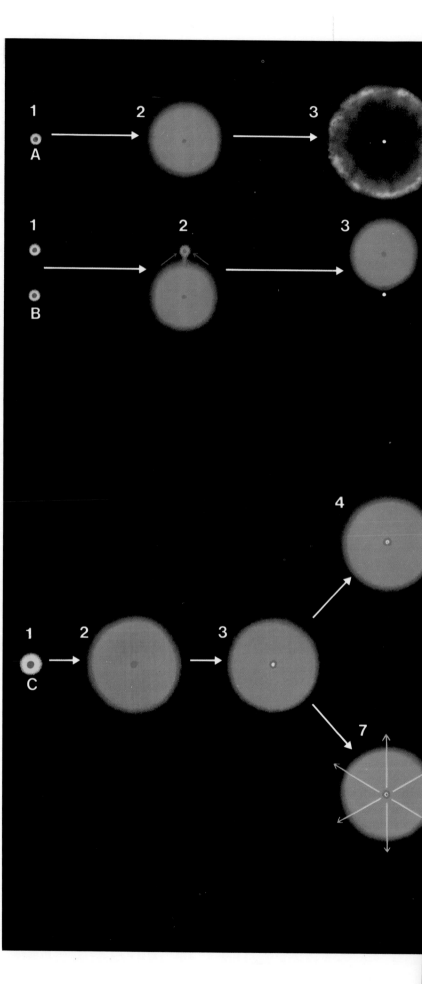

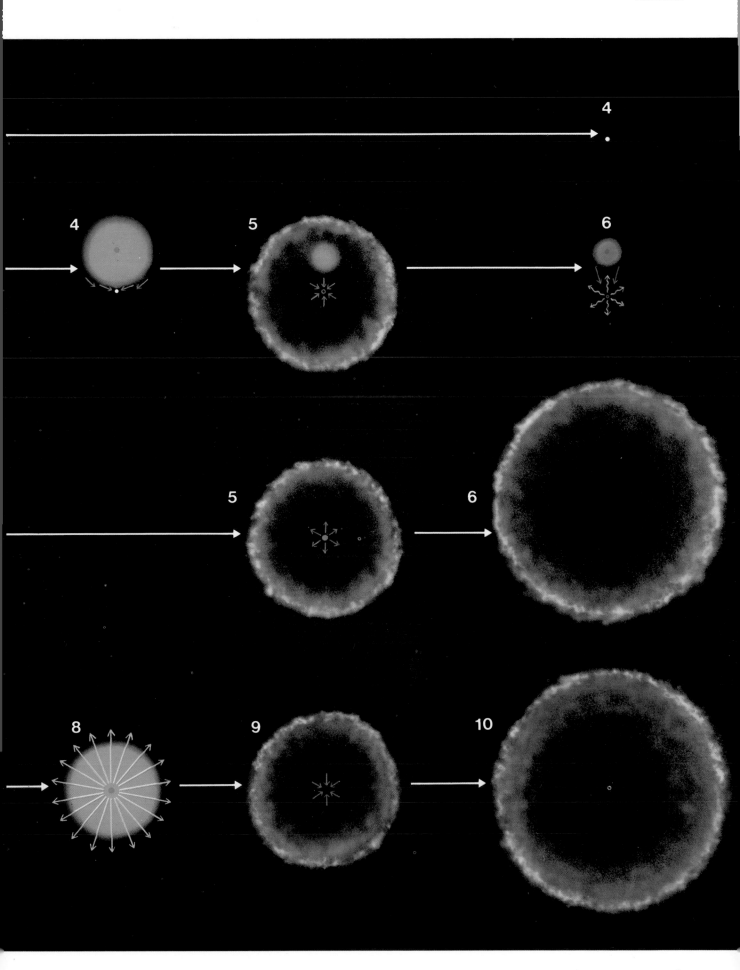

Variable and double stars

passes between Algol and the Earth, Algol is partially obscured causing an apparent fall in its luminosity. What Goodricke had identified was an important class of variable star – the *eclipsing binary*, a double-star system that looks like a single pulsating star.

Eclipsing binaries are the simplest

There are many stars – several that can be seen even with the naked eye – that vary greatly in brightness. The best known is Beta Persei, the second brightest star in the constellation Perseus. Every two days and 21 hours it dims by a full magnitude before quickly regaining its former brightness. Strangely enough, despite the usual care with which ancient astronomers recorded celestial events, no records have been found referring to Beta Persei's fluctuations. Modern astronomers have speculated that the concept of such an obvious exception to the divine order of the heavens was so disturbing that they preferred to ignore it. But the Greeks called Beta Persei the "Demon Star" while the Arabs named it *Algol*, meaning "The Ghoul," so perhaps the early astronomers were ready to acknowledge that there was something strange about the star.

Algol, as it is now popularly called, was first studied in detail in 1782 by the English astronomer John Goodricke. With remarkable insight Goodricke suggested that Algol's variability had nothing to do with the star itself. He suggested instead that it was periodically eclipsed by another, less luminous star, caught in Algol's gravitational field and orbiting it in much the same way as the Earth revolves around the Sun. Each time it

Right: an eclipsing binary – the simplest kind of variable star system. As one star orbits its nearby companion, it periodically passes behind it and its light is eclipsed. Seen from a great distance, the two stars look like a single point of light that varies in intensity in a rhythm corresponding to the orbital movement of one star around the other.

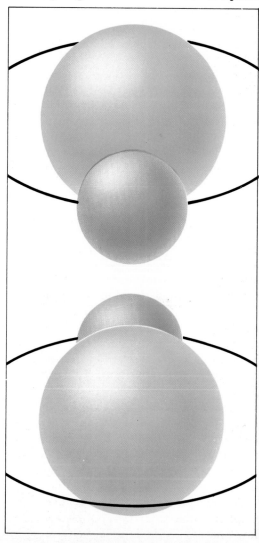

Below: the conspicuous variable star Algol. For about two and a half days it appears to be almost constant in brightness, at magnitude 2.3. Then for five hours it fades to one fourth of its former maximum brilliance. The cause of this is Algol's dark companion, which partially eclipses the brighter member of the pair, cutting off some of its light.

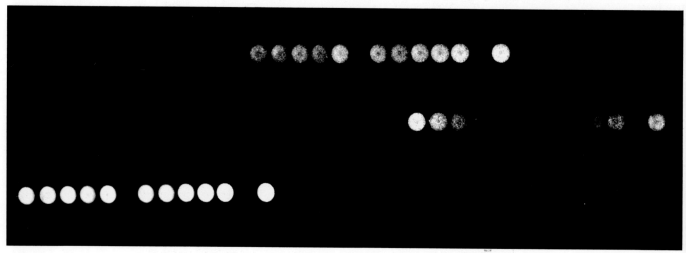

type of variable stars. There are, however, single stars that really do vary in brightness. They are usually old, degenerate, and not very stable giants, whose outer layers periodically collapse in, only to puff out again a short time later. They repeat this routine, over and over, quite regularly, for years.

One of the first true variables to be studied was Omicron Ceti, better known as Mira, meaning "wonderful." When it was discovered in 1596 by the German astronomer David Fabricius, Mira looked like a magnitude 3 star. But within two months it had faded into invisibility. It is now known that Mira's average period is 330 days and during this time the star varies from a maximum magnitude of between 2 and 5 and a minimum of between 8 and 10.

Mira is an example of what is called a long-period variable. Much shorter period stars are known, the most important being a special class of variables known as the Cepheids – named for the first star of the kind to be discovered, Delta Cephei. Cepheid variables have periods ranging from two to 45 days and, according to the researches of the American astronomer Henrietta Leavitt in 1912, are remarkable because their periods are directly linked to their true brightness. This fact was used by astronomers to work out the relative distances of different Cepheids. For example, if they compared two Cepheids with identical periods, they knew that any difference in luminosity must be due simply to distance.

A few years after the discovery of Cepheids, the American astronomer Harlow Shapley worked out how to use them to investigate the true scale of the observable universe. By determining the distance of just one Cepheid he could easily calculate the distance of others by simply comparing their periods. Having done this he could work out the size of the galaxy itself just by observing the distribution of Cepheid stars within it. In later years, similar techniques were used to gain an accurate idea of the nearness or remoteness of giant nebulae such as the Magellanic Clouds and the galaxy M31 in Andromeda.

Double stars are less spectacular than variables but much more common. They are often difficult to detect, however.

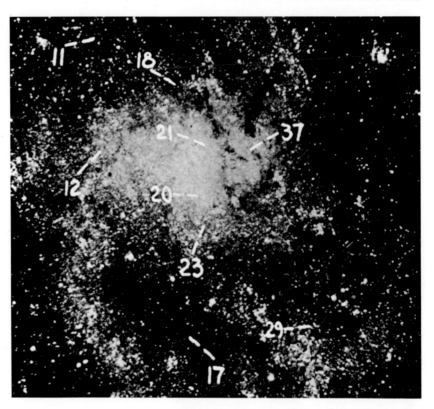

Above: part of a star map of short-period variable stars. The broken bars mark the positions of the variables. These short variable stars, named Cepheids for the first star noted for its variability, have become a yardstick with which to measure the vast distances to the stars.

In the most straightforward cases, such as the eclipsing binary Algol, the companion actually cuts out light from the primary star. But if the stars are very luminous and quite close to one another, variations in brightness of this kind are virtually impossible to detect. In cases where the plane of the orbit of the stars around one another is not completely aligned with Earth, no eclipse can take place anyway.

Today most binary stars are detected by breaking their light into a spectrum. When the two stars are moving so that as one star is approaching the Earth, the other is receding, the spectral lines from the first shift toward the blue end of the spectrum and those belonging to the other shift to the red. The first of these so-called spectroscopic binaries was discovered in 1889 when an American astronomer, Edward C. Pickering found that Mizar, which with Alcor forms a naked-eye double in the constellation of the Big Dipper, was itself actually a double star. Since then several thousand more double stars have been identified. Recent estimates indicate that about one tenth of one percent of all stars are actually star systems consisting of two stars, so close they cannot be visually separated.

Astrology

The origins of astrology are lost in pre-history. Records show that as a body of knowledge, astrology was already important at the dawn of civilization 5000 years ago in Mesopotamia. But the idea that human affairs are influenced by the movement and positions of heavenly bodies must go back much further than this. In earliest times changing seasons could mean the difference between life and death. Because the patterns of lights in the heavens changed with the seasons, it is not surprising that primitive people saw their well-being closely linked to the stars and planets.

In Mesopotamia, astrology was called "the royal art" – practiced only by priests and kings. Their objective was

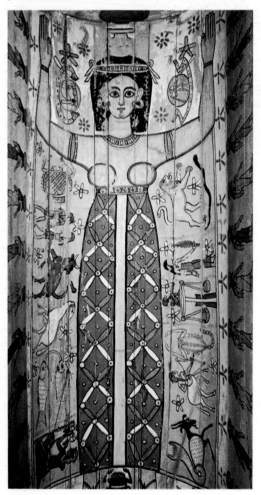

Above: an illustration from a 17th-century Muslim design showing an astrologer taking a reading of the position of a star. Near him is a model of the zodiac, showing the 12 signs. It was the Arabs who kept alive the writings and techniques of astrology after it had been largely discarded in the West with the growth of Christianity.

Left: this Egyptian mummy case dating from the 2nd century AD, depicts the sky goddess Nut surrounded by the signs of the zodiac.

simple – to divine the future. But they worked with only the barest facts about the heavenly bodies and their predictions were usually limited to generalities about the likelihood of good weather, floods, or plague.

By about 1000 BC astrology had spread its influence to Egypt and Greece. The contribution to its development made by the Egyptians was confined largely to inventing a 12-month calendar and a study of the 36 so-called "fixed stars" that later became the basis of the signs of the Zodiac.

More important were the contributions of the Greeks. In particular, philosophers such as Anaximander (611–547 BC) and Hipparchos laid the foundations of the Zodiac by dividing the heavens into 12 segments through which the Sun and planets moved against the background of the stars. These segments – known as houses – were eventually named by linking the shapes of the main stars in them with animal and human identities – the crab, scorpion, ram, lion, archer, water carrier, and so on.

Each of the 12 houses were thought to have influence over a particular aspect of human affairs. The first was the house of life. The second was the house of riches. The third – brotherhood; the fourth – human relationships; the fifth – children; the sixth – health; the seventh – marriage; the eighth – death; the ninth – religion; the tenth – dignity; the eleventh – friend-

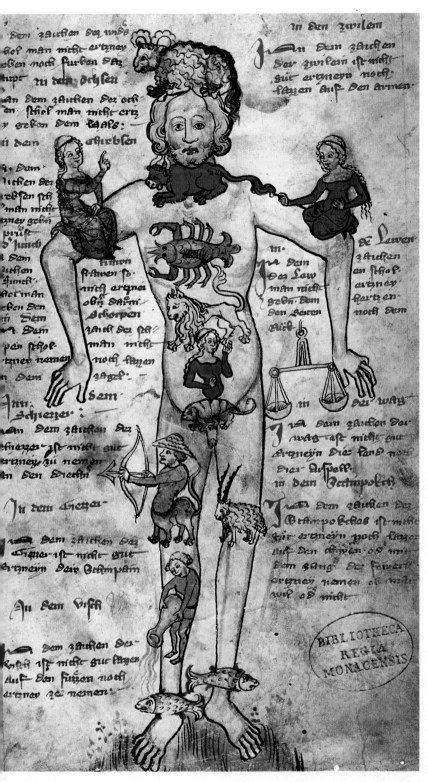

physicians. Illness was held to be closely linked to heavenly influences and doctors claimed that the various houses had specific power over particular parts of the body. Aries, the ram, for example, affected the head whereas Scorpio, the scorpion, influenced the digestive system. Leo, the sign of the lion, held sway over the heart. Treatment was often based on the position of the planets in the houses. Because Mars and Mercury were thought to have different effects on the blood, patients were bled only on Tuesdays and Wednesdays – the days associated with the two planets. Occult medicine was extended further by linking medicinal herbs and metals with the astrological scheme. Early physicians therefore were better trained in celestial mechanics than in the actual workings of the human body.

The scope of astrological prediction was based on the amount of information available on the position and movement of stars and planets. The single most important collection of such data was compiled by the Alexandrian astronomer Ptolemy in the 2nd century AD. Called the *System of Mathematics*, it amounted to a vast catalog of the known heavens. Arab astrologers translated it as *Almagest* – "the great work" – and in numerous translations and editions it was widely used until the Middle Ages. As well as providing a firm basis for astrology, it eventually stimulated detailed studies of the night sky, which evolved into the modern science of astronomy.

Astrology persisted as a widespread system of belief even after more rational studies of the heavens had emerged. In the 1600s, for example, Johannes Kepler made a living by casting horoscopes at the same time as he was working on his basic laws of planetary motion, which are today regarded as the cornerstones of modern astronomy. Even though it is taken less seriously these days, astrology is still pursued throughout the world and almost every popular newspaper has a section devoted to horoscopes. But after thousands of years of influence over human affairs, astrology is today little more than an intriguing amusement. Scientific studies of the heavens have taken over from mysticism, replacing it with a rational system of knowledge that has at last begun to reveal man's true place in the universe.

ship; the twelfth – hatred and ill-feeling. In addition, the known planets were accorded mystic signs and significance so that their position in the various houses at different times could portend a variety of events – from famine and pestilence to the personality traits of a child born on a particular day of the year.

The Zodiac was soon put to use by

Above: an illustration from a German manuscript on astrology in the 14th century. It shows the correlation of the signs of the zodiac with parts of the human body. This is a reflection of the astrological tenet of "As above, so below" in which events in the heavens are matched bv responses on Earth.

65

Above: when the ancient astronomer-priests looked into the night sky, they sought to put some kind of order into the glittering canopy of stars they found there. One group of stars reminded them of a hunter, another seemed to outline the

shape of a mighty bull, so they named the patterns for the heroes of their myths and the gods of their religions. The sky became for them a kind of storybook where the gods and heroes lived forever.

The constellations

For as long as man has looked wonderingly at the night sky, he has traced shapes among the patterns of stars. Among their apparently changeless configurations, early man imagined he could see the figures of animals, hunters, and gods. The earliest records suggest that most of the principal constellations familiar today were known in Mesopotamia about 4000 years ago.

As the ancient civilization of Egypt, Greece, and Rome flourished, a variety of names for the same star patterns emerged. The constellation known to the Greeks and Romans as the Great Bear,

Below: part of the constellation Aquarius, the water carrier, from a Persian star-map manuscript of around AD 1650. The constellations first mapped by the ancients that so neatly divide up the night sky, are still used by modern astronomers as a rough and ready guide to refer to objects in the sky.

for example, was called Sarcophagus by the Egyptians.

Few pictorial records of the early constellations survive, but in about 150 AD, the Greek philosopher Ptolemy compiled a catalog giving the positions of 1022 of the brightest stars, linking them with the 48 constellations then recognized. Some 1400 years later, Ptolemy's great work was used as the basis for two spectacular star maps – one for each hemisphere – drawn and published by the German artist Albrecht Dürer. Using his remarkable talents, Dürer elaborated the maps, drawing figures among the stars to symbolize the various constellations. Since then, Dürer's work has been the model for most subsequent representations of them.

For centuries, a constellation was regarded as relating only to the brightest stars around which the celestial figures were drawn. But this left many less prominent stars unidentified. The first

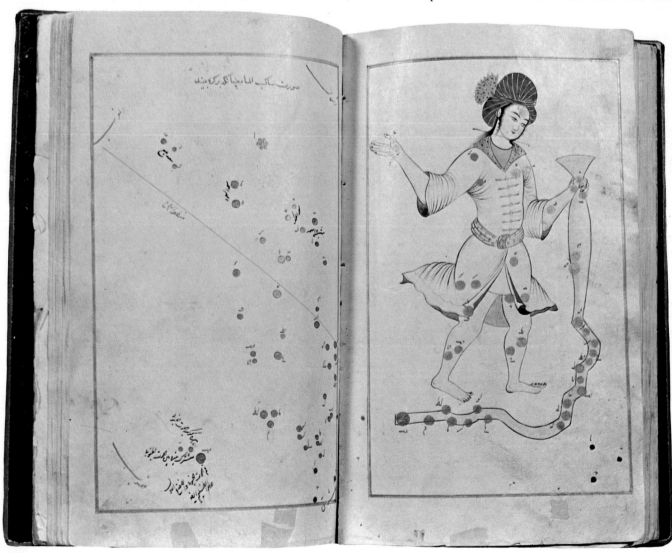

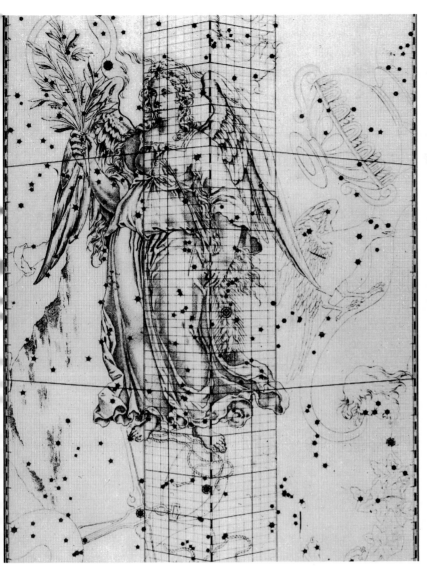

attempt to draw boundaries between the various constellations so that any star could be associated with one, was made in 1801 by a German astronomer, Johann Bode, in his book *Uranographia*. Bode's scheme, refined and modified in subsequent years, laid the foundations of the system of celestial divisions finalized in 1928 by the International Astronomical Union.

Ptolemy's 48 constellations were based on only the brightest stars visible from northerly latitudes. Today 88 are officially recognized covering the skies of both the Northern and Southern hemispheres. Some are themselves divided into subconstellations used commonly because they are so easy to identify. For example, part of the Great Bear is known as the Big Dipper. Similarly, the huge southern constellation Argo Navis is split into

Above: Virgo, the maiden, from an 18th-century celestial atlas. The sheaf of corn in the hand of Virgo ties in with the time of harvest, when the sign occurs.

Above: right: Crux Australis, the Southern Cross, is in the path of the Milky Way and its bright stars form a small but well-defined cross (seen right in this drawing from a description by a Portuguese 15th-century explorer). The upright of the cross points almost to the south celestial pole. The constellation is not visible much north of the equator.

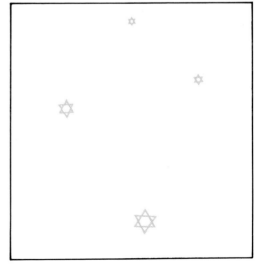

three sub-constellations – the Keel, the Poop, and the Sails.

In northern skies one of the most famous constellations is Orion, the Hunter. It contains two very brilliant stars – the red supergiant Betelgeuse and the blue-white Rigel. It is easy to pick out the Hunter's starry belt and misty sword. The Hunter's belt in particular consists of three prominent stars in an almost straight line. The sword points southward from the belt.

The Hunter's belt is a signpost in the sky. Its three stars point toward the most

brilliant star in the sky – the Dog Star, Sirius. Another obvious celestial signpost is the sub-constellation the Big Dipper, or Plow. Two of its stars – Dubhe and Merak – are known as the pointers because they show the way to the famous North Star, Polaris. Although not among the brightest stars – it is of magnitude 2 – Polaris is actually extremely luminous but at a very great distance from our own Sun.

If Orion's belt is followed in the opposite direction to Sirius, it leads to the constellation of Taurus, the Bull. Most

Below: stars in the far north of the sky. The groups in this portion of the sky include the Great Bear, Little Bear, Cassiopeia, Perseus, and the brilliant stars Capella, Vega, and Deneb. Stars in the inner circle extend to declination 60°N (30° from the celestial pole); those in the outer circle extend to declination 30°N.

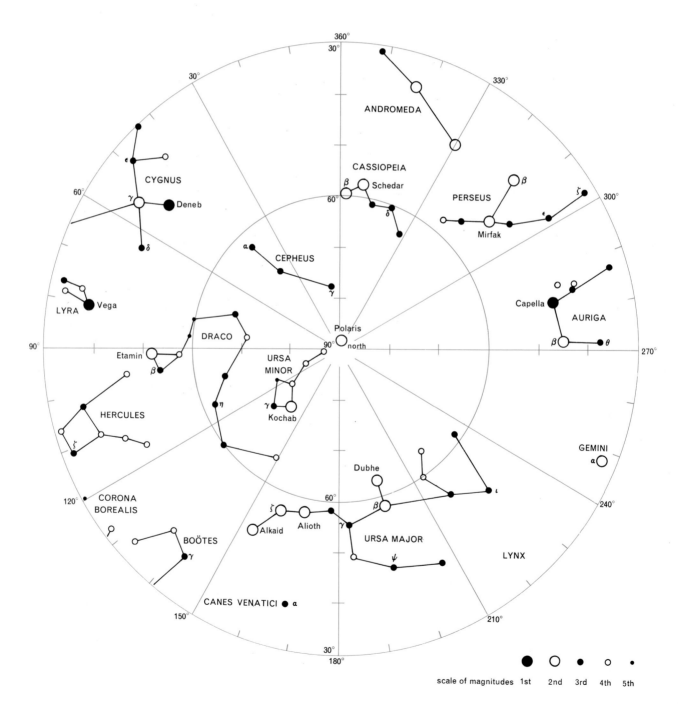

scale of magnitudes 1st 2nd 3rd 4th 5th

Below: stars in the far south of the sky. The south pole is not marked by any conspicuous star, but there are many brilliant groups in the area, notably the Southern Cross, the Centaur, Argo, and parts of the Scorpion and River Eridanus. The scale is the same as for the northern sky, the chart extending to declination 30°S.

prominent of its features is the brilliant orange star, Aldebaran. Its name in Arabic means "the follower (of the Pleiades)". Close to Aldebaran but on the far side from Orion's belt is a faint luminous patch. Seen with even a moderate pair of binoculars, it resolves itself into the beautiful cluster of stars known as the Pleiades or Seven Sisters, also a part of Taurus.

Best known of the constellations in the Southern Hemisphere is Crux, commonly known as the Southern Cross. Close to the cross is the constellation Centaurus,

famous because it contains the star closest to the Sun – Proxima Centauri, just over four light-years away.

Constellations serve a practical purpose, acting as signposts to various parts of the sky and helping to identify star positions by reference to standard segments of the heavens. But they are also a constant reminder of man's enduring fascination with the beautiful patterns visible in the night skies. Their names and shapes evoke the ancient mysticism from which the modern science of astronomy has evolved.

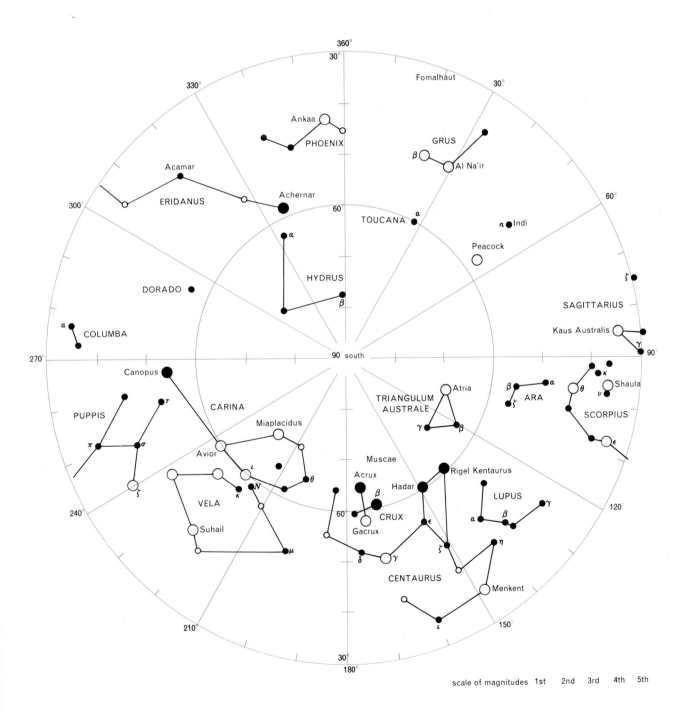

scale of magnitudes 1st 2nd 3rd 4th 5th

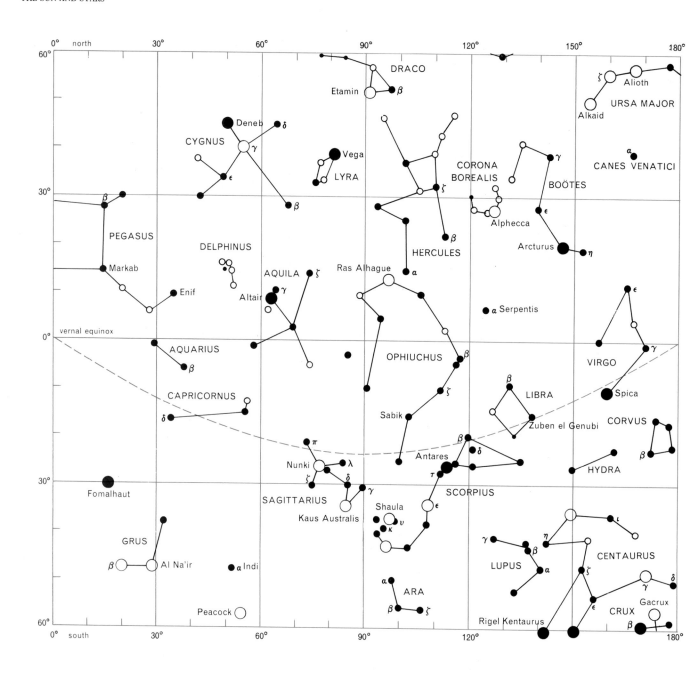

The Twenty Brightest Stars

STAR Constellation	Apparent Magnitude	Colour
SIRIUS Canis Major	—1·43	White
CANOPUS Carina	—0·73	Yellowish
α CENTAURI Centaurus	—0·27	Yellowish
ARCTURUS Boötes	—0·06	Orange
VEGA Lyra	0·04	Bluish-White
CAPELLA Auriga	0·09	Yellowish
RIGEL Orion	0·15	Bluish-White
PROCYON Canis Minor	0·37	Yellowish
ACHERNAR Eridanus	0·53	Bluish-White
BETELGEUSE Orion	variable	Reddish
β CENTAURI Centaurus	0·66	Bluish-White
ALTAIR Aquila	0·80	White
ALDEBARAN Taurus	0·85	Orange

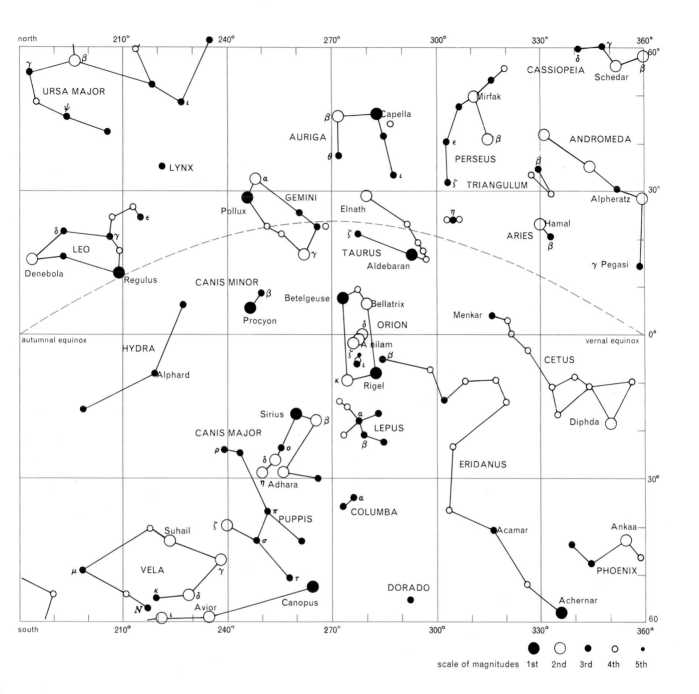

Above: a chart of the constellations, on a Mercator projection; areas close to the celestial poles therefore cannot be shown. Stars are classed into grades of magnitudes of apparent brightness – the smaller the magnitude the brighter the star. The faintest stars seen with the naked eye are of magnitude 6; the chart shows stars down to magnitude 4.

ACRUX Crux	0·87	Bluish-White
ANTARES Scorpio	0·98	Reddish
SPICA Virgo	1·00	Bluish-White
FOMALHAUT Piscis Australis	1·16	White
POLLUX Gemini	1·16	Orange
DENEB Cygnus	1·26	White
β CRUCIS Crux	1·31	Bluish-White

Left: a list of 20 apparently brightest stars, known usually as stars of the first magnitude, though they are by no means of equal brightness.

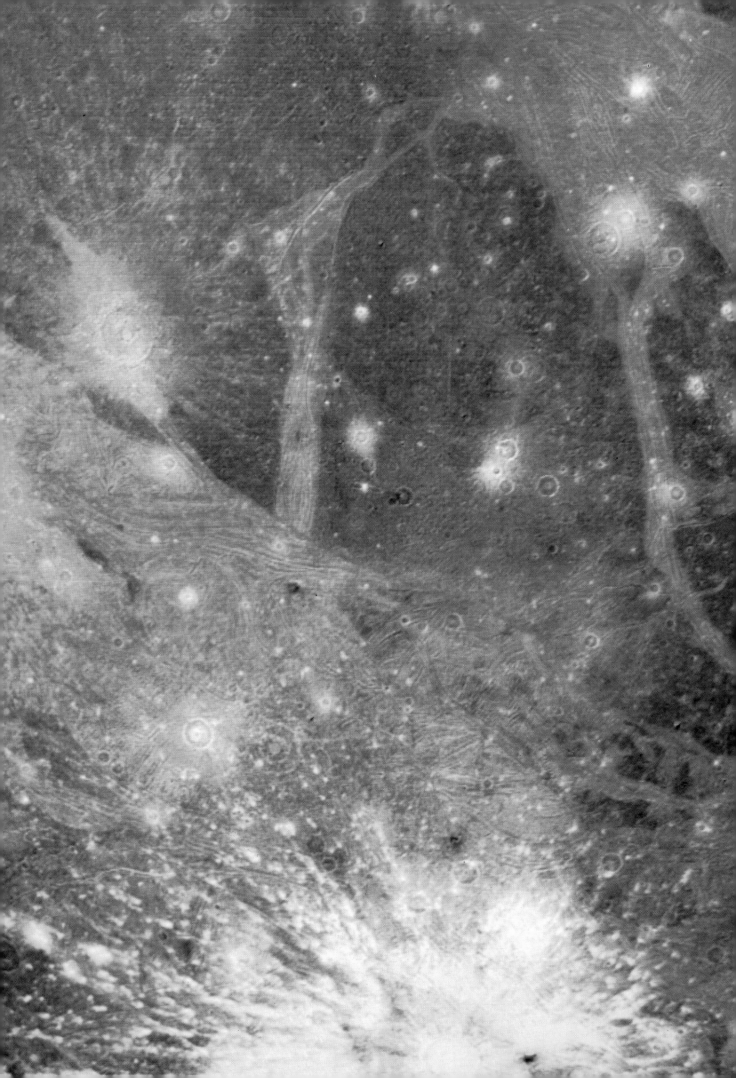

Chapter 3

The Solar System

On a cosmic scale our solar system is totally insignificant – a family of nine planets orbiting a commonplace star, lost in the immensity of interstellar space. But for mankind, the solar system is a rich source of knowledge about the universe. The more that is learned about the planets of the solar system the more extraordinary the Earth appears. The terrifying extremes of temperature and pressure among the other planets, together with atmospheres either too thin to support life or filled with poisonous gases, emphasize the miraculous set of circumstances that has enabled the human race to thrive on Earth.

The dawn of the space age has made it possible to study and examine many of the planets in hitherto unimagined detail. Major new discoveries have been made and the growing knowledge of the solar system and its origins is providing fresh insights into the workings of the universe. But each advance in our knowledge confirms that despite the grandeur of many of the other planets, the Earth remains unique – the sole world on which intelligent life has evolved and prospered.

Opposite: the surface of Ganymede, one of Jupiter's moons, photographed in July 1979 from Voyager 2. At the bottom of the picture is a bright halo impact crater that shows the fresh material thrown out of the crater. The dark background material is the ancient heavily cratered terrain – the oldest material preserved on the Ganymede surface. The 13 moons of Jupiter form a miniature solar system and orbit their giant mother planet much as other planets revolve around the Sun.

Origin and scale of the solar system

For centuries people believed that the Earth was the center of the solar system and that the planets and Sun moved around it. In the 2nd century AD, the Greek philosopher Ptolemy tried to reconcile this idea with the observed movements of the planets. He found the task far from easy. One of the major problems was to explain why a planet sometimes appeared to move toward the Earth while at other times it moved away. Such erratic motion seemed to imply a remarkably complex system of movements. Ptolemy eventually settled on a complicated network of interconnected circles, 40 in all, some drawn inside others.

For centuries Ptolemy's unwieldy plan of the solar system remained unchallenged. Then, between 1507 and his death in 1543, the Polish astronomer Nicolaus Copernicus worked out the revolutionary theory that the Earth, together with other planets, moved around the Sun. The idea removed many of the Ptolemaic complications at a stroke and by the early 1600s the German astronomer Johannes Kepler had verified Coper-

nicus' model of the solar system showing that the path followed by a planet around the Sun is a simple geometrical figure called an ellipse.

Scientists have now plotted the path of these orbits in detail and have a clear idea of why they are elliptical. The way a body – whether it is a planet or a space-

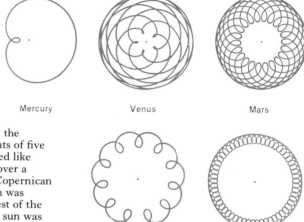

Mercury   Venus   Mars

Jupiter   Saturn

Above and right: the apparent movements of five of the planets looked like this when plotted over a long period. The Copernican view that the Earth was moving with the rest of the planets around the sun was the key to the complicated patterns.

Below: the Sun and the planets of the solar system drawn to scale. The planets are depicted in order of their relative distance from the Sun. At the far left is Mercury, nearest of the planets to the Sun. The others from left to right are Venus, Earth, Mars, Jupiter (the largest), Saturn, Uranus, Neptune, and, farthest away, Pluto. The broad golden arc at the bottom is part of the Sun.

ship – moves in space is largely determined by the force of gravitational attraction that acts between every piece of matter in the universe. The planets therefore can be visualized grouped around the Sun under the influence of the Sun's powerful gravitational field.

The German-born physicist Albert Einstein's general theory of relativity shows that what appears to be the effect of a gravitational force is really a basic

almost as though they were falling through distinct corridors in space, decided by the shape of space itself.

Scientists now believe that the effect of gravity was the most important single factor in the formation of the solar system 4600 million years ago. But before the most recent and widely accepted theory of the solar system's origins was developed, several other hypotheses had been proposed.

One of the most popular ideas was based on some kind of ancient cosmic catastrophe. In 1749 French naturalist Georges Buffon proposed that the planets were formed from the debris of a huge comet that was destroyed in a collision with the Sun. This catastrophe theory was updated in 1913 by an Austrian inventor and engineer, Hans Hörbiger. He held the eccentric view that the entire universe was made up of hot metallic stars and cosmic ice. Hörbiger believed that once a relatively small star had cooled it would eventually become encrusted in cosmic ice. On its journey through space it could have a chance encounter with a large hot star. If the two actually collided, the smaller star might become embedded like an icy bullet in the larger star. But as the ice melted, the large star might become temporarily unstable. A massive explosion would follow and, according to Hörbiger, the cloud of gas and ice ejected from the star would eventually cool, the planets slowly condensing from it as its temperature fell.

Hörbiger carried his bizarre ideas further. He suggested that the various moons in the solar system were thrown off by the Sun in a similar way somewhat later than the formation of the planets. He even suggested that the Earth had captured and lost six previous satellites before grabbing its present Moon. However far-fetched Hörbiger's theories now seem, he had thousands of avid followers in the early 1900s. Many scientific meetings of the day were broken up in disorder with cries of "Out with astronomical orthodoxy! Give us Hörbiger!"

A far more plausible theory was proposed in 1905 by two Americans, a geologist, Thomas Chrowder Chamberlin, and an astronomer, Forest Ray Moulton. They suggested that millions of years ago the Sun and another star had suffered a near collision. The powerful gravita-

Above: a view of the solar system with the Sun in the center. The nine planets all move around the Sun in their own orbit and each takes a different time to complete its cycle around the Sun. The planets are not drawn to scale because the outer planets are in fact much farther from the Sun than the near ones and would be impossible to show in the same diagram. The artist shows clearly, however, how the planets have no light of their own but simply reflect the Sun's light.

quality of space itself. The detailed workings of the theory are extremely difficult to understand but one of its important conclusions is that the actual fabric of space itself can be curved by the presence of anything with mass. The more mass an object has the greater the distortion in space it can cause. If it is imagined that the Sun's huge mass curves space all around it, the masses of the planets modify the overall distortion by varying degrees depending on their own bulk and their distance from the Sun. The result of the bending of space around the Sun is that the planets follow lines of least resistance through curved space. It is

tional attraction between such huge bodies during a near-miss would have drawn off from one of the stars a long streamer of gas. The streamer might then have broken down into several small gas clouds out of which the planets and their moons could have condensed. Although we now know that the chances of a near-miss of this kind are infinitesimal, the principle behind Chamberlin and Moulton's theory was scientifically quite sound.

The modern theory accounting for the formation of the solar system has its origins in a far earlier idea first suggested in 1796 by the German philosopher Immanuel Kant and elaborated by a French mathematician, Pierre de Laplace. Their hypothesis was based on the idea of stars forming out of rotating clouds of dust and gas. They worked out that gravity acting on the core of the cloud would draw more material into the center and make the core contract, so that it became steadily more dense. As the process continues, the gas cloud would rotate at an ever increasing speed until it began to throw off streamers of gas that would form into concentric rings around the dense core. By the time the central region had turned into a star, the planets would have begun condensing from the gas rings surrounding it.

Despite being fashionable for many years the nebular hypothesis as it became known, was dropped when British mathematician James Clerk Maxwell showed in 1859 that it was mathematically impos-

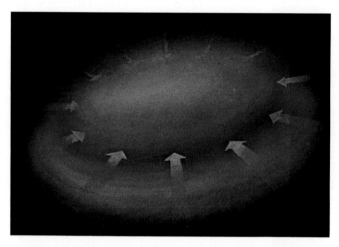

1. The birth of the solar system. A spinning mass of gas and dust contracts under its own gravity. The more it contracts, the faster it rotates and the denser the elements become compacted within its center.

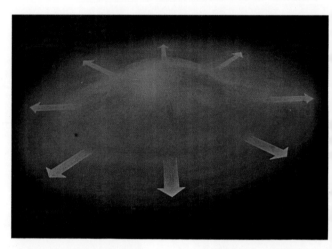

2. Ever-increasing rotation causes the faster-moving particles at the edge of the cloud to be thrown out farther from its center. The stage is now that of a solar nebula, with the ball of gas at the center of a gaseous disk.

5. As the glowing Sun continues to contract and increase in brightness, the residual gas is subjected to increasing radiation, which drives most of the hydrogen and other light elements toward the edge of the disk. The heavier elements remain closer to the central nebula.

6. The disk becomes flatter and more concentrated. Gravitational forces within it cause it to become unstable and to break up into separate clumps of matter – those at the edge of the disk are made up of a higher proportion of lighter elements than those closer to the Sun.

sible for planets to form from a ring of gaseous material. Maxwell found that it would condense only into small particles.

But using modern techniques, the original nebular hypothesis was re-examined in the 1940s by a German physicist, Carl Weizsacker. A new theory emerged which, expanded by an American astronomer, Gerard Kuiper, is now widely accepted as the best description we have of the origin of the Sun and its family of planets.

The modern idea is similar to the Kant-Laplace theory except that the nebula from which the solar system formed is no longer thought to have rotated as a single unit. Instead, turbulent eddies are thought to have been set up in the outer regions – smaller whirl-

All artwork on these two pages by David Hardy, FRAS, from the filmstrip *The Evolution of the Solar System*, published by Visual Publications.

pools in the body of the main whirlpool of gas. As the separate eddies swept close to each other, individual groups of them formed into larger clouds until eventually planets coalesced out of them.

Although the basic idea of the theory is now widely accepted there is still considerable argument over the detailed mechanisms involved. Latest refinements include work by the Swedish astronomer Hannes Alfven and the British astronomer Sir Fred Hoyle, which show how the gaseous components of the embryonic inner planets would be drawn off in the early stages of their formation by the Sun. This is consistent with the inner planets being small rocky bodies whereas the outer planets are giants consisting of a small rocky core enclosed in gas.

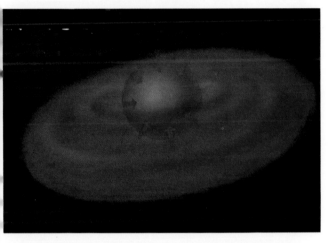

3. The particles at the center of the dust and gas ball are compacted to form a dense core. The closely packed particles are forced into violent collisions, heat is generated, and the protostar begins to glow.

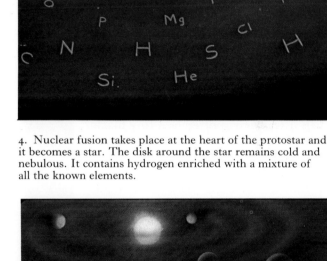

4. Nuclear fusion takes place at the heart of the protostar and it becomes a star. The disk around the star remains cold and nebulous. It contains hydrogen enriched with a mixture of all the known elements.

7. The clumps of matter contract into a solid mass at their centers to become protoplanets. Those nearer to the Sun were the rocky planets – Mercury, Venus, Earth, Moon, Mars, the asteroids and several of the large planets' moons. The outer planets remained mixtures of ice, gas, and rock.

8. Over the next few million years the Sun's increasing radiation – the solar wind – stripped away the coating of gas from the inner planets and much of the residual gas and dust from the disk. Only the outer planets retained their gaseous outer coverings and most of their original size.

Mercury

Mercury is a world of violent extremes. Innermost of the planets, it orbits on average only 36 million miles from the Sun and has virtually no atmosphere to shield it from the fierce solar rays. The daylight of Mercury is a lethal mixture of intense infrared, ultraviolet, and gamma radiation that produce daytime temperatures hot enough to melt lead. By night, temperatures drop to a staggering $-280°$F.

Sunrise on Mercury must be one of the most spectacular sights in the solar system. The Sun itself will be huge in the sky, blindingly bright, dominating the landscape within seconds of its appearance over the horizon. It was long thought that Mercury kept one face permanently fixed toward the Sun. It has since been discovered, however, that it rotates on its axis once every 59 days, and circles the Sun once every 88 days.

Because Mercury is so close to the Sun, it is very difficult to observe from Earth. When it was first spotted as a tiny black dot moving across the great face of the Sun, astronomers almost mistook it for a sunspot. Until the early 1970s, most of our knowledge of the planet was vague and uncertain. Then in 1974 the American space probe Mariner 10 was sent on a close fly-by mission to Mercury. Mariner 10 was the same spacecraft that had already made a close pass with Venus some months earlier. Using the gravitational field of Venus for a kind of slingshot effect, the probe was swung away from Venus on course for Mercury.

Mariner's remarkable discoveries emphasized the enormous value of unmanned space probes in exploring the solar system. During three close passes, Mariner collected data that fundamentally altered many of the views previously held about the planet. In addition, it provided valuable evidence supporting one of the most important theories of planetary evolution.

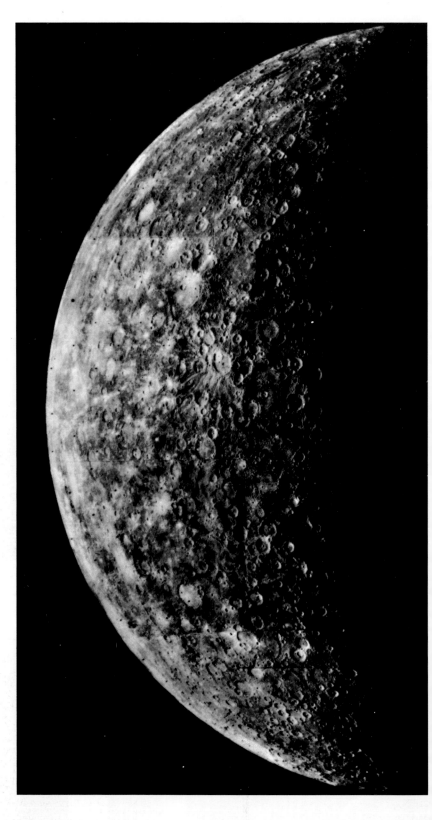

Above: the surface of Mercury, a photomosaic from pictures taken from 124,000 miles away by Mariner 10 in March 1974. Scientists discovered that Mercury has a weak magnetic field – in spite of its slow rotation and non-fluid core.

The first major surprise came when on-board magnetometers revealed a distinct magnetic field surrounding the planet. Most astronomers had thought it impossible for such a small planet – barely 3000 miles in diameter – to generate any magnetism. The discovery immediately revealed to planetary scientists that Mercury must have a dense core consisting of iron and other heavy elements, probably in a molten state much like the core of the Earth.

Other readings from Mariner's sensors showed that Mercury has an atmosphere, tenuous and barely perceptible but definite nonetheless. Astronomer's had always assumed that the intense heat of the Sun would drive off any gases from such a small planet. Probably the faint atmosphere registered by Mariner, consisting of hydrogen, neon, and argon, is vented from fissures in the planet's surface and is formed as the by-product of the decay of underground radioactive deposits.

Infrared scans confirmed the remarkable extremes of heat and cold on the planet and measurements of the rate of cooling once night had fallen provided a useful picture of structure of the surface rocks. The cooling is consistent with an insulating layer of silicate dust – much like that found covering the Moon – broken by large outcroppings of bare rock.

Most spectacular of all Mariner's triumphs was the numerous detailed photographs radioed back to Earth. Considering the technical difficulty of the feat, the quality is astonishingly good. For the first time mankind was able to study the surface features of Mercury. Their appearance reminded most of the scientists seeing them for the first time of an airborne view of the Moon. Bleak and forbidding impact craters among huge laval plains were broken only by mountain ridges and rocky outcrops. One of the most spectacular craters is Caloris Basin, over 840 miles in diameter with a floor covered in concentric ridges and cracks, unlike anything seen before. Around the rim of the basin is a system of subsidiary craters, valleys, and lava flows, obviously linked to whatever event created Caloris. Scientists are still unsure of the exact way in which this region was formed. The most likely idea is that we are now seeing the result of a gigantic impact, extensively modified by subsequent volcanic activity.

Planetary scientists are particularly excited by the evidence of Mercury's high density, about eight times the density of

water. The heavy metal core of the planet lends important support to the best current theory about the origin and evolution of the planets. According to the theory, the most dense planetary bodies "condensed" out of the primeval dust cloud surrounding the young Sun first, ahead of the more gaseous bodies in the outer regions of the solar system. Mercury certainly fits into the expected gradation of densities working from Sun outward.

Studies of Mercury are really only just beginning. It seems certain that new discoveries will await any subsequent space probes visiting the planet. Although it is unlikely that scientists will ever visit Mercury, the growth of knowledge about this tiny, Sun-scorched world may one day provide mankind with major new insights into the evolution of the solar system and its place in our changing universe.

Below: a Mariner 10 photograph of the surface of Mercury. Its heavily cratered surface confirms one theory that all the planets experienced a period of heavy bombardment early in their history and that the craters of Mars and the Moon are not due to localized, isolated bombardments. The wrinkled appearance of the planet's surface has led scientists to speculate that Mercury's large iron core has caused the planet to shrink.

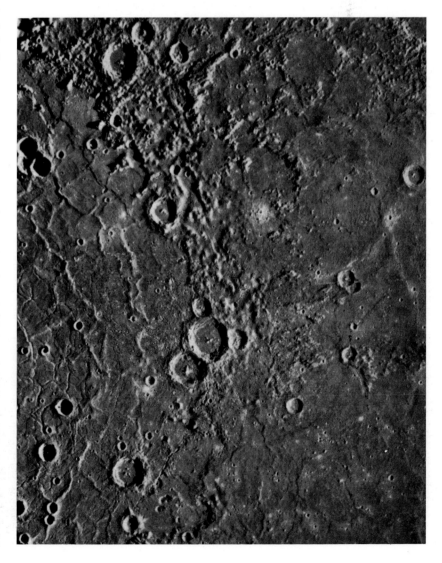

Venus

For centuries Venus has been known as the "morning star." Visible as a bright point of light low in the dawn sky, it is one of the unmistakable sights in the solar system. Its diameter of 7800 miles makes it about Earth's twin in terms of size. Every 19 months it passes within 25 million miles of the Earth, making it our closest planetary neighbor. But despite its nearness to us, Venus has always been a planet of mystery. Its surface is permanently hidden beneath a thick blanket of cloud that not even the most powerful optical telescopes can penetrate.

With no facts to go on, speculations about life on Venus flourished. A popular idea current in the early 1900s was that the surface of Venus was similar to the Earth as it was millions of years ago: a world of steamy jungles teeming with prehistoric life.

But a few years ago scientists realized that <u>the</u> high concentration of carbon dioxide in the Venusian atmosphere could make life on the surface very uncomfortable indeed. On Earth carbon dioxide makes up about one three hundredth of one percent of the atmosphere. The atmosphere of Venus may contain as much as 90 percent. Such a huge concentration would make Venus a kind of planetary heat trap. On Earth gardeners are familiar with the way a greenhouse works. The temperature inside the glass frame is always higher than the external air temperature. This is because the

Below: the planet Venus at five different phases. When "new," Venus is between the Earth and the Sun, but usually it cannot be seen at all, as its dark side faces Earth. When "full," Venus is almost behind the Sun. It is brightest when a crescent.

Sun's heat can pass through the glass but cannot get out again. Venus imitates this so-called "greenhouse effect" on a massive scale. The Sun's heat penetrates the carbon dioxide atmosphere and strikes the planet's surface, heating it. Some is reflected or re-radiated at a slightly different wavelength. But because the wavelength has changed, the heat can no longer penetrate the atmosphere. The carbon dioxide acts as a one-way barrier, allowing heat in but trapping it and preventing it leaking back into space. The result is a steadily rising temperature at the surface, leveling off at about 900° Fahrenheit, hot enough to melt lead.

The greenhouse effect put an end to the idea of lush jungle vegetation on the planet. Instead, astronomers pictured Venus as a world of endless deserts, swelteringly hot and without life. Others suggested the planet might be completely covered by an immense ocean with the possibility of a rich variety of marine life in its depths.

The first unmanned space probes to visit the planet put an end to all speculations. Venus was revealed as one of the most inhospitable planets in the solar system. All hopes of finding life were completely dashed. Far from being Earth's twin, a landing on Venus would be like a descent into hell.

American and Soviet probes in the 1960s confirmed the blistering temperatures on the planet's surface. Even hotter than Mercury, some parts of Venus must be literally red hot. The atmosphere was found to be extremely dense – the surface pressure is over 90 times greater than it is on Earth.

But the most startling feature of the environment on Venus is the cloud blanket covering its surface. The cloud

1910, September 10.

1910, June 10.

1927, October 24.

1919, September 25.

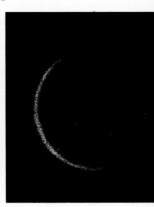

1964, June 19.

begins at an altitude of about 40 miles and has a well-defined structure. The uppermost layer consists mainly of sulfuric acid droplets for a depth of 3 miles. This yellow, corrosive cloud is responsible for reflecting up to 75 percent of all the sunlight that reaches Venus, making the planet one of the brightest objects in our night sky. From the bottom of this initial cloud layer, the Sun can barely be seen and the atmosphere has the appearance of a dimly lit fog. A second cloud layer extends from 35 to 32 miles, even more

Above: the planet Venus with its cover of swirling cloud, photographed by Mariner 10. Venus is Earth's nearest planetary neighbor and possesses a size, mass, and density only slightly less than those of Earth.

noxious than the first. In addition to droplets of sulfuric acid, it is also thick with particles of solid and liquid sulfur. Visibility in the cloud is less than a mile and the gloom is even more pronounced than in the layers above.

Below the second layer is a clear space followed by a third layer similar to those above but less dense and extending to about 30 miles above the surface. Below this the atmosphere is relatively free of cloud although probably filled with a light haze.

83

What is Venus like below the deadly clouds of acid? The thick atmosphere will cause some strange effects. Anyone standing on the surface would find it difficult to see much beyond a mile or two. The weak sunlight that penetrates the cloud will be scattered at the surface so that everything is bathed in a deep red glow. Some scientists suggest that the thickness of the atmosphere might also create bizarre optical illusions. On Earth the atmosphere can often bend light by a few degrees. The most common effect of this distortion is the creation of mirages in hot regions such as deserts. But on Venus the atmosphere is so dense that rays of light could be bent by more than 90 degrees. If this happened, an observer standing on Venus could theoretically see the entire surface of the planet rising up around him like the walls of huge circular cliffs. Fortunately the poor visibility at the surface would protect him from the full effect of such an illusion which might otherwise drive him insane. But if the lurid twilight could ever be penetrated, Venus might present one of the strangest spectacles in the solar system.

We have already had a glimpse of the surface of Venus. In October 1975 two Soviet probes touched down on the planet. Despite their rugged construction they remained in working order for only an hour. But during that time they radioed the first photographs of the Venusian landscape. Although the panoramas were restricted by the camera angles, the visibility was remarkably good. The high temperatures and corrosive atmosphere of Venus had led scientists to expect a smooth, almost polished surface. Instead, they saw a stone-covered slope, covered with loose angular rocks. The lack of any obvious erosion may mean that the rocks are quite young and that Venus is therefore still a geologically active planet. This idea is strengthened by the shelflike appearance of the landing site of one of the probes, which may be the weathered surface of a comparatively recent lava flow.

At about the same time that the first pictures were being transmitted to Earth, American astronomers were making their own survey of the surface of Venus using radar beams from the huge radio telescope at Arecibo in Puerto Rico. By

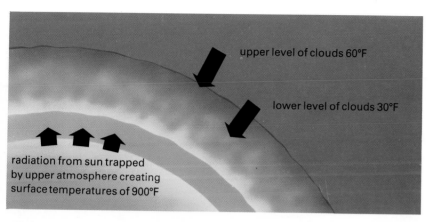

upper level of clouds 60°F

lower level of clouds 30°F

radiation from sun trapped by upper atmosphere creating surface temperatures of 900°F

Above: diagram showing how Venus' atmosphere shuts in the Sun's heat, producing the ground temperature of 900°F, as recorded by Mariner 2.

Left: an artist's impression of the surface of Venus. Above the surface swirl clouds of liquid and solid sulfur, and others that produce a rain of sulfuric acid. Although the atmosphere at the surface is clearer, it is extremely inhospitable with temperatures of 900°F and jetstream winds of up to 400 miles an hour. Tension in the planet's crust appears to have pulled some of its surface apart, according to radar pictures from the Pioneer Venus orbiter circling the planet. It detected a great rift valley 900 miles long, 175 miles wide, and three miles deep.

Below: the rocky surface of Venus, photographed by the Soviet probe Venera 9. The probe was chilled before landing so as to protect it from the intense heat. The probe transmitted for one hour.

analyzing the echos bounced from Venus they could construct a reasonably accurate picture of a large region in the Northern Hemisphere of the planet. The most striking feature of the landscape is a large pear-shaped basin more than 100 miles across. The floor of the basin seems to be comparatively smooth suggesting the action of lava flows at some stage in the past. The basin appears to be surrounded by a system of mountainous ridges.

More detailed mapping of the surface began late in 1978 when NASA placed the probe Pioneer 2 in orbit around the planet. Although only a tiny part of the planet had been mapped by mid-1979 some amazing discoveries had been made, such as a mountain – named Mount Maxwell – 8000 feet taller than Earth's Mount Everest; a plateau twice as large as Tibet's; and the largest canyon ever seen. The surface of Venus also appears to have been heavily battered by giant meteorites.

The atmosphere of Venus is not only poisonous. It is wracked by high-altitude wind storms. Photographs of the cloud layers show rapid swirling movements with clouds spiraling outward from the equator toward the poles at speeds of up to 225 miles an hour. In fact the entire upper atmosphere is rotating much more rapidly than the planet itself.

Venus rotates very slowly about its axis, so that one Venusian day lasts the equivalent of 243 Earth days. The planet is also unusual because it turns on its axis in a counterclockwise direction. The Sun – if it were visible from the surface – would therefore rise in the West and set in the East. Most other planets and moons turn in a clockwise direction and there is no satisfactory explanation why Venus should be different. The planet takes about 225 Earth days to complete one orbit around the Sun, compared with the Earth's 365.

A recent discovery about the composition of the Venusian atmosphere has cast doubt on the most popular theory accounting for the formation of the solar system. The theory suggests that the Sun and planets formed at the same time. The innermost planets – Mercury, Venus, Earth, and Mars – are thought to be small and rocky because the Sun drew their lighter constituents away. If this idea is correct, the closer a planet is to the Sun the less likely there is to be lighter gases in the atmosphere. But in the atmosphere of Venus, the opposite appears to be true. In particular, there seems to be 500 times as much argon gas and 2700 times as much neon as in the atmosphere of Earth.

So far scientists cannot explain why these gases were not drawn away from the planet during the birth of the solar system. Further discoveries about Venus may soon force a revision of the most basic ideas about how the Sun and planets were formed.

Earth

The planet Earth is the only planet in the solar system that can support advanced forms of life. The composition of its atmosphere and its distance from the Sun combine to create a generally temperate climate with enough vital ingredients – such as oxygen, carbon dioxide, and water – to support life. But the balance of factors essential to all plant and animal life is extremely delicate. Even minute changes in the Earth's orbit or fluctuations in the composition of the atmosphere could cause a total catastrophe, destroying life on Earth forever. Planet Earth is like a giant spaceship, protected from the dangers of space by complex life-support systems. Just as any astronaut monitors and maintains the systems on his spacecraft, so must mankind be constantly alert to ensure that, as civilization expands, its growing power to change the environment does not at the same time interfere with Earth's own life-support systems.

The Earth as seen from space is a magnificent sight. Predominantly blue in color because of its vast oceans and seas, its surface is crossed with layers of white cloud. It is the third planet from the Sun, on average about 93 million miles away and completes an orbit around it once every 365.25 days. Roughly spherical in shape, the Earth has a

Above: the small blue planet Earth from 22,000 miles above its surface. North·America lies in the northwest of the globe, Africa in the southeast.

Below: Clouds are a visible reminder of the ocean of air that enshrouds Earth. Not only does air – the atmosphere – nourish life on Earth, it also acts as a barrier to screen out the harmful radiation that comes from the Sun.

diameter of 7899 miles from pole to pole with a slight bulge around its equator giving an equatorial diameter of 7929.

The thin envelope of gases surrounding the Earth is called the atmosphere. It is made of a mixture of gases we usually call "air." The main constituents are nitrogen, making up about 78 percent of the air, and oxygen, which makes up 21 percent. The remaining proportion consists largely of argon gas – about 0.93 percent – and carbon dioxide – 0.03 percent. There are also traces of rare gases such as neon and krypton and a variable amount of water vapor.

The atmosphere performs two vital roles. It provides an ocean of air that enables living creatures to breathe and it also acts as a shield against some of the worst hazards of space.

Although everyone can immediately appreciate the importance of air to breathe it is more difficult to imagine the canopy of the atmosphere as an important roof over people's heads. But at every moment the atmosphere is shielding us from lethal radiation from the Sun. Seen from the Earth, the Sun is a provider of warmth and light. Families spend large sums of money to travel to sunny regions of the world just to be able to enjoy its

health-giving rays. But beyond the atmosphere, the Sun is a powerhouse of deadly radiations. In addition to ordinary visible light, it produces vast amounts of ultraviolet radiation. On Earth, the ultraviolet in sunlight helps to give us a healthy tan. But in space, it is so intense it quickly destroys life. The atmosphere acts as a huge ultraviolet filter with the most intense rays penetrating no farther than an altitude of about 30 miles.

But the atmosphere does not just block invisible radiation. It also shields Earth against collisions with meteors. Space is filled with countless fragments of rock and metal, some no bigger than a grain of sand, others boulder-sized or even a mile or more across. These fragments hurtle through space at several miles a second, occasionally being attracted by a planet's gravitational field. The effect of impacts by meteors can be seen on the face of Earth's nearest neighbor, the Moon. Over millions of years, its surface has become pitted and scarred by thousands of violent collisions. If this had been happening on Earth, life would have had no chance to evolve as quickly as it has done. Fortunately,

Below left: the table shows the relative percentage of x-rays, ultraviolet, visible light (subdivided into the colors of the spectrum), and infrared rays radiated by the Sun.

Below: the diagram illustrates absorption, scattering, and reflection of incoming solar radiation by gases, clouds, and particles in the atmosphere and by the Earth's surface. Less than half of all incoming radiation reaches the Earth. Even so, this fraction, transformed to heat, is enough to warm the sea, land, and lower atmosphere. By making water evaporate and air move about, this radiation helps create weather.

however, when meteors plunge into the Earth's atmosphere, the friction between their surfaces and the molecules of the air, causes them to heat up. Eventually they become red hot, disintegrating in a plume of flame and incandescent fragments. On a clear night it is possible to see several meteor tracks and as they blaze across the sky, it is easy to understand why for centuries they have been called "shooting stars."

The atmosphere is made up of distinct layers. The lowest is called the *troposphere*. It extends to an altitude of around 10 miles and contains about 75 percent of all the gases in the atmosphere. It is the region in which all the world's weather occurs. Clouds, winds, and air currents affect the troposphere more than any other region. Clouds are formed by evaporating water vapor while winds and currents are caused by unequal heating of the air. The temperature is highest at the Earth's surface, falling to about −140°F at the upper limit of the troposphere. This difference in temperature keeps the air in constant motion.

Above the troposphere is the *tropopause*, a region mainly free from the

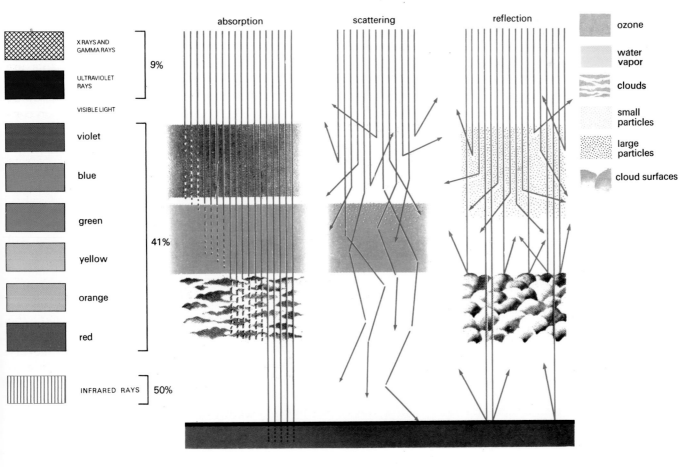

turbulence of lower levels. It is really an intermediary layer that quickly gives way to the next important region – the stratosphere. The tropopause is remarkable, however, because of a layer of extremely high winds that race horizontally along its upper reaches. Known as the jet stream, it is quite common for wind speeds to reach 100 miles an hour and velocities of nearly 200 miles an hour have been recorded.

The stratosphere extends above the troposphere to an altitude of about 30 miles. Air in the stratosphere is thin and dry and almost weather-free. Because of the presence of an oxygen-like gas known as ozone, the temperature of the stratosphere actually rises with altitude and at the upper limit of the region, it is almost as warm as at sea level. The ozone warms the air by absorbing ultraviolet radiation from sunlight, providing the vital filtering agent that protects life on the Earth's surface.

Above the stratosphere are various other layers with the outermost, the *exosphere*, starting at an altitude of over 300 miles and gradually fading into the vacuum of space itself. The zone extending from about 50 to 375 miles is known as the *ionosphere*. It takes its name from the large number of electrically charged atoms – *ions* – that are trapped

in this part of the atmosphere. The ionosphere is especially important to modern life because it reflects radio waves, enabling radio transmissions to be literally bounced from the sky so that they can reach receivers all over the Earth.

The structure of the Earth itself is just as complex as the atmosphere surrounding it. In fact, less is known about the processes taking place deep inside the Earth simply because there is no direct means of investigating what is taking place. Knowledge of the composition and workings of the interior has been built up from studies of seismic waves from earthquakes and man-made explosions. Sound waves traveling through the various layers of the Earth move at different speeds and are reflected and bent by the different rock structures they encounter. Analysis of the way such vibrations behave, therefore, tells scientists a great deal about the nature of the world beneath our feet.

Life exists on a thin skin of solid rock known as the *crust*. Under the oceans the crust is an average of only three miles thick whereas in continental regions it has an average thickness of about 20 miles. Beneath the crust is a solid area of *mantle* going down 60 miles. The crust and solid mantle float on a semiliquid, viscous layer of upper mantle extending

Above: mountains are a common feature of the Earth's surface and are created by slow, large-scale movements of the crust. Because of different reactions of the Earth's surface to these movements there are three main ways in which mountains are formed – by pressures within the crust creating uplift of the surface layers, by volcanic activity, and by erosion.

Right: diagram of a cross section of the Earth illustrating how new crust is formed. New material wells up from the mantle at the oceanic ridges, spreads over the ocean floor, and descends again into the mantle at the oceanic trenches.

downward to about 150 miles. Below this is the mantle proper, which at a depth of 1800 miles gives way to the liquid outer core. At a depth of 3000 miles the solid, iron-rich core of the Earth extends for the final 800 miles to the very center of the planet.

The immense stresses from within the Earth play a continuous role in changing the face of the surface layer. One of the most remarkable ideas explaining how the land masses of the world reached their present configuration is known as the theory of continental drift. The origin of the theory was a simple and obvious observation. In 1911 a German meteorologist, Alfred Wegener, pointed out that the continental regions of Africa and South America appeared to fit together like the pieces of a gigantic jigsaw puzzle. He suggested that the continents may once have been joined and that at some point in the remote past they broke away from one another and drifted apart. At first Wegener's ideas were ridiculed. Who could possibly believe that the solid and obviously unchanging continents could drift about like rafts on a sea? But some years later the mechanism by which such movements could take place was discovered and the theory of continental drift became an established part of geophysical science.

Above: distorted sedimentary rock layers at Lulworth Cove in Dorset, England. They furnish an example, on a miniature scale, of the process of mountain-building. The folding and buckling action clearly shown here is also accompanied by shearing of the rock layers – a process which takes place over hundreds of miles of rock.

The basic idea of the drift theory is that the Earth's solid crust and mantle region is cracked like an eggshell, forming six large segments and several smaller ones, and known as *plates*. The plates are floating on the viscous upper mantle just like rafts on a sea. The plates drift under the influence of convection currents in the mantle, sometimes pulling away from one another and at other times rubbing against one another. The continents ride on the plates, being carried over the face of the Earth on journeys determined by the convection currents in the mantle. The movement is infinitesimally slow, only perceptible over hundreds of millions of years. But its effect can be sudden and dramatic. Along lines of stress where two plates press together, earthquakes and volcanic activity may be common. Slower kinds of change may also take place. Where one plate rides under another deep trenches in the Earth's surface are formed. Alternatively, the crust may be buckled upward slowly forming new mountain ranges. The Himalayas – the largest mountain range in the world – were formed by the collision between the edges of two plates.

Mountains are among the most spectacular features on the Earth's surface. There are three main kinds – dissected mountains, formed by erosion; structural mountains formed by some kind of uplift from inside the Earth; and volcanic cones.

Dissected mountains are cut from an original flat landscape by the action of water and weather. The most common process is for a river to cut a drainage channel into the land. As its valley

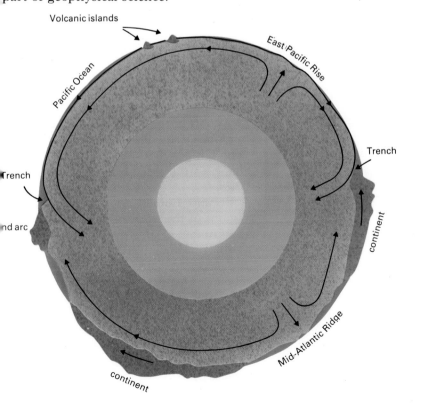

Volcanic islands

East Pacific Rise

Pacific Ocean

Trench

Trench

nd arc

continent

Mid-Atlantic Ridge

continent

ridges high above the surrounding terrain.

Volcanos are common in all areas where mountain building has taken place. In most regions of the Earth, the mountains were formed in the remote past and obvious evidence of volcanic cones has been eroded away while actual volcanic activity has long since ceased. But in areas of fairly young mountains or where mountain building is still taking place, active volcanoes are a spectacular part of the landscape.

As uplift takes place and mountains rise, many faults and cracks occur in the surrounding terrain. These sometimes fill with molten rock – a fearsome, glowing liquid called magma – forced up under pressure from the mantle. In most cases, the magma cools and solidifies before it reaches the surface. But sometimes it reaches the surface molten, erupting in a cloud of dust, rock, and ash and producing flows of magma, known as lava.

The shape of a volcano depends on the kind of lava it ejects. Free-flowing lava such as basalt tends to run quickly away from the vent hole, hardening and building up a low, squat mountain with a wide base. Because of their shape they are sometimes called shield volcanoes and some of the best examples are those common in the Hawaiian islands. Volcanoes that give out thick lavas, which stop flowing and harden quickly, form into domes – relatively small features with steep sides. The most spectacular volcano is the type that has hot gas mixed in its lava. When the lava reaches the surface, it is literally blown out of the vent. The lava cools in the air forming small pieces of cinder that collect around the original vent. Cinder cones are the classic volcano shape – steep sided but with fairly broad bases.

Because humans live only on the land regions of the world, it is often forgotten that many of the geological processes visible on land are also taking place under the sea. In fact, although the most advanced forms of life are land-dwelling, the world is really a water planet. About two thirds of its surface is covered with lakes, seas, and oceans and much of the world's climate depends on an important water-based process known as the hydrological cycle.

Water is continuously moving from

deepens, the forces of wind, frost, sun, and rain work on the valley sides forming the rough faces of hills and mountains. The Mesas of central North America are an example of dissected mountains.

Most major mountain ranges, however, are structural in origin. The exact way in which they form varies from place to place. Most are the result of huge folds in the crust, which cause powerful uplifts that raise rocky peaks and connecting

Above: molten rock jets high into the sky against a background of ash and smoke emitted by an Icelandic volcano in 1973. Iceland is directly over one of the divisions between the large plates of crust that float on the magma layer of the Earth. As the plates diverge, hot magma wells up between them (mostly in oceanic areas) and intense volcanic activity occurs.

oceans to land and back to the oceans. The cycle consists of water evaporating from the Earth's surface and rising as vapor into the atmosphere. As it rises and cools, it condenses to produce droplets of water that fall back to the Earth in the form of rain, snow, sleet, or hail. Rivers then return most of this to the lakes and seas. The pump that keeps the cycle going is the Sun. Without its heat, the energy would not be available to keep the water circulating.

The ocean is the largest unexplored region of the Earth. Startling discoveries are still being made. One of the most remarkable finds in the history of science was uncovered recently when deep-sea probes found strange plant and animal life at the bottom of the ocean in the vicinity of vents that allow heat from below the Earth's crust to bubble through. It has always been accepted that sunlight was a given in the evolution of life, and yet no sunlight ever penetrates to the depths where these creatures dwell. These deep-sea life forms, sustained by thermal energy, have evolved in ways that were previously unimagined.

Life evolved from complex chemicals first synthesized in the ocean, billions of years ago, and mankind is now slowly returning to the environment from which its remote ancestors emerged. New developments in diving and submersible technology and off-shore engineering are at last making it possible for scientists to utilize

Above: an oil rig burning off natural gas in the North Sea. The oceans are one of the few unexplored regions of the Earth, though the latest advances in technology are enabling scientists to utilize more of the ocean's resources.

Below: Tokyo, typical of the busy, noisy, overcrowded cities of much of the world. Overpopulation places heavy strains on the world's resources.

some of the ocean's resources. Most important in recent years has been the discovery of undersea oil fields and some ocean scientists argue that efforts should be made to establish permanent undersea communities that could mine the oil and mineral rich seabed.

What is the future of planet Earth? It already had a history of 4600 million years and for about 3700 million years, it has successfully supported life. But within the last 100 years a rapidly expanding technological civilization has begun to strain the resources of the planet. More disturbing even than this is the growing pollution of the sea, land, and air by the effluents that are the by-products of developing society. The real danger of the situation is that it is often difficult to predict the long-term effects of technological activity on the environment. Mankind cannot fully consider the implications of its actions because there is no way of working out what those actions may be – the possibilities are too complex for anything but approximate predictions of what may happen.

Even though more and more people are becoming aware of the dangers of overexploiting and polluting their world, only time will tell whether the Earth will have a long-term future as the cradle of life or whether it may one day be reduced to a lifeless and barren world like so many of the other planets in the solar system.

Moon

Our closest neighbor in space is the Moon – Earth's only natural satellite. On average it orbits the Earth at a distance of 238,900 miles, close enough for many of its surface features to be seen with the naked eye. Although only the sixth largest in the solar system, Earth's moon is probably larger than any other in comparison with its mother planet. Its diameter of about 2200 miles is roughly a quarter of the Earth's.

Even before the first manned expeditions of Project Apollo, astronomers knew that the Moon was a very different world from Earth. The main reason is its mass. Although a fairly large body compared to the Earth its mass is only about 1.2 percent of the Earth's. Because gravity depends on mass, the Moon's gravity is relatively low – only 16 percent that of Earth. This means that the Moon cannot hold onto an atmosphere and any envelope of gases that may have once surrounded it has long since leaked away into space. And without an atmospheric pressure acting on it, any surface water has evaporated and also drifted away into space. The Moon therefore is an airless, waterless world. But the absence of an atmosphere also means that its surface is not protected from the full glare of radiation from the Sun. Many of the deadly emissions, such as ultraviolet and x-rays, that are shielded from the Earth by its atmosphere are a normal part of sunlight on the Moon. To make things worse, meteors that normally burn up in a planetary atmosphere, fall unhindered onto the Moon, often producing huge impact craters.

The Moon differs from the Earth in another important way. In the course of a complete orbit around Earth, it rotates only once on its axis. A day on the Moon therefore lasts for two Earth weeks. And because of the way the Moon spins in its orbit, the same hemisphere is always turned away from the Earth. The so-called far side of the Moon remained a

Above: a painting of about 1700 by Donato Creti showing the Moon as it looked when observed through a Galilean telescope. The invention of the telescope in 1608 opened the way for the development of astronomy, and the Moon as Earth's nearest neighbor began to come under close scrutiny.

mystery to Earthbound astronomers until 1959 when the Soviet space probe Lunik 3 made the first lunar circumnavigation. The historic pictures it radioed back to Earth showed a similar moonscape to the known hemisphere but less cratered and with fewer large plains.

How was the Moon formed? Even today astronomers are not sure of the answer. Three main theories have been proposed. The earliest was suggested by a British mathematician, George Darwin, in the late 1800s. His idea was that the Moon was once a part of the Earth and somehow broke away to form an independent planet. The modern version of this theory suggests an ancient cataclysm that rocked the Earth during the very process of its creation.

Some 4,600 million years ago the Earth was still semimolten. As the heaviest elements sank through the liquid rocks toward the center of the Earth the change in the planet's center of mass made it rotate on its axis more quickly. Eventually when the Earth was spinning so fast that a day was little more than 2 hours 30 minutes long, a huge bulge in he Earth's crust developed, thrown out by the tremendous forces caused by the rapid spinning motion. As soon as the bulge formed, the whole shape of the Earth became unstable, passing quickly through a number of distortions until a huge fragment of the planet actually broke away. This dramatic picture of the Moon's formation is supported by the similarity between the density of lunar rock and material in the outer layers of the Earth's crust. But the chemical composition is quite different. So far no one has been able to reconcile this anomaly with the theory.

The second major theory suggests that the Moon is a kind of sister planet to the Earth, formed independently but at about the same time and in roughly the same place. To explain why the chemical composition of the Moon differs from the Earth, supporters of the theory argue that the Earth formed from the primeval cloud of dust and gas first and the Moon was created soon afterward from the chemically different leftover material.

Perhaps the most remarkable hypothesis of all is the idea that the Moon comes from another part of the Universe altogether. This theory suggests that the Moon was once a wandering minor planet captured by the Earth's gravitational field millions of years ago.

Although the mystery of the Moon's origin remains unsolved, the evidence brought back to Earth by the Apollo lunar expeditions has revealed most of the Moon's more recent history and has at last given us a detailed insight into its structure and composition.

The Moon has had a violent past during which a hail of meteor impacts have pitted and cratered its barren surface. One of the most striking features of the lunar landscape is the large areas of plains that astronomers call *maria*, the Latin word meaning "seas." Most of these were formed between 3000 and 4000 million years ago when the Moon's gravitational field attracted some large objects – probably small asteroids. These plowed into the lunar surface leaving huge impact basins as scars. The largest of these spectacular features on the near side of the Moon is the Mare Imbrium, the Sea of Rains. Over 650 miles across, the huge basin was formed when an asteroid about 50 miles in diameter col-

Above: two of the Moon's phases. At the top is the last crescent before the new Moon; at the bottom is the full Moon. The dark, waterless "seas" can be seen in the photograph.

Below: a nearly vertical photograph of the far side of the lunar surface. It was taken with a telephoto lens during the Apollo 8 mission. The whole of the lunar surface is pockmarked, the various craters originating in volcanic activity or meteorite impact.

lided with the Moon.

Although the seas were initially formed by impacts they have subsequently been affected by volcanic action. Heat generated deep within the Moon by the decay of various radioactive elements produced large subterranean pockets of molten lava. The lava forced its way to the surface wherever the crust was cracked, which was mostly in the regions that had suffered major impacts. The great basins therefore were filled with smooth flows of lava producing even and firm rock floors that 1000 million years later would provide excellent landing sites for the first manned expeditions.

As well as the huge lunar plains, many individual craters are easily seen from Earth. Some, like the crater Tycho, were formed as recently as 100 million years ago. A characteristic of most young craters is a system of prominent rays, plumes of light-colored surface material thrown outward by the force of the original meteor impact. Tycho itself is surrounded by rays extending over 50 miles in every direction. Older craters lose their rays over the millennia as the soil surrounding them is pounded and churned up by the countless impacts of micrometeors – tiny particles of rock often no larger than a grain of sand. The huge regions of uplands that include many impressive peaks are in striking contrast to the flat lunar seas. On Earth it would take millions of years for such rugged hills and mountains to form. But on the Moon none of the usual processes of mountain building can take place. The

huge peaks and ridges are the walls of craters and basins, thrown up in minutes following the impact of a meteor or asteroid. On the southeastern side of the Mare Imbrium for example is a mountain range known as the Apennines – stretching 120 miles from the ancient crater Eratosthenes in the south to the 6000-foot Mount Hadley in the east.

The region around Mount Hadley was chosen as a landing site for one of the Apollo missions. The main reason was to investigate an area of upland for the first time. But scientists were also intrigued by a nearby surface feature spotted previously from lunar orbit. Known as the Hadley Rille, a sinuous channel over 80 miles long, many people believed it was evidence of the action of water millions of years ago. But although it looks like a dried-up river bed, close study revealed that it was in fact the remains of an ancient lava flow. Similar rilles have been seen in other regions, while on the outskirts of some of the larger mare, rille-like systems of ridges, sometimes shaped like concentric circles, have been observed. They were also formed by lava but instead of the molten rock etching the surface into a steep-sided valley, ridges were thrown up as lava cooled under the lunar surface causing sudden contractions in the crust.

All the scientific observations and measurements show the Moon to be a bleak and forbidding place.

Above: a lunar rover, used by the Apollo 15 crew in 1971. It enabled Irwin, seen at the vehicle, and Scott to collect the richest scientific haul of the whole program.

Below: a piece of Moon rock. Later analysis of the samples brought back to Earth provided scientists with much valuable and interesting information. Some of the samples contained large numbers of "glass marbles" which are believed to have been ejected from the interior of craters. Many of the samples contained elements similar to those found on Earth – though with differences in chemical make-up, with a larger proportion of the higher melting-point elements such as chronium, titanium, yttrium, and zirconium.

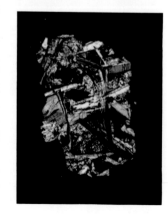

No one seriously expected to find any living organisms there, even though astronauts taking part in the Apollo lunar landing missions went through a period of quarantine after their return to Earth just in case there were some virulent bacteria on the Moon that might threaten the health of people on Earth. Scientists are still trying to decide whether or not the Moon is alive volcanically, however.

Much of the lava action seen on the Moon is extremely ancient. But from time to time since the 1700s astronomers have reported seeing flashes of light and strange glows from certain regions. Although none of the Apollo missions discovered evidence of volcanic processes taking place today, some areas of natural radioactivity were discovered. These are probably due to radioactive gases being vented through the surface from the Moon's interior and could well cause fleeting glows to appear, visible even from the Earth.

There now seems little doubt that the surface of the Moon is geologically dead. Instruments such as seismic detectors, left by Apollo astronauts, have so far failed to reveal a single moonquake. In July 1972, however, a seismograph picked up vibrations from a large meteor that had struck the far side of the Moon. The shock waves had passed through the Moon and from their behavior scientists were able for the first time to work out what the Moon's interior must be like. Their conclusion was remarkable. Despite its bleak, unchanging surface, the Moon's core is hot and partially molten – much like the innermost core of the Earth.

Despite the vivid and detailed picture of the Moon we can now piece together, for most people its most important feature is its appearance in the night sky. Like any planet, the Moon shines only by reflected light but because of its closeness to us, a full Moon is a spectacular sight. The various phases of the Moon are formed by its position in relation to the Sun as seen from Earth. The new Moon – sometimes called the dark of the Moon – occurs when the Moon passes either above or below the Sun. The first crescent appears a day or two later. As the angle between the Sun and Moon increases, the crescent grows first to a quarter when the angle is 90°

to full Moon at 180°. Then as the angle begins to decrease and the slanting rays illuminate less and less of the Moon's face, the phases move back toward the dark of the Moon. The complete cycle – usually called lunation – takes on average 29½ days.

A less familiar sight is the spectacle of a lunar eclipse. Like any other solid body, the Earth casts a shadow. Occasionally when the Moon moves directly into the shadow so that Sun, Earth, and Moon are exactly aligned with the Earth in the middle, the source of the Moon's light is cut off. When this happens the Moon is said to be eclipsed by the Earth and plunges for several minutes into semi-darkness. Usually it does not disappear completely. The Earth's atmosphere bends some of the light from its Sunward

Below: an Apollo 11 photograph of a full-disk Moon. The lack of atmosphere means that there is no veiling of the surface features. The dark area just right of center is the Sea of Crises (Mare Crisium), which covers about 66,000 square miles. To the lower left is the Sea of Tranquillity (Mare Tranquillitatis). It covers about 110,000 square miles. In this photograph North is at the top, whereas most photographs – and all the early drawings – show South at the top because astronomical telescopes give an inverted image.

side onto the lunar surface and the Moon, even in the depths of an eclipse, can be seen as a dim, shadowy object, often with a slightly reddish tint.

The breakthrough in mankind's knowledge of the Moon came with the first manned expeditions. More has been learned since the first astronaut stepped onto the lunar surface in 1969 than in all the previous 2000 years. Before the end of this century the first permanent colonies may well have been established and for the first time in over 4600 million years, life will have come to the Moon's forbidding surface.

The Moon, our closest neighbor and the most beautiful sight in the night sky, may one day be a stepping stone from which mankind can begin an exploration of the solar system and the stars beyond.

Mars

Above: a photograph taken before one of the Viking landings on Mars in 1976. Previous Soviet and United States missions had found a lifeless looking planet. The Viking landings provided far more detailed information and even a suggestion that there might be some form of primitive life on the planet.

In 1877 Mars was making one of its close approaches to Earth. Astronomers estimated that at one point during the year, the planet would be less than 40 million miles away. All over Europe the most advanced telescopes available were being readied for a detailed examination of the so-called red planet.

In Milan, Giovanni Schiaparelli was among the luckiest observers. Taking advantage of a few brief moments of exceptional clarity in the Earth's atmosphere he made some observations of Mars that were to start a legend. To his astonishment he made out a series of regular lines apparently lacing the Martian surface. He completed a few rough sketches of them and announced his discovery to the world. He called the lines *canali*, the Italian word for "channels." But before he could contradict them, other astronomers translated Schiaparelli's word into "canals." Soon the idea was established that the face of Mars was covered with a remarkable system of artificial waterways.

A few years later, an American astronomer, Percival Lowell, working from his own private observatory in Flagstaff, Arizona, confirmed Schiaparelli's observations and elaborated on the origins of the waterways. According to Lowell, the great canal system was a relic of a past civilization. Climatic changes on Mars had made temperatures rise dangerously and had led to planet-wide drought. To survive the new conditions Martian engineers had attempted to irrigate their world by drawing water from the melting polar ice caps. Obviously they had only

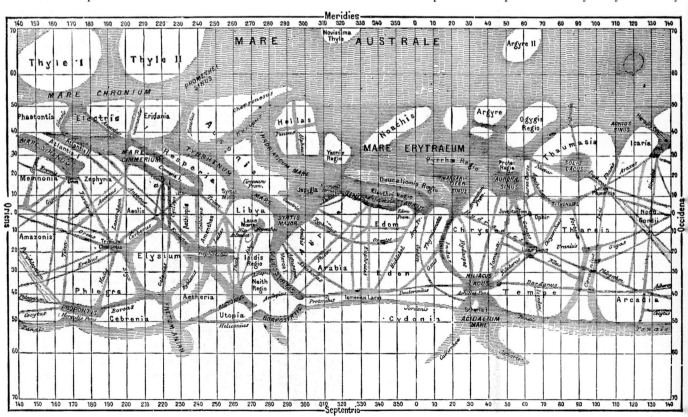

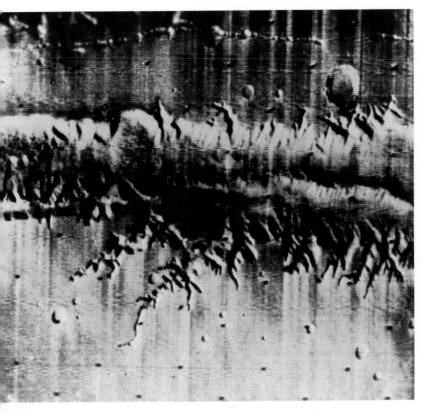

Above: a vast Martian canyon photographed by Mariner 9. It is part of the Tithonius Lacus Mosaic, which has a total length of 3500 miles, with a width of up to 250 miles.

Left: the canals of Mars, drawn by the Italian astronomer Giovanni Schiaparelli in 1877.

Below: Percival Lowell, the 19th-century American astronomer who carried out an intensive study of Mars and came to the conclusion that the canals of Mars were actually irrigation channels used by intelligent beings to carry water from the polar icecaps to more arid regions.

put off the eventual extinction of their race leaving the canals as a lasting relic of their efforts, so huge in scale as to be visible from over 40 million miles.

Today we know that Mars probably has undergone massive climatic upheavals in its past. Perhaps it was in fact once the home of an advanced civilization. But close examination of the Martian surface by unmanned space probes both in orbit and on the surface itself has revealed no trace of either the canals or the race supposed to have built them. The intricate markings seen by Schiaparelli and Lowell were almost certainly no more than a trick of the Earth's atmosphere.

But even if we have found no traces of past civilizations as yet, Mars may still provide us with our first contact with extraterrestrial life. Experiments performed by the American Viking landers in the 1970s designed to spot any life forms, have been tantilizingly inconclusive. What is known about surface conditions, however, suggests that any Martians that are discovered will be very rudimentary creatures, perhaps bacteria or green algae. Certainly more advanced creatures could not survive the hostile environment.

Mars is smaller and less dense than the Earth. The closest similarity it has with Earth is the length of its day, only 23 minutes longer than a day on Earth. The atmosphere of Mars is extremely tenuous, far too thin to sustain human life. There also appears to be little free oxygen, the Martian air being mainly composed of carbon dioxide and nitrogen. Because Mars is on average 142 million miles from the Sun, it receives about half as much radiant energy as Earth. Its thin atmosphere creates great extremes of temperatures, ranging from about 90°F at noon on the equator to −50°F just before dawn. At the poles temperatures as low as −193°F have been recorded.

One of the greatest puzzles about Mars is whether there is any water on the planet. Photographs taken from orbit by the American Mariner and Viking spacecraft, show geological features indicating the action of water at some time in the past. One particularly spectacular feature is the great rift valley named Coprates. The valley runs 3000 miles east to west, near the Martian equator. In places it is about 50 miles wide and more than a mile deep. Running roughly perpendicularly away from Coprates are innumerable sinuous channels, meandering through the highlands around the valley and often joined by other tributary channels. These and other similar discoveries have led scientists to speculate that Mars may have suffered several major changes in its climate during past millennia. The present-day American astronomer Carl Sagan has suggested that these changes take place in cycles linked with small variations in the planet's orbit around the Sun. Sagan believes that every few thousand years, water returns to the surface of Mars, the atmosphere thickens and the once dead planet becomes capable of sustaining life again. Perhaps, together with the water and regenerated atmosphere, a long-dormant form of life reappears to flourish briefly in the interval before the next great climatic reversal.

Much of the Martian countryside appears to be dominated by immense dust-covered plains. Dune systems have been spotted from orbit similar to those found in deserts on Earth. From time to time huge dust storms race across the planet. The ferocity of these storms can scarcely be imagined. In 1971, for ex-

ample, Mariner 9 arrived to find the entire hemisphere of the planet below enveloped in a raging cloud of dust. The speed at which the storms spread suggest wind speeds of over 300 mph.

Mars is far from being a featureless dusty plain. Some spectacular mountainous features have been seen from orbit. Best known of these is Mount Olympus, a giant lava cone 300 miles across at its base and approximately 87,000 feet high. Numerous other volcanic peaks and craters have been mapped along with other features that are clearly the result of meteorite impacts. One region known as Hellas is a single vast impact basin probably created when a small asteroid collided with Mars sometime in the remote past. Over 1100 miles across and three miles deep, the floor of the basin shows signs of complex lava flows that may have been caused by tremendous heating of the Martian soil during the impact itself.

Above: this color photograph from the Viking 1 probe of July 1976 provides proof that Mars is in fact a red planet. The whole of the Martian surface appears to be covered in rocks and dust, and it is as heavily pitted with meteorite craters as the Moon. Huge volcanoes also feature in the landscape – some, like Olympus Mons, are more than three times the height of Mount Everest.

Right: a computer-enhanced picture of Deimos, the smaller of the two satellites of Mars. It was created by a pair of images of Deimos taken from Orbiter 1. One was through an orange filter, another through the camera's violet filter. The images were combined into a single picture to search for color differences on the surface of Deimos.

Even the early observations of Mars revealed the existence of polar regions. The regular seasonal advance of ice caps originally encouraged the belief that a ready source of water was available on Mars to sustain life. More recently it has become clear that the Martian poles are not covered with ordinary ice. Much of it is solidified carbon dioxide frozen out of the atmosphere by the bitter polar climate. However, detailed analysis seems

to show a good deal of water ice mixed with the frozen gas and other evidence suggests a layer of permafrost extending about a mile below the surface of the near-polar regions. Judging by the thick fog banks that have been observed from time to time, at least some of this moisture evaporates in the relative warmth of the Martian day.

When the first astronauts step onto the surface of Mars, the landscape will seem rugged and forbidding. For most of their time they will be gathering soil samples and studying rock formations. But one of the most exciting views they will have will come when they look away from the planet's surface and scan the Martian sky. There above them will be the twin moons of Mars, Phobos (fear) and Deimos (terror).

Until recently the origins of the moons were surrounded in mystery. Some observations suggested that Phobos, in particular, was in a slowly decaying orbit. If this was the case, the moon was being slowed by the drag of the upper atmosphere of Mars. In 1960, a Soviet astronomer, I. S. Shklovskii, worked out that to be affected by atmospheric drag at an altitude of over 6000 miles above the Martian surface, the moon's density would have to be one thousandth that of water. Since no natural substance has such a low density, Shklovskii concluded that the moon was hollow. A vast hollow object nearly 10 miles across could hardly be a natural phenomenon and Shklovskii suggested that perhaps both

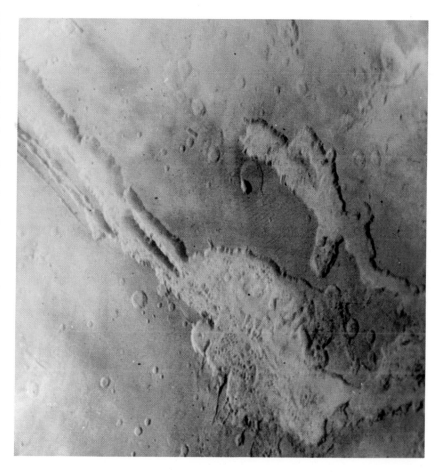

Above: this huge canyon, Valles Marineris, is part of a vast, highly fractured area south of the Martian equator.

Below: sand dunes in the Planitia Basin of Chryse, seen from the Viking 1 lander.

Phobos and Deimos were actually artificial satellites launched in the distant past by an advanced civilization.

The exciting hypothesis was dramatically shattered by the first close-up photographs of Phobos. Taken by Mariner 9 at a distance of less than 4000 miles, the moon appeared unmistakably natural in origin. Instead of a gleaming space vehicle, Phobos turned out to be a large, irregular hulk of rock, pitted by impact craters. Photographs obtained later of Deimos told a similar story.

Although our knowledge about Mars has grown rapidly since the planet was visited by the unmanned space probes Mariner and Viking, many unanswered questions remain. Perhaps the mystery scientists most want to resolve is whether any form of life exists in the Martian soil. Further remote probes will no doubt produce vital new information about the possibilities. But only when the first manned expeditions touch down will the real exploration of Mars begin. Once intensive planet-wide surveys can be undertaken guided by human judgment and intelligence, the Martian world will at last give up its remaining secrets.

Jupiter

normal liquid hydrogen 15,000 miles deep.

The atmosphere above the surface of Jupiter is a complex mixture of hydrogen, helium, ammonia, methane, and water. The lower regions of the 600-mile-thick belt of gases is probably layered with ice crystals. The atmosphere is filled with turbulent weather systems and great convection currents of rising and falling gases. The planet's rapid rate of rotation – the Jovian day lasts nine hours five minutes, the shortest in the solar system – creates ferocious easterly and westerly wind systems that mix the bands of variously colored gases into remarkable loops, swirls, and plumes.

The most striking feature of Jupiter's spectacular multicolored atmosphere is the Great Red Spot. It was first noted in 1664 by a British astronomer, Robert Hooke, so we know it is at least 300 years old. Today scientists think it may have existed for thousands of years. Since it was first observed, its shape and color have varied and the latest close-up photographs even show a complex structure inside the Spot. At present it is nearly 7000 miles wide and 25,000 miles long. Scientists believe it to be a gigantic storm zone consisting of a mixture of gases caught in a violent and ceaseless vortex.

Scientists sometimes describe Jupiter as a failed star. Certainly it bears little resemblance to the inner planets of the solar system. Heat trapped deep inside it since its formation, or generated by the decay of radioactive elements has raised the temperature of the core to a staggering 60,000 degrees Fahrenheit – over three times the temperature on the surface of the Sun (18,000°F). Overall, Jupiter emits twice as much radiation as it receives and if it were more massive it could quite soon turn into a thermonuclear furnace that would transform the planet into a minor star.

Between the intensely hot core and the bitterly cold upper atmosphere, Jupiter has a bewildering range of temperatures and pressures. Almost certainly there are some fairly temperate zones. Scientists are now seriously considering the possibility that life may have evolved in these regions. The idea is not as far-fetched as it sounds. The Jovian atmosphere appears to contain all the

Jupiter, giant of the solar system, is 300 times more massive than the Earth. It has a diameter of 89,250 miles and never comes closer to Earth than 370 million miles. Unlike the smaller planets, Jupiter has retained most of the hydrogen and helium that went into its making during the birth of the solar system. Because of this the planet may have only a small rocky core, surrounded by a thick shell of solid and liquid gases.

The exact structure of Jupiter remains a mystery but recent fly-by missions by the American Pioneer and Voyager probes have made it possible to work out a rough model of the planet's interior. At its center an Earth-sized core of iron-rich minerals is surrounded by a shell of dense liquid hydrogen thousands of miles thick. The hydrogen is so densely packed that it behaves almost like a metal, and electrical currents generated in the shell are probably the source of Jupiter's powerful magnetic field. Enclosing the layer of metallic hydrogen is an ocean of

Above: Jupiter, largest planet in the solar system, photographed by Voyager 1 from a distance of 29 million miles. It was during the Voyager mission that a ring was discovered around the planet that had not been detected by Earth telescopes.

basic ingredients from which complex organic molecules could be synthesized. It is also known that the atmosphere is constantly blasted by violent lightning storms. Powerful electrical discharges in a mixture of hydrogen, methane, ammonia, and water could well lead to the formation of basic amino acids – the building blocks of life. Fantastic as it may seem, living organisms could well have evolved drifting in Jupiter's turbulent atmosphere.

Jupiter has 13 moons orbiting at average distances ranging from 108,000 miles to nearly 14 million miles. The four principal moons – called Galilean moons after Italian astronomer Galileo who first discovered them – are Io, Europa, Ganymede, and Callisto. Early in 1979 American space probe Voyager I sent back superb close-up photographs of each of them. Io – second closest to Jupiter – turned out to be the most

Below Jupiter's great red spot, photographed from Voyager 1. The spot, some 30,000 miles long by 8000 miles wide, was found to be the greatest of a number of rotating storms, which behave like the hurricanes on Earth. The planet also has a very powerful magnetic field surrounded by zones of intense radiation over the equatorial regions.

remarkable. It has a yellowish brown surface unmarked by the usual impact craters but with broad plains crossed by cliffs, channels, fault lines, and basins. Most startling of all was the most likely reason why no craters were visible. At least eight active volcanoes were spotted erupting in great sulfurous clouds. Io's surface is therefore being constantly remade by layers of volcanic rock and lava ejected from deep within the tiny planet.

An even more exciting discovery made by the cameras of Voyager I was a ring around Jupiter similar to those encircling Saturn and Uranus. It lies well inside the orbit of Amalthea, innermost of the Jovian moons, less than 7000 miles above Jupiter's swirling cloudtops. Apparently made of rocky debris, Jupiter's one ring could well be the broken remains of another moon or perhaps the leftovers from Jupiter's formation 4600 million years ago.

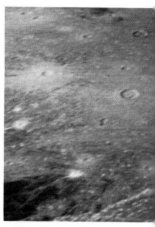

Left: Ganymede from 1,600,000 miles, and (above) from about 400 miles. Ganymede, the half rock, half water Jovian major satellite or moon, is larger than Mercury. Jupiter has 13 satellites, possibly 14, all but four of them are believed to be captured asteroids.

Below: Amalthea, the tiny, red, innermost satellite of Jupiter, photographed from 255,000 miles away by Voyager 1. It is only 70,000 miles away from Jupiter, and would make an excellent site for future astronauts to observe the planet were it not for the rain of deadly radiation striking its surface.

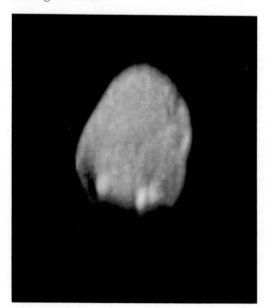

Right: Callisto, photographed from 121,000 miles, has been described as a ball of slush with a rock center. It is larger than the Earth's Moon, and has the oldest surface in the solar system.

Below: Io, closest of Jupiter's moons, and the most volcanic body in the solar system. It is thought to be covered by large numbers of continuously erupting volcanoes, more violent than any found on Earth. It is also believed to be connected in some way with radio emissions of particular wavelengths, which occur only when Io crosses the plane of Jupiter's magnetic field.

Below right: Europa, Jupiter's smallest moon, is slightly smaller than the Earth's Moon. Its surface of cracked ice has led some to theorize that Europa may be covered by a vast ocean, frozen on the surface but kept liquid below by its internal heat. Perhaps life, much like that which lives near heat vents at the bottom of Earth's oceans, may exist in Europa's seas.

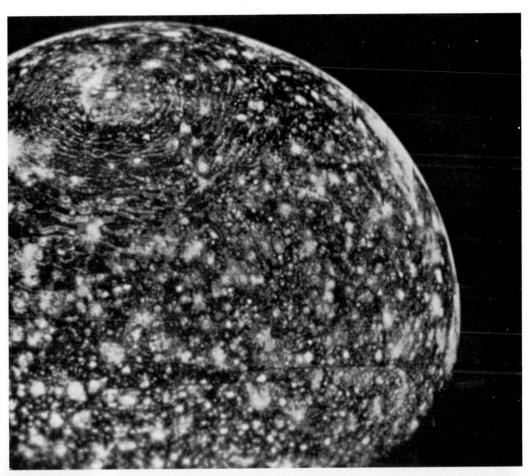

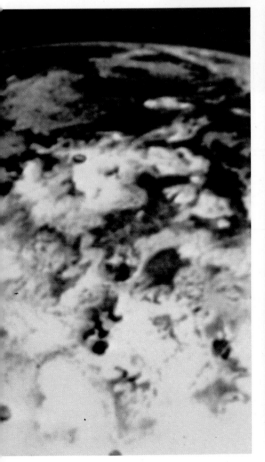

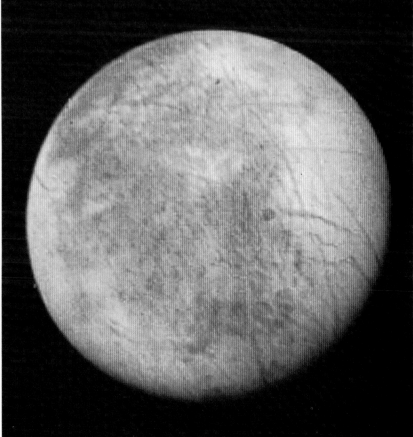

Saturn

the patterns constantly change although the poles always appear somewhat darker than the rest of the planet.

Saturn is a huge body, over 75,000 miles in diameter at the equator. It moves slowly in its remote orbit so that a year on Saturn – one circuit of the Sun – is the equivalent of 29.46 Earth years.

Saturn, most distant of all the planets visible with the naked eye, orbits the Sun at an average distance of 886 million miles. Like its nearest planetary neighbor Jupiter, it has an extremely hot, rocky core. But this does not mean the planet is a warm place. The core is encased in a superdense shell of solid hydrogen. Around this is an outer layer of liquid gas and ice. In the cloudy atmosphere above its frozen surface, temperatures probably fall as low as −290°F.

The atmosphere itself is a choking mixture of hydrogen, helium, ammonia, and methane. Although not as turbulent as the atmosphere of Jupiter, Saturn's rapid rate of rotation about its axis – a day on Saturn lasts only 10 hours 14 minutes – produces multicolored gas bands across its surface. At the equator the planet is a whitish-yellow, flanked by darker colored bands. In other regions

Above: the planet Saturn, from a computer-processed photograph by Pioneer 11. The slight distortion of the planet is due to the uneven spacing of the photographing device, which builds up pictures line by line, somewhat like a television camera. Saturn's moon, Rhea, can be seen as a speck of light below the planet. It is the sixth from the center of Saturn's 10 moons.

Right: Saturn with four of its satellite moons – Tethys, Dione, Rhea, and Titan.

Despite its vast size, however, Saturn is so light that it could float in water.

Seen with the naked eye, Saturn is a faint point of light dimly glimpsed against the background of stars. But with even a medium-power telescope, the planet's true magnificence becomes apparent. Saturn is surrounded at its equator by a system of beautiful concentric rings. The rings extend several hundred thousand miles from the surface and are composed primarily of chunks of ice, some the size of large boulders, others mere specks. The rings are thought to be between 10 to 50 meters thick.

There are three main rings in the system. Ring A is the outermost. It is 10,000 miles wide and is separated from Ring B by a 1,700 mile gap called Cassini's division for astronomer Jean Cassini who first discovered it in 1675. Ring B is 17,500 miles wide and is the most clearly visible of all Saturn's rings. Next comes the faint, barely discernible Ring C, about 10,700 miles wide. Of the tiny rings inside Ring C, the most famous is the F Ring, discovered by Pioneer 11 as it flew by in September 1979. The F Ring is only 800 miles wide, and it encircles the planet only 8,000 miles above the cloud tops. When the Voyager spaceprobes flew by in 1981, an even more startling discovery was made: The F Ring consists of three "ringlets" – and two of them are braided together! This interweaving of the ringlets would seem to defy the laws of orbital mechanics, but then it was discovered that the F-Ring ringlets are "shepherded" into this form by two tiny moons. The Voyagers also discovered that there are "spokes" of microscopic charged particles cutting across the rings, and that there are dozens of tiny moons, moonlets and huge rocks scattered throughout the ring system.

Of the larger moons, the innermost is Janus, orbiting, on average, just under 100,000 miles from the surface. Farthest out is Phoebe, at 8 million miles. Because of its extreme distance and retrograde motion, some suggest Phoebe is a captured satellite and didn't develop with the rest of the Saturnian system. The most fascinating moon, however, is Titan, which may be the only satellite in the solar system to have its own atmosphere.

Discovered in 1655 by Dutch astronomer Christian Huygens, Titan has a diameter of 3,600 miles, and is therefore larger than the planet Mercury. Its atmosphere, which is twice the density of Earth's, is composed primarily of nitrogen, with some ethane, acetylene and hydrogen cyanide thrown in. Although this sounds unpleasant, it is remarkably like the hypothesized atmosphere of primeval Earth, although in the case of Titan, the oxygen and carbon dioxide have been frozen out by the moon's −290°F surface temperature. This low temperature, however, is the only reason this small body can retain an atmosphere – if it were much warmer the gases would become agitated and would start leaking off into space. As it is, a little bit of ionized gas does escape, but it remains in Titan's orbit, forming a "torus" (donut) around Saturn.

Although Voyager 1 came very close to Titan, its cameras couldn't penetrate the thick mass of orange cloud cover, and so the composition and surface conditions of

Below: an image of Saturn's ring system taken by Pioneer 11 in September 1979. It shows details never before seen, including another faint, hitherto undiscovered ring. Saturn's rings – its most distinctive and beautiful feature – are either particles of ice or ice-covered rock.

the moon remain a mystery. According to one theory, a rocky core is surrounded by a mantle of rock particles mixed with water. Above this is a layer of water and ammonia locked inside a crust of ice and frozen methane. It has been suggested that because of the very low surface temperature, it might rain methane on Titan. The rain would gather into rivers and lakes of natural gas, which would empty into huge methane oceans, and all under an orange sky. That would be a very strange sight indeed.

Uranus

Far beyond the orbit of Saturn, moving on average 1783 million miles from the Sun, is Uranus. Discovered in 1781 by William Herschel, Uranus was for some years called the Georgian planet, for Herschel's patron, King George III of Britain.

Because of its remoteness from Earth knowledge of Uranus is sketchy. It is the third largest planet in the solar system, and is about half the size of Saturn but a good deal more dense. It probably has a hot rocky core enclosed in an icy mantle. Its atmosphere is a mixture of hydrogen, helium, ammonia, and methane and although no surface features can be distinguished the planet has an overall greenish hue and will probably be covered with distinctive belts of cloud similar to those visible on Jupiter and Saturn.

The most remarkable thing about Uranus is that it orbits the Sun almost on its side. The axis of the planet is tilted at an angle of 98 degrees. This means that the seasons of the Uranian year are extremely unusual. Although the planet's period of rotation is not known exactly, it is thought to be between 17 and 23 hours. But because of the extreme tilt of the planet's axis, the length of a day will vary widely from place to place on the planet during its circuit of the Sun – which it takes 84.01 years to complete. First, much of the Northern Hemisphere, and then much of the Southern Hemisphere will be darkness for periods of up to 20 Earth years.

A major discovery about Uranus was made in March 1977. A team of astronomers were planning to make detailed observations of the planet as it passed in front of a particular star. They hoped to learn more about the atmosphere of Uranus and by carefully measuring the time Uranus took to eclipse the star, they were expecting to make some accurate calculations of the planet's diameter. Instead, the researchers were astonished to find that just before Uranus

passed in front of it, the star faded and brightened five times. The same sequence took place soon after the star emerged from behind the planet. The only explanation for the effect was that Uranus has a system of at least five narrow rings surrounding it (recent research indicates there could actually be nine or more rings).

Later observations have confirmed the discovery and scientists now know that the rings lie in the planet's equatorial plane in a band between 12,500 and 15,500 miles from its surface. So far only one photograph has been successfully taken showing details of the rings. It seems certain that they are much more tenuous and thinner than the rings of Saturn, perhaps only a few hundred yards deep in places. The first close-up glimpses of Uranus' rings will have to wait until about 1986 when an American Voyager space probe is due to fly past the planet on a long and lonely course that will eventually take it out of the solar system.

Uranus has a family of five moons. The two outermost moons – Oberon and Titania – were identified by Uranus'

Below: Uranus, with its satellite system of five moons, photographed in 1948 with an 82-inch reflector telescope at McDonald Observatory in Texas. The moons, from the left, are Oberon; Ariel (lower left on the photographic halation ring); Uranus itself; the tiny, most recently discovered moon, Miranda, within the ring; Umbriel; and farthest right, Titania.

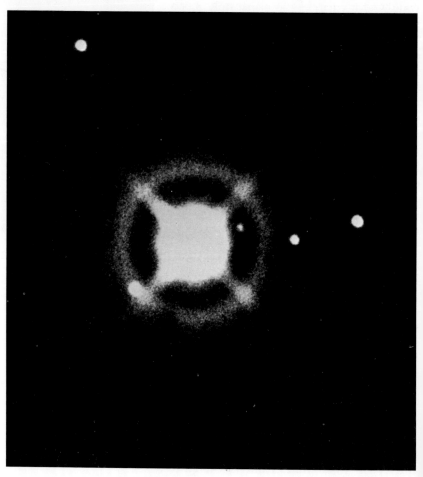

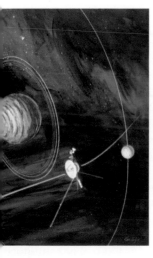

Above: Voyager 2 flies past the planet Uranus – an artist's impression of an event to come in 1986. The planet has a severe axial tilt of 98 degrees.

Right: ringed Uranus, viewed from an icy asteroid. The artist has sought to capture the impression of the curiously tilted planet – a feature not shared by any other of the solar planets. By the time the Voyager 2 probe flies past the planet, one of its poles will be pointing toward the sun so that Uranus will appear – as in the painting here – to be a half-planet.

discoverer, William Herschell. Oberon is about 1500 miles across and orbits an average 366,000 miles above Uranus. Titania is about the same size and moves at an average orbital distance of 274,000 miles. The next two inner moons, Umbriel and Ariel, were not discovered until 1851 when they were observed by the British astronomer William Lassell. The tiny innermost moon, Miranda, was spotted by the American astronomer Gerard Kuiper in 1948. Orbiting at an average 82,000 miles from Uranus, Miranda is barely 70 miles across.

Soon after the rings of Uranus were discovered scientists realized that the relationship between the position of the rings and the family of known moons, suggests that there could be a sixth moon waiting to be discovered.

Neptune

Neptune, outermost of the giant planets, was discovered by mathematics before it was ever seen with a telescope. The 19th-century British astronomer John Couch Adams was deeply interested in the mechanics of the solar system. He had made a detailed study of the way in which the gravitational fields of the planets interact with each other causing slight changes in their orbits around the Sun. In 1841 he began examining reported discrepancies between the observed orbit of Uranus and the path that it had been calculated it should follow. By 1845 Adams was convinced that the motion of Uranus was being disturbed by another more distant and hitherto undiscovered planet. Adams was even able to predict where in the night sky the new planet should be visible. Eventually in 1846 Neptune was spotted by two German astronomers, Johann Galle and Heinrich d'Arrest.

On average Neptune is 2793 million miles from the Sun, so far in fact that the sun would seem from the planet's surface no more than an extremely bright star. At present Neptune is the most distant planet in the Sun's family. Some time in 1979 Pluto – normally the outermost planet – swept inside the orbit of Neptune where it will remain until 1999. Although the orbits of Neptune and Pluto cross each other, the orbital planes are inclined so that there is never any possibility of a collision.

Even using a powerful telescope, Neptune appears small and extremely faint. It appears to have a bluish tinge and some recent observations suggest that the planet is crossed with light and dark bands of cloud, similar to the other giant planets. Neptune is roughly the same size as Uranus with an estimated diameter of 31,000 miles. Its composition is probably also similar – a hot rocky core estimated to be about 12,000 miles in diameter, and enclosed in ice. Its atmosphere, roughly 2000 miles deep,

probably consists of hydrogen, helium, ammonia, and methane. Average temperatures in Neptune's clouds are thought to be as low as $-400°$F.

Neptune does not share Uranus' exaggerated axial tilt. The $29°$ inclination means that a day on Neptune is roughly the same as the planet's period of rotation – somewhere between 19 and 23 hours. The long circuit around the Sun takes Neptune 165 Earth years. This means that a year on Neptune consists of nearly 60,000 Earth days.

Neptune has two known moons – Triton and Nereid. Triton has an estimated diameter of over 2300 miles – which makes it one of the largest moons in the solar system. Some observations seem to indicate that it is quite dense. If this is the case, it may well have a sufficiently strong gravitational field for it to retain some kind of atmosphere.

It moves in a circular orbit around 220,000 miles from the cloud tops of its

mother planet. It takes just under six Earth days to complete one orbit but because it travels in a direction opposite to the direction in which Neptune turns on its axis, the surface features of the giant planet will seem to pass below it at dizzying speed.

Neptune's only other moon, the tiny Nereid, was discovered in 1949 by American astronomer Gerard Kuiper. With a diameter of little more than 187 miles, Nereid describes a wildly eccentric orbit. Its distance from Neptune swings from as little as 867,000 miles to well over 6 million miles.

There is little hope of discovering much more about Neptune from the Earth. But by about 1989 an American Voyager space probe is due to pass fairly close to the planet. Assuming that the craft's cameras and transmitters are still working by then, we will be able to see if this gas giant, like the others, has rings and many more moons.

Above: Neptune, discovered in 1846, with its two satellites Triton and Nereid, photographed with the 82-inch reflector telescope at McDonald Observatory. Triton is larger than the Earth's Moon, but Nereid (arrowed) is very small and has a highly irregular orbit around Neptune. The fuzzy spots in the photograph are distant stars and galaxies.

Left: an artist's impression of Neptune viewed from its major satellite Triton. The lack of atmosphere on Triton makes the sky appear black, and the red and yellow light-absorbing methane atmosphere of Neptune gives the planet a bluish tinge. The Sun's light will appear feeble at this great distance, but the artist has worked out a lighting scale based on the rapid acclimatization of the human eye to a reduced light source.

Pluto

Toward the end of the last century the American astronomer Percival Lowell worked out that the orbits of the outer planets Uranus and Neptune were being affected by a still more distant and hitherto undiscovered planet. Working in his private observatory in Flagstaff, Arizona, Lowell made a painstaking search for this new "Planet X." He worked for 10 fruitless years until his death in 1916. Then, 14 years later, working from the same observatory but using improved equipment, the astronomer Clyde Tombaugh found the new planet – named Pluto for the Greek god of the underworld.

It is not surprising that Lowell failed in his search. At the time Pluto was the most distant planet in the solar system, visible only as a faint point of light even through powerful telescopes. The latest research shows Pluto's orbit to be highly eccentric carrying it as far from the Sun as 4500 million miles and as close as 2750 million miles. This means that for about 20 years of its 248-year journey around the Sun, Pluto is actually inside the orbit of Neptune.

Pluto has been the center of controversy ever since its discovery. Instead of being a giant planet like Jupiter, Saturn, Uranus, and Neptune, Pluto appears to be extremely small – an estimate made in 1978 suggests a diameter of less than 2000 miles. But this presents an immediate problem. How could such a tiny planet have caused the gravitational effects that suggested its existence in the first place. A body of such tiny dimensions could surely have no measurable effect on such giants as Uranus and Neptune. And yet it was by these very effects that Pluto was tracked down.

Several theories have been proposed to account for the mystery. One idea was that Pluto was extremely dense. But even if it were made from solid iron its gravitational field would not be strong enough to disturb the orbits of the other planets. Another suggestion was that Pluto was actually much larger than it appears and that all we can observe is a bright region of a giant planet, most of which is hidden

Above: an imagined view from the bleak, inhospitable Pluto. The sky appears black in the nonexistent atmosphere, the Sun merely a bright star (seen in the center of the picture), its light mirrored in a lake of frozen methane. The tiny planet – smaller than Earth or Mars – has not as yet been explored by space probes.

in darkness. Latest measurements however are based not on visible light but on infrared radiation from the planet and its tiny diameter seems beyond question.

The most exciting possibility is that another undiscovered planet exists somewhere beyond the orbit of Pluto. Fantastic as the idea sounds it seems the best explanation of the discrepancies in the orbits of Uranus and Neptune. Some astronomers have even tried to work out the characteristics of the mysterious planet. Latest calculations show it could be orbiting on average over 5700 million miles from the Sun. At this distance it would need to have a diameter of about 10,000 miles and a mass six times as great as the Earth.

In July 1978 another remarkable discovery was made from an observatory near Lowell's home in Flagstaff. An American naval astronomer, James Christy, was trying to refine details of Pluto's orbit when he noticed an apparent elongation in the image of the planet on a photographic plate. Subsequent studies have tended to confirm Christy's conclusion that the elongation is due to the existence of a moon orbiting close to Pluto. Christy has called it Charon for the mythological boatman who ferried souls into the underworld.

Charon is an extremely unusual satellite. It appears to have a diameter about 40 percent that of Pluto, making it larger in proportion to its mother planet than any other satellite in the solar system. In fact, this proportion has led some to suggest that Pluto-Charon should really be considered a double planet system (the same can al-

Below: two photographs taken by American astronomer Clyde Tombaugh at Lowell Observatory, Arizona, in 1930, show the planet Pluto (arrowed) for the first time. It had been predicted as early as 1905, when the American astronomer Percival Lowell showed that there had to be a ninth planet to account for small irregularities in the orbit of the recently discovered Neptune. Pluto had evaded earlier detection because it was unexpectedly faint. The planet remains largely a mystery. A moon was discovered around the remote planet in 1978, which led to a reassessment of its size – it is now thought to be both smaller and lighter than at first predicted.

most be said of the Earth and the Moon). Even more remarkable is the fact that Charon orbits Pluto in 6.4 days, exactly the same time as it takes Pluto to complete a rotation on its axis. This means that Charon must hang stationary over one point on Pluto's surface, a phenomenon unique in the solar system.

Some astronomers even question whether Pluto is a true planet at all. Its similarity to some of the larger moons of the giant planets suggest it may actually have once been a moon of Neptune torn loose by some ancient cataclysm, perhaps even by the intrusion of the mysterious tenth planet, although recent research indicates that this is unlikely.

At the moment scientists can only speculate about conditions on the surface of Pluto. Surprisingly, it is believed that it may indeed have an atmosphere, composed of methane and something heavier. It is believed that the rest of the planet is a mix of frozen methane and ice. The Sun would appear as a bright star but its light would hardly be strong enough to illuminate Pluto's bleak landscape.

Pluto is a lonely outpost world. Beyond its remote orbit lies only the unresolved mystery of the tenth planet and then the vast gulf of interstellar space.

Asteroids and meteors

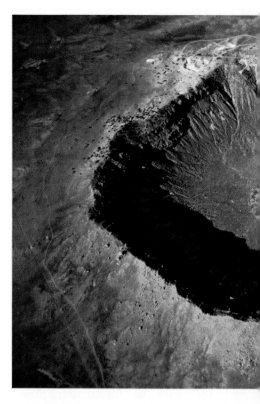

Left: part of the total asteroid population of an estimated 45,000 that have their orbits in the solar system. More than 1500 asteroids have now had their orbits worked out – a difficult task, for even in large telescopes they appear as mere points of light.

Right: a gigantic crater, nearly a mile wide, near Winslow, Arizona. It was caused by a meteor that fell some 30,000 years ago.

Below right: the orbits of some of the exceptional asteroids. Most asteroids are found inside the wide gap between the orbits of Mars and Jupiter, but some swing closer to the Sun, moving in eccentric orbits, and may pass relatively close to Earth. Hermes, for instance, may come to within 500,000 miles of Earth. In June 1968 Icarus passed the Earth at 4 million miles distance. Hidalgo, on the other hand, has a path that carries it out almost as far as Saturn.

Between the orbits of Mars and Jupiter lies a belt of dwarf worlds. Several thousand are known and their total number may be as high as 100,000. Known as *asteroids* (starlike) or – more correctly since they are not stars – minor planets, most of them are tiny irregular shaped bodies. But a few are extremely large. The largest of all is Ceres, which at almost 650 miles across was the first of the asteroids to be discovered.

Astronomers in the 18th century were convinced that the arrangement of the planets followed a numerical rule known as Bode's Law. By fairly arbitrary manipulation of numbers it seemed possible to work out a good approximation to the orbits of the known planets. But according to the rule there should have been a planet somewhere between Mars and Jupiter. Astronomers all over Europe searched for the missing planet until in 1801 an Italian astronomer, Giuseppe

Piazzi, announced the discovery of a starlike object in roughly the position predicted by Bode's Law. He named the object Ceres.

But soon astronomers found that Ceres was not alone. A whole belt of minor planets seemed to exist. Most lie in a main group with average orbits ranging from 205 million to 304 million miles from the Sun. Some follow highly eccentric orbits that carry them well away from the main group.

Best known of these is Icarus, which

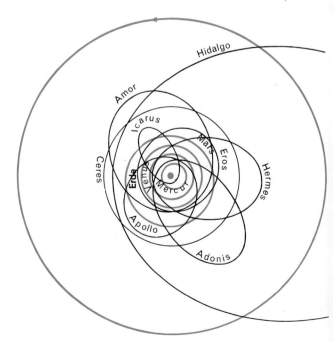

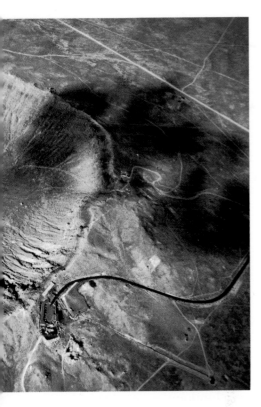

in the course of about 200 days moves from a distance of 183 million miles – beyond the orbit of Mars – to within 17 million miles of the Sun, well inside the orbit of Mercury. Because it spins on its axis once every 80 minutes, its entire surface must bear the scorching heat of the Sun and at its closest approach its surface rocks probably become literally red hot.

Another well-known asteroid is Eros, discovered in 1889. At its closest approach to Earth, it passes within 15 million miles of Earth. Careful observations of the way its brightness varies, suggest that Eros is cigar-shaped, about 18 miles long and four miles wide.

Another group of asteroids known as the Trojans lie actually in the path of Jupiter's orbit. They are gathered in two groups ahead and behind the giant planet. An asteroid beyond Jupiter has also been spotted. In November 1977 the American astronomer Charles Kowal discovered Chiron, a lump of rock about 100 miles across that is currently moving somewhere between Saturn and Uranus. Latest calculations however show that Chiron is not in a stable orbit. Within a few million years it will almost certainly be ejected from the solar system forever.

Where did the asteroids come from? One of the most popular theories suggest

Below: an artist's vision of a future method of mining the metal-rich asteroids to replace the rapidly decreasing supplies of some of these materials here on Earth. Scientists are looking into the possibility – given the technological advances – of using the larger asteroids as remote research stations, where instruments could be assembled that could send back to Earth vast quantities of invaluable information.

they are the debris of a giant planet that once existed between Mars and Jupiter. At some stage in the remote past the planet is supposed to have exploded leaving in its wake the fragments of rock we now know as the asteroids. A more likely suggestion is that they are formed from the primordial dust and rock left over when the planets of the solar system were formed 4600 million years ago.

The largest asteroids may be rich in metal and minerals. Some space scientists have suggested that it may one day become possible to mine them. A more practical proposition would be to use asteroids as remote research stations. Instruments deployed on Icarus, for example, could radio back to Earth fantastic close-ups of the Sun's surface. It is doubtful, however, whether any equipment that could be built with present-day technology would survive such a close encounter with the Sun. The only hope would be to put the instruments in insulated chambers installed under the surface of the asteroid and to operate them by remote control either from Earth or satellite. Other asteroids such as Eros could provide an excellent laboratory for teams of astronomers. Secure in a permanent base built while the asteroid is still close to Earth, scientists could enjoy a free ride through the asteroid belt.

Comets

Halley's comet, which flies by every 76 years, is the most famous of all the comets. When it next passes around the sun in 1985–86, many of the techniques of modern science (including three space probes) will be brought to bear on it, for today, comets remain one of the unresolved mysteries of space. For centuries people believed they were omens of war or natural disaster. The sight of a comet in the night sky therefore was often a matter of deep anxiety. Halley's comet appeared in 1066 just before the Norman invasion of Britain. The same comet reappeared in 1456 when Pope Calixtus III actually excommunicated it.

The first attempt to provide a scientific explanation of comets was made by Aristotle in about 350 BC. He taught that both comets and meteors were caused by hot, dry vapors rising from the ground into the upper air. This idea of comets as a phenomenon of Earth's upper atmosphere persisted for centuries.

Today it is known that comets are visitors from space swinging on long, eccentric orbits around the Sun. But what are comets made of? They are spec-

Above: the famous Bayeux Tapestry showing Halley's Comet, observed in the year 1066. In the tapestry, the English King Harold is so alarmed at the news of the comet that he sways on his throne.

Below: an illustration from the French astronomer Camille Flammarion's *Astronomie Populaire* of 1879. It shows the path of the great comet of 1843 as it passed around the Sun in February of that year. Other brilliant visitors were seen in 1843, 1858, 1861, and two in 1882.

tacular enough to observe – some have been bright enough to see even by day. Their head or nucleus appears as a compact luminous region from which streams a brilliant tail. This tail can be immense. The Great Comet of 1843 had one over 200 million miles in length. But despite their grandeur, comets are composed of very little solid material. One astronomer described a comet as the nearest approach to nothing that can still be anything.

Although no one is sure of the exact composition of comets, scientists now know that in the main they consist of small particles of metal and minerals in a cloud of gas. The head of a comet is called the *coma*. Inside it is thought to be a nucleus containing most of the comet's mass – mainly in the form of small fragments of iron, magnesium, and nickel mixed with particles of frozen water, ammonia, and methane. Although the nucleus is quite small – rarely more than a few miles in diameter – the coma of which it is a part may be huge. In 1811 a comet was spotted with a coma 1,250,000 miles across, the largest ever seen, much bigger even than the Sun.

The splendor of a comet lies in its tail. The long, glowing plumes extending from the coma are probably made of an extremely fine haze of dust and gas. Tails vary enormously in size and shape, however. Some comets have no tails at all whereas others trail magnificent streamers of gas millions of miles long.

A comet shines with light reflected

from the Sun. The tail itself only develops as the comet approaches the Sun – when icy materials begin to evaporate and particles of gas and dust are driven out of the coma. Once a comet begins to recede from the Sun its tail vanishes, and the once brilliant coma fades into invisibility. A feature of the tail that has for long intrigued astronomers is the way it always seems to orientate itself so that it is pointed away from the Sun. For years this was thought to be caused by a physical pressure exerted by sunlight on the tenuous material in the tail. Today astronomers think it is more likely to be an effect caused by the solar wind – the more or less nonstop stream of electrically charged particles that are emitted by the Sun.

Where do comets come from? Several

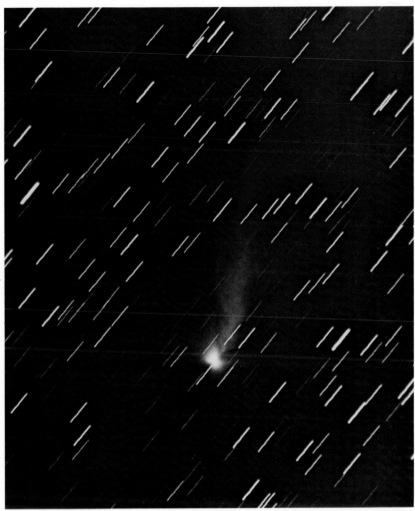

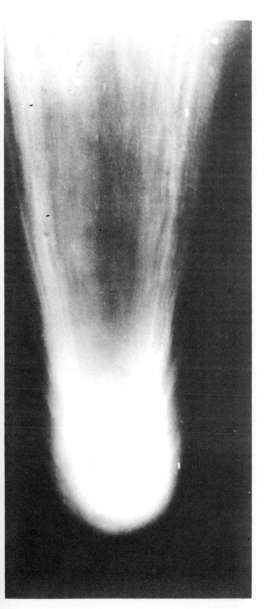

Above: the Comet Humason, discovered in 1961 by the American astronomer Milton Humason with a 48-inch Schmidt telescope. When comets are discovered they are identified by the name (or, in the case of many discoverers, the first three observers to report it) of its discoverer and by an alphabetical letter that follows a strict sequence in any one year. Humason's comet was designated Comet Humason 1961e – that is, Humason's comet was the fifth comet discovered in 1961.

Left: Haley's Comet, which will next pass around the Sun 1985–1986.

theories have been proposed to explain their origins but no astronomers are sure of the real answer. One of the most widely accepted theories was proposed by Dutch astronomer Jan Hendrik Oort in 1950. According to Oort comets are made from the debris left over from the formation of the solar system. This in itself is a fairly obvious idea but Oort went on to suggest that far beyond the range of the most powerful telescopes, a vast cloud of comet nuclei is orbiting the Sun. The few comets observed from Earth are just those thrown out of the main swarm by a collision or some other astronomical event that has sent them into an orbit much closer to the Sun. Today many astronomers are beginning to accept this remarkable idea and current estimates suggest there may be as many as 100,000 million comets orbiting at distances from the Sun ranging from nearly 3 million miles to over 9 million million miles. This puts the middle of the swarm about one light-year from the Sun.

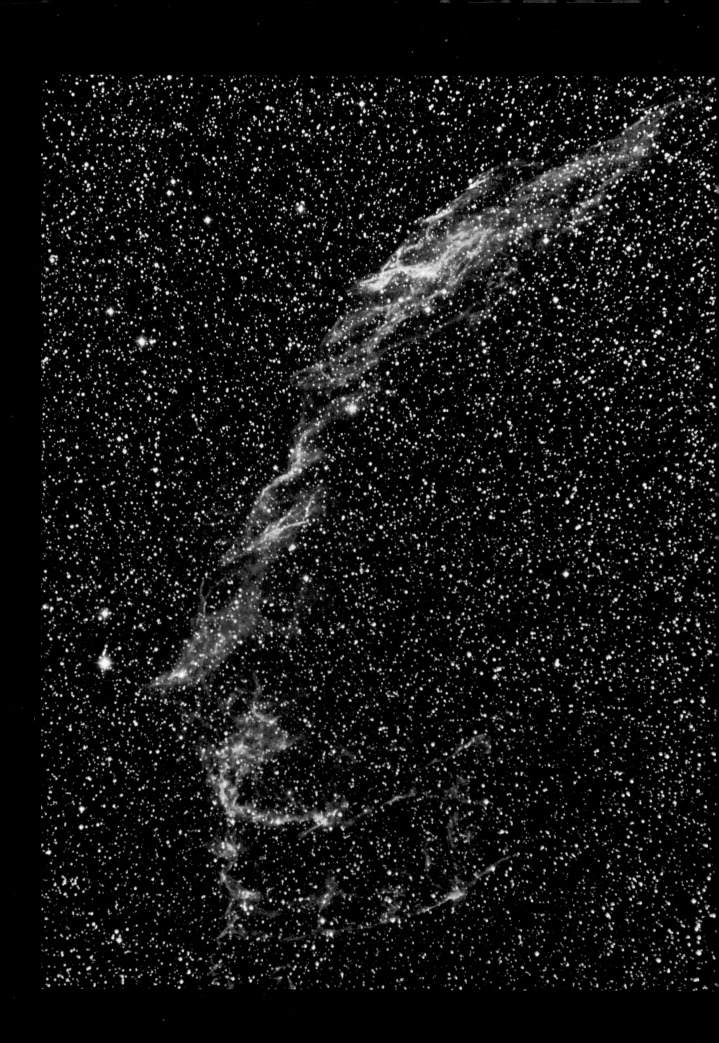

Chapter 4

Galaxies

For thousands of years people's vision of the universe was limited to the stars and planets they could see in the night sky. The most advanced thinkers believed the stars existed in a single galaxy, perhaps infinite in extent, and distributed in some manner around the great band of celestial light known as the Milky Way.

The truth was much more awesome. During the early part of this century astronomers at last realized that beyond the Milky Way are countless other galaxies, vast island universes made up of billions of stars drifting through the void of intergalactic space.

The discovery of the galaxies revealed an exciting new perspective on the universe and the Earth's place in it. More remarkable still are some of the mysteries of galactic evolution which astronomers are now investigating at the outer limits of the observable universe.

Opposite: the Veil Nebula in the constellation Cygnus. It probably originated in the supernova explosion of a star and consists of bright areas of still-expanding gas and dust spread thinly over a wide area. Other nebulae have been found to be island universes, independent galaxies consisting of billions of stars at vast distances from our own Milky Way.

The Milky Way and beyond

nomer William Herschel made a systematic survey of the heavens. He mapped out 683 regions scattered over the sky and counted the stars visible in each region through his telescope. He found that the distribution of stars rose steadily approaching the Milky Way, reached a sharp maximum in the plane

Stretching across the night sky through the constellations of Orion, Perseus, Cassiopeia, Cygnus, Aquila, Sagittarius, Centaurus, and Carina is a softly glowing band of light. Seen on a clear night, it is a beautiful and awesome sight. The ancient Greeks called it *galaxias kyklos*, meaning the "milky circle," and believed that it was a road to heaven. The Romans called it *via lactea* – the Milky Way.

In 1610 the Italian scientist Galileo Galilei looked at the Milky Way using one of the earliest astronomical telescopes. Even with such a primitive instrument, the uniform band of light resolved itself into a myriad of glittering stars, too numerous and closely packed to be distinguished with the naked eye.

For more than 150 years astronomers puzzled over Galileo's discovery. In past centuries most people had believed that the universe extended outward from the Earth in all directions, infinite in size and filled with a roughly uniform distribution of stars. But Galileo's observations showed quite clearly that there were many more stars in the Milky Way than in any other direction.

In 1784 the celebrated English astro-

Above: the Milky Way. Astronomers are reasonably sure that some nearby stars have planetary systems. Statistically, therefore, in the immensity of our galaxy, there must be millions of planets – worlds on which alien life may well have evolved.

nomer William Herschel made a systematic survey of the heavens. He mapped out 683 regions scattered over the sky and counted the stars visible in each region through his telescope. He found that the distribution of stars rose steadily approaching the Milky Way, reached a sharp maximum in the plane

of the Milky Way, and fell to a minimum at right angles to it.

To explain his findings Herschel suggested that the stars surrounding Earth were grouped in a volume of space shaped like a lens with the Sun positioned near its center. Herschel pointed out that on

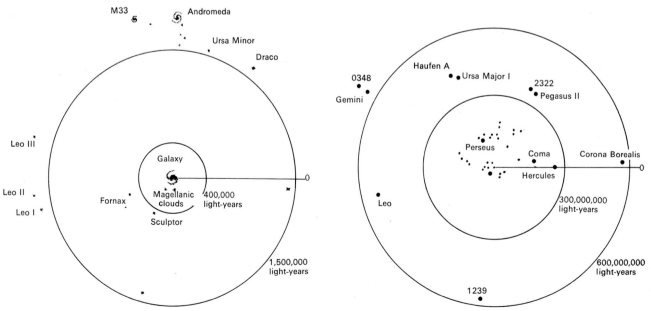

sighting along the long diameter of the lens, there would appear a number of bright close stars with vast numbers of more distant stars behind them. These distant accumulations of stars would produce the pale luminosity we know as the Milky Way. Looking away from the long diameter of the lens, only the nearby brighter stars would be visible with relatively few more distant stars behind. Herschel even went as far as to estimate the size of the Milky Way. He thought it was about 8000 light-years across its long diameter and 1500 light-years across its short.

While Herschel's idea of the galaxy's shape was remarkably accurate, his estimate of its size was much too low. By the early 1930s astronomers had worked out that the Milky Way galaxy is about 100,000 light-years across with the Sun 27,000 light-years from the center. At the center itself, the galaxy is 16,000

Below: the Dumb Bell Nebula in the constellation Vulpecula. It is an example of a planetary nebula in which during the final stage of a star's life (in the Dumb Bell's case, two stars), the outer layers of a star completely separate from the core and slowly expand into space. After a short period of time – some 50,000 years – the gas is dispersed into interstellar space. Scientists therefore know that planetary nebulae are very recent phenomena – always under 50,000 years old.

light-years thick, thinning to 3000 light-years at the Sun's position.

The next logical question was – how many stars are there in the galaxy? The Dutch astronomer Jan Oort worked out the answer in the 1920s, showing at the same time that the entire galaxy is turning in space in a huge and complex rotation. Oort found that the stars closest to the galactic center were revolving around the center more rapidly than the outer stars. By observing the relative motions of stars seen from the Earth, he showed that the Sun was itself moving in a roughly circular orbit at a speed of about 140 miles per second relative to the center. At this rate, the Sun completes a single orbit of the Milky Way galaxy once every 230,000,000 years.

Oort then worked out the strength of the gravitational attraction of the inner regions of the galaxy needed to drive the

Opposite far left: distances of some of the galaxies in our local group. The Milky Way is shown in the center; the circles are of radii 400,000 and 1,500,000 light-years respectively. The most important members of the local group are the Milky Way, the Andromeda spiral (M31), the Triangulum Spiral (M33), and the Greater and the Smaller Magellanic Clouds. The other members are small irregular galaxies.

Opposite left: distances of some of the clusters of the galaxies; the radii of the circles are 300,000,000 and 600,000,000 light-years respectively. The Coma cluster is particularly rich, and is also comparatively close to our galaxy, but on this scale our local group would be reduced to a small point, so that clearly it is not possible to see the detailed forms of individual members of a cluster of galaxies.

Sun around in its orbit. The galactic center, he found, would need to have a mass 90,000 million times the mass of the Sun. Assuming that about nine tenths of the galaxy's mass is in its central regions, its total mass would be about 100,000 million times that of the Sun. So, if the Sun is a fairly average star, the galaxy must contain about 100,000 million other stars.

After Galileo first used a telescope to study the skies, astronomers became increasingly aware that stars are not the only brightly glowing objects in space.

There are also faint, misty patches of light like clouds. Best known of these is in the constellation Andromeda, a pale smudge of light just visible to the naked eye. Arab astronomers included it in star charts drawn centuries ago. In modern times, the first to observe it was the German astronomer Simon Marius, who in 1612 used a small telescope to observe the wispy, luminous cloud which he likened to the flame of a candle.

As telescopes become more widely used, increasing numbers of these strange objects were detected, though none ap-

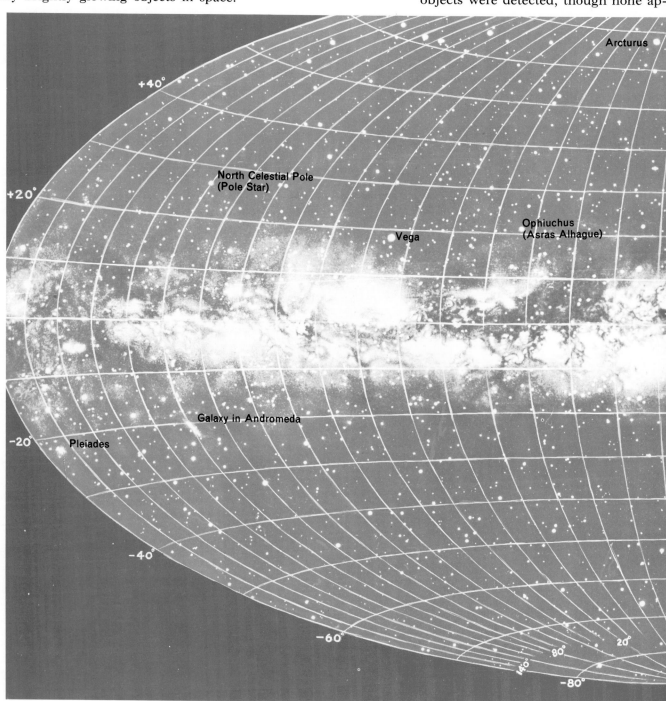

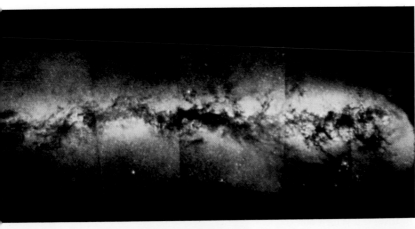

Left: a mosaic of the Milky Way from the constellation of Sagittarius to Cassiopeia. The Milky Way – the galaxy containing our Sun – is believed to be shaped like a flattened disk that swells to a hub in the center. The Solar system is some 30,000 light-years from the center, while the whole disk is some 100,000 light-years across.

Below: a composite picture of the Milky Way. It is thought that the Milky Way was formed from a very large cloud of gas hundreds of thousands of light-years in diameter. Most of the gas reached the disklike form before stars were formed, and then star formation could have happened very quickly. The gas that remained uncondensed now forms the interstellar gas.

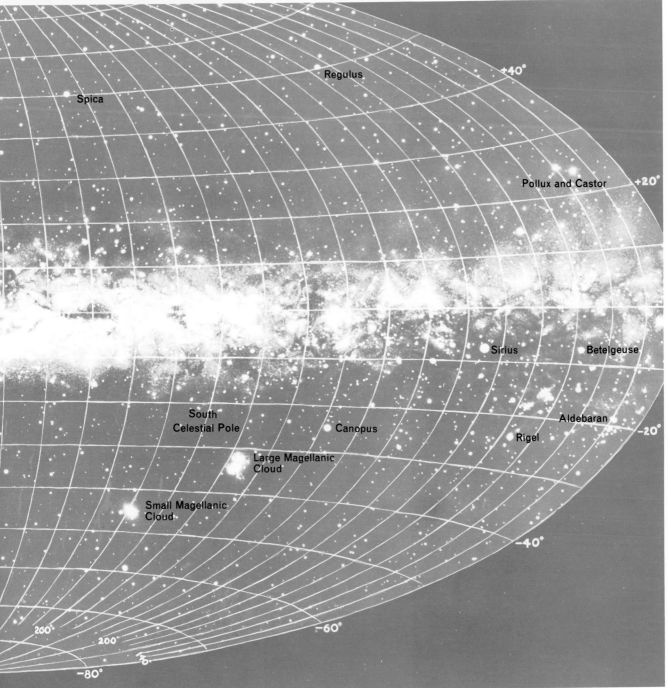

peared as large and bright as the one in Andromeda. Astronomers called them *nebulae*, the Latin word meaning "clouds." In 1781 French astronomer Charles Messier began cataloging them. His reasons for doing so had nothing to do with any real interest in nebulae themselves. Messier was a comet-hunter, and nebulae were easily confused with comets. To make things easier, Messier decided to compile a list of nebulae, mapping their position so that he and other comethunters would no longer be misled. His initial list included about 40 nebulae but within a few years the number had been extended to 100. The great nebula in Andromeda, for example, is known as M31 because it is number 31 on Messier's

Right: the American astronomer Edwin Hubble. His investigations gave scientists a new sense of the vastness of the universe. Hubble is seen here with the 100-inch Mount Wilson telescope, through which he discovered that the nearest and largest of the faint misty patches of light was in fact an island universe, a galaxy separate from our own, about 1.5 million light-years away from the Earth.

Below: the Whirlpool galaxy as seen through a modern telescope. Its spiraling arms can be seen here to contain many highly luminous stars.

list.

Messier's catalog is still used but the nebulae it identifies are now infinitely more important to astronomy than the comets that interested Messier so much.

Some astronomers explained the existence of nebulae by supposing that they were the gas clouds from which stars condensed to form solar systems complete with a family of planets. The French mathematician and astronomer Pierre-Simon de Laplace, a contemporary of Messier, suggested that the Andromeda nebula was a solar system in the process of forming. By this argument, it was a fairly small object quite close to the Sun. But the greatest German philosopher of that time, Immanuel Kant, rejected Laplace's idea proposing instead that the nebula was an immensely large group of stars which only appeared as a pale cloud because of its extreme remoteness from the Sun. Kant went further and suggested that many of the known nebula, including the one in Andromeda, were huge "island universes" − independent galaxies drifting in space at a great distance.

Ironically, Laplace's inaccurate views

were based on observational data while Kant relied on intuition. Because of this, Kant's remarkable insight was rejected as worthless speculation and Laplace's concept – known as the nebular hypothesis – became generally accepted.

Vindication of Kant's idea of island universes had to wait until the early 1900s. The real breakthrough came in 1924 on the slopes of Mount Wilson northeast of Pasadena, California. A few years earlier the observatory on Mount Wilson had received a massive donation enabling it to install a reflecting telescope with a mirror 100 inches in diameter. At the time and for another generation, that instrument was the most powerful telescope in the world. Its remarkable capabilities were put to use by the American astronomer Edwin Hubble. The telescope's immense resolving power enabled him to scrutinize the Andromeda nebula in unprecedented detail. In a series of historic observations, Hubble swept aside Laplace's nebular hypothesis and gave mankind a totally new perspective on the universe.

Through the Mount Wilson telescope the outskirts of the luminous cloud could be resolved into a vast number of individual stars. The apparently featureless cloud was, as Kant had suspected, a vast galaxy of stars. By observing special kinds of variable stars visible in the nebula, Hubble made the first realistic estimate of its distance from the Sun. His results were a dramatic confirmation of the concept of remote island universes. The Andromeda nebula was over 800,000 light-years away – obviously far outside the confines of the Milky Way.

Hubble called Andromeda the first extragalactic nebula. It is known to be about 1.5 million light-years away, and is the Milky Way's closest galactic neighbor. The most distant now known are thousands of millions of light-years away – on the very edge of the observable universe.

Modern astronomers set no real limit on the total number of galaxies. Some estimate that our best instruments can detect about 100,000,000,000 of them, each an island universe containing countless millions of individual stars. By contrast with this fantastic perspective, early speculations that the Milky Way represented the entire universe seem pitifully

naive. The scale of the universe is far more immense than early astronomers imagined – so vast that it is virtually beyond the grasp of the human mind.

Confirmation that many nebula were in fact distant galaxies was only the beginning of a series of discoveries changing the ideas about the universe and Earth's place in it. Many nebulae were discovered to be exactly what they appeared – huge clouds of faintly luminous gas, at great distances but inside the Milky Way. These are called galactic nebulae. One of the best known is the Great Nebula in Orion, 1500 light-years away yet visible with the naked eye as a misty patch in Orion's "sword." It has been shown that Laplace was not entirely

Above: a galaxy in the southern constellation Sculptor. The galaxy is one of many that are millions of times farther away than the farthest star that we can see. With the naked eye we cannot even see more than a small part of our own galaxy, but with a large radio or optical telescope we can see thousands of galaxies at great distances.

wrong in his nebular hypothesis, because clouds such as the one in Orion do generate stars – not singly as Laplace believed, but in their millions.

Other beautiful galactic nebulae appear in many of the constellations in a great variety of shapes and colors. The Lagoon Nebula in Sagittarius is a beautiful swirl of luminous cloud, and the Veil Nebula in Cygnus is a thin feathery glow against the bright canopy of stars.

There is a distinct class of galactic nebulae that do not glow, but their presence is obvious by the way they

block the light from stars behind them. Two of the most dramatic of these dark nebulae are the Horsehead Nebula of Orion and the Coal Sack, a region of intense darkness near the Southern Cross.

While galaxies are the largest known accumulations of matter, they are themselves drawn into groups by the mutual attraction of their gravitational fields. Just as planets revolve around stars and stars are drawn together into galaxies, the galaxies themselves move through space in clusters, bound together by the universal force of gravity. The Milky Way belongs to a group of about 19 scattered over an area about three million light-years across. The only other large galaxy in this cluster, known as the Local Group, is the Andromeda Galaxy, M31.

Above and below: the Magellanic Clouds, closest of the important external galaxies. In the Large Magellanic Cloud (below) is the largest known gaseous nebula, the vast Looped Nebula, and also the most luminous star detected, Doradus, which has a brightness calculated to be 300,000 times that of the Sun. The small Magellanic Cloud (above) is reproduced from one of American astronomer Henrietta Leavitt's original negatives used in her classic investigations into the short-period variable Cepheid stars.

Right: the Lagoon Nebula in the constellation Sagittarius, photographed with the 200-inch Hale reflector at Mount Palomar.

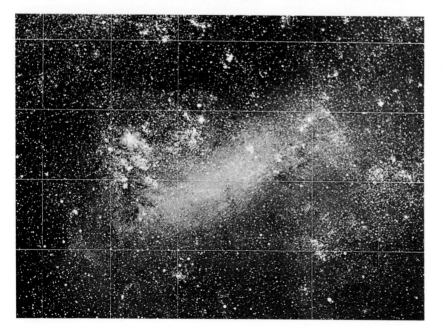

Others are comparatively small. Closest to the Milky Way are the two very small galaxies first noticed by the explorer Ferdinand Magellan during his round-the-world voyage. They are called the Greater and Smaller Magellanic Clouds. The large cloud is about 155,000 light-years away and the smaller about 165,000 light-years. There is evidence that they may have been even closer in the past. In particular, about 500 million years ago the Greater Magellanic Cloud may have actually brushed the Milky Way, coming within 70,000 light-years of its center.

Clusters vary widely in size. Giants such as the one in Virgo that contains about 1000 galaxies are themselves dwarfed by supergiants like the remote Coma Berenices cluster – over 10,000 galaxies at a distance of some 400 million light-years.

Clusters of galaxies are still not the largest groupings of matter in nature. Evidence has recently been discovered that the Local Group of galaxies is in fact a small part of an elliptically shaped local supercluster, formed from about 50 conventional clusters. This super-cluster measures over 100 million light-years across and is about 30 million light-years thick at its center.

Because of the scale of superclusters, it is extremely difficult to be sure of identifying them. But most scientists now agree that the local supercluster is not a chance event. Almost certainly, many of the distant groups of galaxies are only

Above: a cluster of galaxies in Coma Berenices.

Below: the Horse's Head Nebula is one of many nebulae that appear as cloudlike wisps of luminous gas, often assuming individual forms that give rise to the names they bear (Horse's Head, Veil, North America, Trifid).

single elements in other superclusters.

The question still remains whether superclusters are the ultimate form of natural order in the universe. Could superclusters themselves be a part of still more gigantic groupings? Perhaps recognition of such an immense scale is simply beyond the power of human observation and comprehension.

The structure of galaxies

When Edwin Hubble discovered that the Andromeda nebula was actually a huge island universe far outside the Milky Way, astronomy entered a new era of development. For centuries, observations had been confined to the nearer stars and planets. The discovery of galaxies beyond our own gave astronomers a vast new field to explore. For the first time, the true immensity of space was revealed and the study of galaxies soon provided humanity with important insights into the structure and origins of the universe itself.

Hubble was the pioneer of galactic research. Using the 100-inch reflecting telescope at Mount Wilson in California, he made a careful study of a large number of extragalactic nebulae. However, Hubble was not the first to examine nebulae.

Left: a drawing of the Whirlpool galaxy made in 1850 by the Irish astronomer Lord Rosse. He detected the spiral structure of the Whirlpool – and many other galaxies – with the 72-inch reflector telescope he built.

In the mid-1800s, long before nebulae were regarded as independent galaxies, the Irish astronomer William Parsons, Third Earl of Rosse, used his self-built 72-inch telescope to examine them. He noted that many looked like tiny whirlpools of light and suggested that most were made from spirals of glowing gas.

By 1900 other observers had confirmed that about three quarters of nebulae had some kind of spiral structure. Using the extremely powerful 100-inch telescope Hubble made studies of only the extragalactic nebulae. He found that spirals were in fact most common but that a great variety of structures existed when examined in detail. Hubble eventually refined his observations into the first modern theory of galactic evolution based on the different structures he discovered.

Hubble identified three classes of galaxy – irregulars, spirals, and ellipsoidals. Within these groups, he distinguished special subdivisions based on detailed differences between members of the same general family. For example, Hubble found that ellipsoidal galaxies varied in flatness from almost exact spheres to thin disks. He labeled them E0 to E7, the higher numbers indicating increasing degrees of flatness. Spiral galaxies came in two types – ordinary spirals, denoted by the letter S, and those with a "bar" running through them, denoted by the symbol SB.

Refining the descriptions still more, Hubble added the letters a, b, or c to indicate how tightly the spiral arms were wrapped around the galactic center. A galaxy classified as Sa, for example, is an ordinary spiral with tightly wrapped arms. Another, classified SBc is a barred spiral with much more loosely formed arms. The Milky Way is of the Sb type, very much like the neighboring galaxy in Andromeda.

A few galaxies such as the Magellanic Clouds have no well-defined shape. Hubble called them "irregulars" and believed that they represented the earliest stage in the evolution of a galaxy. His idea was that gravitational forces within an irregular galaxy caused it to gradually coalesce into a rotating ball. Under the influence of its rotation the sphere would begin to flatten, passing through the seven subclassifications of the ellipsoidal type and eventually flinging out arms and evolving into a spiral.

Hubble's ideas were graphic and plausible. Unfortunately they did not stand up to the test of actual observations. In 1942, working with the same telescope with which Hubble pioneered galactic research, the German-born astronomer

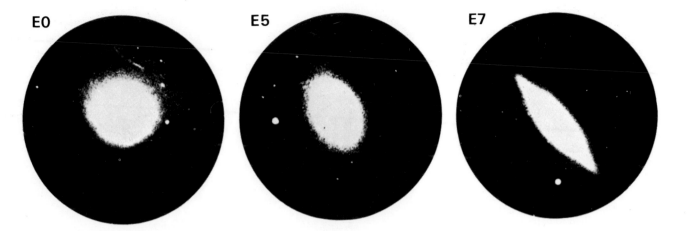

E0

E5

E7

Walter Baade turned Hubble's theory on its head. Baade made his observations during wartime when most cities had their brightest lights turned off and visibility was therefore exceptionally good. Baade made the most detailed studies yet of the Andromeda galaxy. He was able to distinguish individual stars not only in the spiral arms of the great galaxy, but also of its center. He immediately noticed an important difference between them. Those near the heart of the galaxy were reddish in color. Those on the outskirts were bluish and

Above and below: these photographs of nine galaxies taken with the 60-inch telescope at Mount Wilson Observatory are labeled according to the classification system invented by the American astronomer Edwin Hubble. Starting at *E0*, the ellipticals become increasingly flattened through *E5* and *E7*. Spirals are more easily identifiable where the galaxy nucleus is small. With a few exceptions, all the galaxies observed by Hubble fitted into his classification.

much brighter. He also saw that the outer regions of the galaxy were thick with interstellar dust while the interior was relatively free of it. Baade classified the bluish stars as Population I and the red stars at the center as Population II.

Baade continued his studies using the mighty 200-inch reflector that had recently been built at Mount Palomar. With the much greater resolving power of the giant instrument, he was able to examine other galaxies in almost as much detail. His results spelled the end of Hubble's theory.

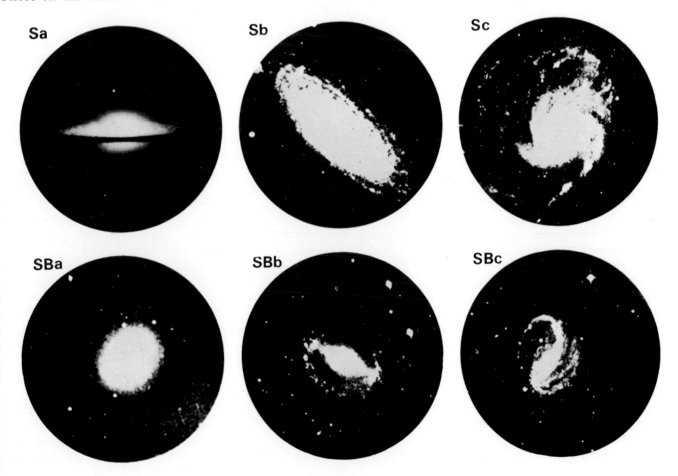

Sa

Sb

Sc

SBa

SBb

SBc

Right: classification of galaxies. The illustration on the far right shows how the galaxies appear in the sky. The key, top left, shows the galaxies in diagrammatic form, labeled according to the method laid down by the famous American astronomer Edwin Hubble. In the handle of the fork are the ellipticals, denoted E. Ellipticals range in shape from globelike to the purely elliptical. The roundest ones are referred to as Eo galaxies, and then the numbers go down in sequence to more oval ones, the most elliptical being $E7$ galaxies. The prongs of the fork are divided into spirals, denoted S, and barred spirals, SB. On the left fork, the degrees of tightness of the spirals are lettered a, b, and c so that a galaxy can be designated from this type as Sa, Sb, or Sc. The same lettered designation is applied to barred spirals also, so that they are labeled SBa, SBb, or SBc. The normal and barred spiral galaxies differ considerably in size. The smaller ones are around 30,000 light-years in diameter and the larger ones appear to have a diameter of about 120,000 light-years.

Baade knew that Population II stars were mainly aging red giants while Population I were young hot stars, forming out of clouds of interstellar dust. Hubble had thought that ellipsoidal galaxies were an early stage in the general evolutionary picture. But Baade showed that the ellipsoidals tend to be dominated by old and dying Population II stars, clearly not a starting point for any kind of galactic evolution. In fact, Baade's results showed that most spherical or near-spherical groupings of stars seemed to be dominated by huge old red stars. Baade's work suggested that the youngest galaxies are the spiral ones. New stars condense from the dust clouds in the spiral arms and, as the galaxy ages, its arms tighten around the galactic center. In this way a loose spiral evolves into a flattened ellipsoidal galaxy which, as it ages, grows increasingly spherical.

Spiral galaxies are characterized by so-called globular clusters of stars, remarkable compact balls consisting of hundreds of thousands of very old stars. Astronomers believe that these were among the first stars to appear in the universe, forming when the galaxies themselves began to separate from the primeval gas clouds over 10,000 million years ago. The clusters are usually found in a kind of halo around the flattened disk of the galactic center. In the Milky Way, about 100 globular clusters are known, set in a huge circle around it.

Above: the globular cluster known as M13 in the constellation Hercules, photographed with the 200-inch Hale reflector telescope at Palomar. It is in fact a compact swarm of thousands of stars in a tightly knit sphere and is the only such cluster visible to the naked eye from Europe – where it is seen as a faint hazy star.

Left: the Pleiades, or Seven Sisters, in the constellation Taurus, is perhaps the most famous open star cluster visible to the naked eye. The cluster is actually made up of over 200 stars, and is 490 light-years away from Earth. (Photo copyright is owned by The California Institute of Technology and the Carnegie Institution of Washington.)

The globular clusters are one of the most spectacular components of a spiral galaxy. Varying in diameter from almost 50 to 300 light-years, the density of stars near their centers must be thousands of times greater than in the vicinity of the Sun. The night sky seen from the heart of a globular cluster would be a dazzling sight, filled with a myriad of brilliant red lights. One such cluster in the constellation of Hercules can just be seen with the naked eye. Visible as a faint patch of light, it might almost be taken for an ordinary star. In a sense it is like seeing into the remote past – looking at stars already ancient 1000 million years before the Sun was formed.

Today, astronomers see the pattern of galactic evolution as more complex than Baade's findings suggest. The spiral shape is now regarded as a secondary stage in an evolution toward an ellipsoidal galaxy. The initial stage is the actual formation of what astronomers call the protogalaxy – a kind of coalescence of stellar matter from the primeval gas cloud produced during the creation of the universe itself. The first sign of a recognizable galaxy, therefore, is a rotating sphere of young stars and luminous dust. But almost as soon as it forms, gravitational attraction dragging material toward its center starts a contraction into an even denser sphere.

As the protogalaxy contracts, it spins faster and faster until streamers of galactic material, mainly dust and gas, are flung outward from the center. This forms a disk-shaped inner region around the slowly collapsing core of older stars. From the disk, long spiral arms emerge, curling outward into space. The disk and arms become the main breeding grounds of new generations of stars. The arms in particular, rich in hydrogen gas and interstellar dust, contain the youngest stars and many of the huge nebulae from which they emerge.

Over many millions of years, the material from which the new stars evolve is used up. As the galaxy ages, its outermost stars begin to redden and the spiral arms close around the galactic disk. Eventually the spiral galaxy disappears, transformed into a flat ellipsoid. With the passing of more time, it evolves with infinite slowness into a huge sphere of ancient and dying stars.

131

Galaxies in chaos

Not long after Edwin Hubble's historic discovery that the Andromeda Nebula was in reality an independent galaxy far from the Milky Way, evidence was found that certain galaxies were going through some kind of titanic upheaval.

In 1943 American astronomer Carl Seyfert discovered a rare class of galaxies whose inner cores seemed to be in a state of total chaos. Because galaxies are usually huge accumulations of stars, the light emitted by them is much like the light from stars. Seyfert found that in about one percent of all spiral galaxies, the central regions were extremely luminous, giving out the kind of light normally expected from a cloud of incandescent gas, but there was no evidence of any stars at all.

In the years following Seyfert's remarkable discovery, radio and optical astronomers found that the strange superactive galaxies were even more mysterious than had first been imagined.

Optical astronomers made detailed studies of the spectrum of light from the core regions. Their findings suggested that there were huge dust clouds at temperatures of up to 36,000°F, and swirling about at speeds of up to 16 million miles per hour, at times shooting out in jets to disrupt the outer reaches of a galaxy. This is what could be expected if a vast explosion were shaking the core of the galaxy. But the power of the blast would be unimaginable, dwarfing even the might of a supernova.

The visible region of a Seyfert galaxy's spectrum is only one aspect of the puzzle they represent. When other parts of the electromagnetic spectrum are examined, the mystery deepens. For example, startling results have followed measurements of the infrared radiation they emit. Infrared radiation is invisible to the human eye, its wavelengths being too long to create any sensation on the retina.

Below: a Seyfert-type galaxy (named for their discoverer, the American astronomer Carl Seyfert). Seyfert galaxies are those with a relatively small bright nuclei of high-temperature gases. This is NGC4151, which emits both radio and x-rays. Seyferts are thought to be galaxies where inner regions are undergoing upheavals so violent that even the vast power of a supernova explosion is dwarfed by comparison.

It is a kind of radiated heat characteristic of any hot or incandescent body. But Seyfert galaxies generate it on a fantastic scale. Results show that a single Seyfert galaxy sends out more energy in the form of radiant heat than 500 galaxies like the Milky Way produce over the whole range of the spectrum. More amazing still is the fact that this huge output varies noticeably over periods as short as a few weeks. Such intense fluctuations limit the size of the source of energy to the time it would take light to travel across the source. For example, if an object changed dramatically in a day, then it could be no larger than a light-day across. This means that the Seyfert cores are no more than a few light-weeks across, and in some cases may be no bigger than the size of a few solar systems.

One way of accounting for the powerful infrared emission is to imagine a compact x-ray source embedded in the galactic nucleus. As x-rays pour out from the center they warm up the surrounding envelope of gases causing it to give out huge quantities of heat. This idea received sudden dramatic support in 1971 when the first x-ray telescope to be placed in orbit around the Earth was turned

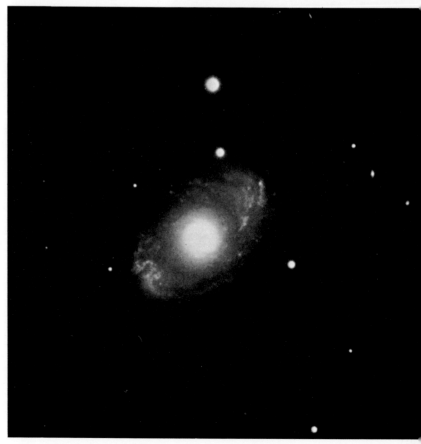

toward selected Seyfert galaxies. Results showed that they were as luminous in terms of x-rays as they were in the infrared region of the spectrum.

The x-ray hypothesis really only replaces one mystery by another. What kind of object could possibly be such a powerful source of x-rays? One suggestion is that in the heart of each galaxy is a black hole – a region of space with a gravitational field so strong that not even light can escape it. Dust and gas around such an object would constantly be drawn into its depths, making the hole ever larger and more voracious. As far as Seyfert galaxies are concerned, the important point about black holes is that as material falls into them, an "accretion disk" of high energy matter would form around the hole, and it would give off intense bursts of x-rays. These bursts could account for the fluctuations from the nuclei of Seyfert galaxies.

Radioastronomy has also revealed startling information about these chaotic galaxies. The best-known discovery took place in 1946, only a few years after Seyfert made his observations. A British team using a telescope made of surplus radar equipment picked up a strong source of radio waves in the constellation Cygnus. They thought it was no more than a particularly powerful radio star and paid it no special attention. But in 1951 the English astronomer F. Graham Smith made some detailed studies of the source – by then called Cygnus A – and worked out that far from being a nearby star, it was in fact a good-sized galaxy 550 million light-years away from us.

The discovery rocked the scientific world. Despite the fact that it is so far away, Cygnus A is one of the strongest sources of radio waves known. To account for this, it would have to be a million times more luminous in terms of radio waves than the Milky Way. What kind of physical processes could lead to such a tremendous output?

The first photographs of Cygnus A taken with the 200-inch reflector at Mount Palomar showed a curious double structure. It looked as if two roughly spherical galaxies had collided. Perhaps the complicated results of such a mighty impact could produce the radio emissions. But collisions between galaxies are extremely rare, and before long the

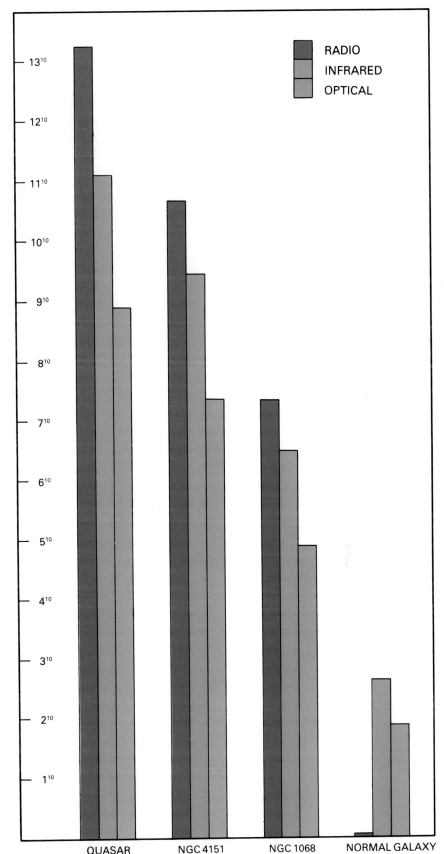

Above: diagram shows comparisons of wavelength distribution of energy emitted by Seyfert galaxies, a quasistellar source (quasar), and a normal galaxy. The three sources are placed on the scale so that the greatest emitters of energy are at the top – Quasar 3C 273, closely followed by the Seyfert galaxies NGC 4151, and NGC 1068 and the least energetic, the normal galaxy, at the bottom.

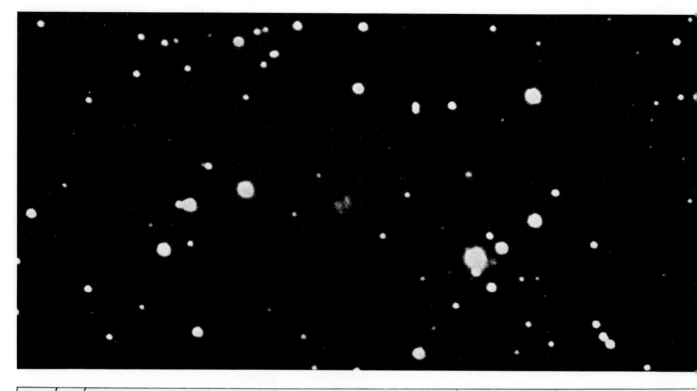

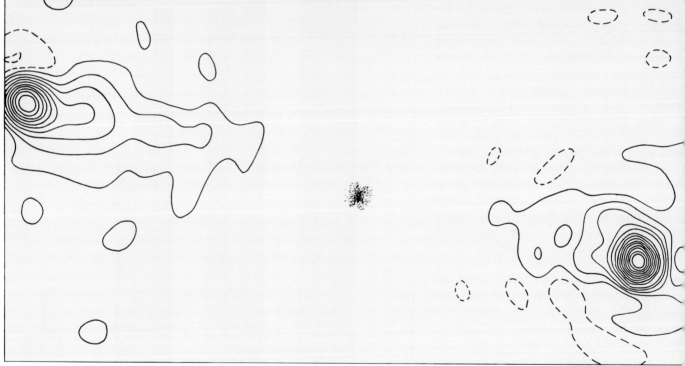

collision theory had to be abandoned because so many radio galaxies had been discovered.

With the development of increasingly sophisticated radiotelescopes, a wealth of new detail was revealed about Cygnus A. Most important of all was the discovery that it was made up of three distinct radio sources, two outlying bright lobes and a dim central component between

Above: a contour map of the strong radio emissions from Cygnus *A*. The emissions are concentrated in two lobes some 500,000 light-years on either side of the parent galaxy. Scientists interpret the map as suggesting that electrically charged particles have been violently ejected from the galaxy in the center.

them. This structure became the basis for the most widely accepted explanation of radio galaxies. It appears that they formed from violent explosions in their central regions. Electrically charged atoms and subatomic particles are shot outward at high speeds and meet powerful magnetic fields surrounding the galactic nucleus. The result is powerful bursts of radio waves. The predicted

were astonishing. BL Lac showed no spectral lines at all – just a featureless band of radiation stretching from the ultraviolet to the infrared. Not a single bright or dark line was visible.

Although the spectrum was featureless, measurements of its intensity gave accurate information about the amount of energy emitted by BL Lac. The output was enormous – far greater than any known Seyfert galaxy. More detailed optical studies showed that BL Lac objects are made up of a brilliant nucleus, giving the blank spectrum, surrounded by a faint hazy cloud. Some researchers believe they have successfully picked out some spectral information from the nebulous outer regions. If they are right, their findings suggest BL Lac objects are extremely remote. BL Lac itself, for example, may be over 1000 million light-years away.

Whatever is going on inside BL Lac objects must be similar to events inside a normal Seyfert galaxy. The difference is one of scale. BL Lac objects are far brighter. Like Seyferts their energy output varies (at factors of up to 100), but over much smaller periods of time, sometimes a matter of a few hours. Again, this implies that the source of their energy must be rather small, perhaps even less than the solar system in size. The black hole theory is again offered as an explanation of what is going on at the center of the BL Lacs.

The most mysterious type of celestial object in chaos is the quasi-stellar source – quasar for short – which, as the name suggests, were first thought to be stars, but are now believed to be galaxies. Quasars are brighter than BL Lacs and about the same size, but are a great deal farther away. In fact, at up to 10 billion light-years, they are the most distant objects we can see, and as it has taken their light that long to reach us, they are also the oldest. Some theorize that giant black holes may also be at the heart of quasars. The most striking recent idea is that quasars, BL Lacs and Seyferts are all related. Perhaps quasars are very young giant elliptical galaxies and BL Lacs are slightly older ones as they are closer and dimmer. Seyferts, the closest, may be simply young spiral galaxies, for as some have noted, one in ten spiral galaxies are Seyferts; it could be that all spiral galaxies spend one-tenth of their time, early on, as Seyferts. Even the Milky Way.

Above: the Copernicus satellite, launched in 1972. Copernicus uses an array of x-ray telescopes mounted like a Gatling gun, whose "barrels" provide a range of different kinds of x-ray imaging down to pinpoint focus.

Left, above: Cygnus *A*, the most powerful radio source known, in the constellation Cygnus. It is the fuzzy spot just left of center in the photograph. At first the double image was thought to show two galaxies in collision. The latest theory is that it is a single galaxy with a band of obscuring dust through the center.

shape of the magnetic fields fits well with the actual pattern of the radio emissions. The weak central source is believed to be the simmering remains of the original explosion.

Today the term Seyfert is used to describe any galaxy in a state of upheaval. There is a type of Seyfert, however, that is so remarkable it has a name to distinguish it from all others. It is named for the first of the class to be discovered, BL Lacertae, usually abbreviated to BL Lac.

Although technically a kind of Seyfert galaxy, BL Lac objects show their own distinct and puzzling characteristics. The first was spotted in 1968. Like many superactive galaxies, it was initially mistaken for a star. But close examination showed that it was a very strange object, certainly not a star and yet like no other galaxy known. As usual with any new discovery, astronomers started by studying the spectrum of its light. Their results

Chapter 5

Origin of the Universe

How was the universe created? Was there ever a beginning or has it always existed much as it is today? These are some of the most fundamental questions we can ask about the world around us. For centuries only philosophers and theologians attempted answers but now the new science of cosmology is beginning to penetrate these ultimate mysteries.

The Ancients restricted themselves to pondering on the origin of the Earth. But the perspective changed as soon as astronomy revealed the true immensity of the universe and Earth's relatively insignificant place in it. By the early 1900s cosmologists were searching for the key not only to the beginning of a planet but also to time and space itself.

Exciting advances have now made it possible to reconstruct the first moments of the universe's existence. What has been learned is hard to comprehend – a fantastic holocaust of energy and matter, a primeval fireball out of which came the building blocks of the present-day universe.

Opposite: the Orion Nebula, some 1500 light-years away, is a region where new stars in our galaxy are being formed. Much of the gaseous material that makes up a nebula consists not only of the raw material out of which new stars are formed, but also matter that was once in the interior of stars and has since been returned to interstellar space. This cycle of birth, death, and rebirth seems to occur throughout the universe. Has it always been like this?

The shape of the universe

By the early 1800s the distances to many of the stars had been calculated with a fair degree of accuracy. Astronomers were just beginning to turn their attention to a much more fundamental question: how far do stars extend? Do they go on forever throughout space or are they contained in a huge but finite volume? Allowing just for the 6000 stars visible to the naked eye, it has been calculated that they lie in a sphere roughly 330 light-years across – a diameter of about 2000 million million miles. By human standards this is an enormous distance, but today it is known that the sphere of visible stars is a micro-

scopically small element in the immensity of the universe as a whole.

Speculations about the true scale of the universe inevitably lead to the question of what shape it is. In the early 1900s a remarkable answer was worked out by the German-born scientist Albert Einstein. He was not interested merely in the way stars and galaxies were arranged in space. His idea of "shape" was more basic: he believed that there was a shape to space itself. This is a difficult idea, often incomprehensible to nonscientists. After all, in everyday language space is emptiness, and to talk of it as having shape seems totally meaningless. But using sophisticated mathematics, Einstein showed that even completely empty

Below left: medieval cosmology, based on Aristotle's theories, decreed that space was finite and had a definite edge. Here an astronomer looks beyond the edge to gaze at the mechanics that operate the ordered universe.

Right: the geometry of space. *A*, space which conforms to the laws of Euclidean geometry, the same familiar geometrical laws taught in school classrooms throughout the world. Up to the beginning of the 20th-century space was thought to be Euclidean. In it the shortest path between any two points should be a straight line and the angles of any triangle traced out in space should add up to 180 degrees. But Einstein showed that space was in reality much more complex than this. *B*, curved or non-Euclidean space which Einstein's general theory of relativity shows is the true nature of real space. Here the angles of a triangle measured on the surface of a sphere representing the curvature of space, add up to more than 180°. Moreover, the shortest path between two points is a curve, similar to the Great Circles that are the shortest routes from one place to another on the surface of the Earth. *C* illustrates the effect of a strong gravitational field in space, causing a localized extra curvature. A beam of light passing through the affected region is noticeably deflected from its otherwise direct path. This relationship between gravity and the fabric of space was first predicted by Einstein and verified by British astronomer Arthur Eddington who showed in a series of historic photographs that the light from stars is bent as it passes close to the sun, a deflection caused by the Sun's powerful gravitational field.

space – such as a vacuum – has a definite shape and structure.

Einstein's work was based on the idea that space and time exist as a single entity called space-time. This entity has some very special properties. Most important is that it has a kind of elasticity – it can stretch, bend, and distort like a sheet of rubber. From this theory Einstein made some far-reaching suggestions about the shape of the universe itself.

The question he tried to answer was, what kind of geometry holds true in space? If it was the kind taught in most schools, which dates back to the Greek mathematician Euclid, then space could be described as Euclidean and the rules of classical geometry would then apply to it. For example, on the two-dimensional page of a book, the three angles of a triangle always add up to 360° and the shortest distance between two points is a straight line. If rules like these apply to space, a ray of light traveling through the universe must move in a straight line

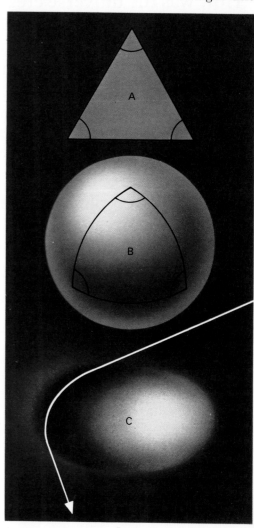

forever. We might call this a "flat universe," just as we would describe the Earth as flat if we could walk in a straight line along its surface indefinitely without returning to the same place. A flat universe is therefore also an infinite one, extending in all directions forever.

Einstein's findings about the structure of space suggested that it was not so simple. His work revealed a non-Euclidean universe, one in which space is curved equally in all directions; but because space is really a four-dimensional amalgam of space and time, the shape of the universe is not an ordinary sphere. Since humans are three-dimensional creatures, they have no words to describe the shape fully. The best that can be done is to describe it by the mathematical term of hypersphere.

The result of Einstein's idea is clear enough. A light ray traveling through the curved universe will eventually return to its starting point. Theoretically, therefore, if people's vision were super-

Below: An attempt by the Dutch artist Murits Escher to portray a universe that is both infinite and bounded. He represents a three-dimensional sphere as a circle filled with blue devils and white angels interlocked and extending to the outermost limits of the circle. But as the figures approach the boundary they become steadily smaller. At the extreme limits of Escher's pseudo-universe, the figures are infinitely small. Within the boundary of the circle, an infinite number of devils and angels can therefore coexist.

naturally sharp, they could stand in space and see an image of the back of their own head in the remote distance. In other words, the universe is closed in upon itself and yet is infinite in extent. Think of it in terms of people walking on the surface of the Earth: they feel as if they are walking in a straight line when in fact they are taking a curved path around a sphere. They can walk forever without reaching an end to the Earth but they will be endlessly repeating their path.

Today the idea of a universe infinite yet bounded is widely accepted, although it is no easier to understand now than when Einstein first formulated the concept. However, we now have observable evidence supporting his theory.

Even Einstein had not discovered the complete truth. He imagined the universe to be static and unchanging hypersphere. It is actually undergoing a steady expansion, carrying the galaxies farther away from each other as though fragmented by some mighty explosion.

Above: if we try to imagine ourselves as an observer inside Escher's sphere we find that the universe looks much the same wherever we stand. If we stand at *A* (top circle) we seem to be at the center of the universe with *B* at its observable edge. But if we travel to *B* (circle above) point *A* looks as if it is at the edge and *B* seems to be the center.

139

The expanding universe

In the early 1920s American astronomer Edwin Hubble made the first systematic studies of galaxies beyond ours. His work was of overwhelming importance, and much of the present knowledge about the origins and evolution of the universe is based on his findings.

Hubble's historic discoveries forced astronomers to accept that there are innumerable galaxies beside this one. Just as the Sun is an insignificant member of a vast family of stars in the Milky Way, so this galaxy is lost among the countless millions of other galaxies spreading in all directions as far as the most powerful telescopes can see. Hubble also made some of the earliest realistic estimates of the distance to the galaxies. He worked out that some were about 500 million light-years away, and that the nearest was more than 800,000 light-years distant.

Perhaps his most far-reaching discovery came when he examined the light emitted by the galaxies. Using a spectroscope he split the light into a rainbow of different wavelengths and examined the spectral lines revealed. Normally the different bright and dark lines indicate the presence of well-known chemical elements such as carbon and hydrogen, and lines characterizing the different elements occur in positions in a spectrum determined by their wavelengths. Hubble discovered a remarkable distortion in the spectra he examined. The familiar patterns of lines appeared to be consistently displaced toward the red end of the spectrum. Broadly speaking, Hubble found that the light from the faintest galaxies tended to be more red than that from bright galaxies. This is known as the *red shift*.

The importance of the red shift in the spectrum is best understood by comparing it with a similar effect familiar in sound waves. Most people have noticed at one time or another that the sound of an approaching car rises in pitch as it ap-

Right: the Doppler Effect applied to light. Named for the Austrian physicist Christian Doppler, who first applied it to sound, the diagram shows a light source (dot inside the smallest circle) moving toward an observer. As successive wave crests reach the observer, they are closer together (shorter wavelengths) than if the source were stationary. Because of this light appears bluer than it is in reality. To someone observing the light source moving away, the light appears redder (longer wavelength).

Below: one practical test of the Doppler Effect is to listen to the sound of cars as they approach, pass, and recede on a motorway. As they approach, their sound rises in pitch. As they pass and begin moving away, the pitch falls.

proaches. As soon as it passes, the pitch drops. The change in pitch is caused by an apparent change in the wavelength of the sound waves as their source moves. The explanation for this phenomenon was first worked out by the Austrian physicist Christian Johann Doppler in 1842. The easiest way to understand it is

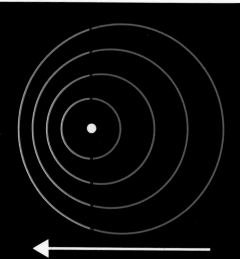

to imagine a ship at sea. If the ship is at rest, waves will be striking its bows at a rate of perhaps one wave per second. If the ship gets under way and sails directly into the waves, the frequency with which they strike the bows will increase. To the crew of the ship it will seem as if the distance between the crest of each consecutive wave – a distance known as wavelength – will have become shorter. If the ship were sailing away from the prevailing direction of the waves, the opposite effect would take place and the ship's crew would see an apparent increase in the wavelength. If this phenomenon is imagined in reverse so that the

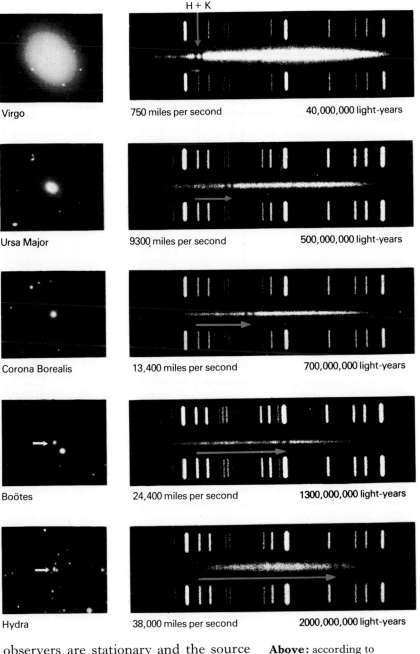

H + K

| | |
Virgo | 750 miles per second | 40,000,000 light-years

Ursa Major | 9300 miles per second | 500,000,000 light-years

Corona Borealis | 13,400 miles per second | 700,000,000 light-years

Boötes | 24,400 miles per second | 1300,000,000 light-years

Hydra | 38,000 miles per second | 2000,000,000 light-years

cornerstones of modern cosmology. It states that the apparent speed of a receding galaxy is proportional to its distance. In other words, galaxies twice as far away recede twice as fast.

Hubble's Law was the first real indication that the universe was not static, as Albert Einstein had believed. According to the measured red shifts of those galaxies observed, it was in a state of overall change. One of the most significant features of Hubble's discovery is that he found few galaxies with a blue shift in their spectrum. Virtually every galaxy he examined appeared to be moving away. This could suggest that the Ancients may have been right to imagine the Earth as the center of the universe, but Hubble's Law makes it possible to deduce the movement of the galaxies not only as viewed from the Earth but also as it would appear from any point in the universe. It then turns out that the galaxies are receding from each other at the same rate throughout the universe, so that at every point in space, the observer would always appear to be in the center of the overall movement.

The uniform expansion of the universe revealed by Hubble's Law was the starting point for modern cosmology. For the first time evidence had been uncovered showing that the universe as a whole was undergoing some common dynamic process. Astronomers and physicists were presented with an exciting new picture of the universe, one that showed an obvious evolution and therefore suggested the likelihood of a birth and perhaps an eventual death.

Hubble believed that the galaxies were moving apart under their own impulsion like the fragments from some gigantic explosion, but the true picture is more complicated. Einstein had shown that space has a physical structure and can twist and stretch like a sheet of rubber. If space is imagined as a balloon instead of as a flat rubber sheet, it makes an approximate model for the universe as a whole. It is important to remember that only the rubber membrane of the balloon represents space, not the volume either inside or outside it. The galaxies can be imagined as tiny dots evenly distributed over the baloon's surface. The expansion Hubble observed can be compared to the

observers are stationary and the source of the waves is moving, the wavelength will seem longer if the source moves away and shorter if it approaches the observers.

This phenomenon – now called the *Doppler Effect* – is true of all kinds of waves including light. This means that if a source of light is receding, its spectrum is reddened by an amount depending upon its speed measured in relation to the Earth.

Hubble soon realized that galaxies with the biggest red shifts were also the most distant. Once he had worked out a few of the distances involved he could formulate his discovery as a principle that is now called *Hubble's Law*, one of the

Above: according to Hubble's Law (first formulated by the American astronomer Edwin Hubble) the recessional velocity of a galaxy increases with its distance from us. In the lefthand column are clusters of galaxies lying at various distances. In the righthand column are their spectra. Arrows indicate the amount of red shift of the lines when measured from a specific baseline – in this case the *H* and *K* lines of the calcium atom. The remotest galaxies have the largest red shifts indicating that they are receding from us at the highest relative velocities.

inflation of the balloon. As it expands, the membrane stretches and the dots move farther away from one another.

The expanding universe is not simply an explosion of galaxies from a common point in space. There is no edge of the universe from which the most distant galaxies hurtle outward into the empty space beyond them. Instead the galaxies must be pictured as spread uniformly throughout space. It is only the distance between the galaxies that grows as space itself expands.

The most difficult part to understand of this theory is how space can expand if there is nothing into which it can expand. It is important to remember that space is not spreading out into some pre-existing region as it expands. Its expansion is best looked at as a kind of stretching

Right: Hubble's Law – the *apparent speed of a receding galaxy is proportional to its distance from Earth.* In simple terms this means that the farther a galaxy is away from us, the higher is its speed of recession from us.

Left: the balloon analogy illustrating the concept of an expanding universe. In particular it shows why all galaxies seem to be moving away from us. As the surface of the balloon expands any two points on its surface move farther apart. Points that are already far apart, such as A and C in the illustration, show a bigger increase in distance than points that are close – A and B.

Below left: a two-dimensional model of the universe as viewed by two observers (A and B) at different points in the universe. The circles represent the observable "horizons" for each of the observers. The dots indicate galaxies, which from both observation posts will appear to go on for ever. The choice of observation posts is random, the picture would be the same from any other points in the universe. The observer always appears to himself to be in the center.

Below right: this two-dimensional model of an expanding universe shows the same system as the model on the left but with every distance between the galaxies expanded 1.5 times.

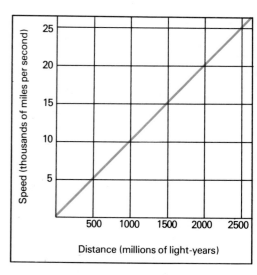

which changes the scale of all distances measured in space. The Earth itself does not stretch with space, so the effect of the expansion of distance can be gauged by the way all other objects in the universe are involved in an overall recession of one from another.

Although matter is not stretched by the expansion, light waves are. This causes the wavelength to increase and the light to redden. Hubble's red shift therefore is not really caused by the movement of the galaxies themselves, which can be regarded as being at rest in a universe of expanding space.

The idea of expanding space still suggests that there was an aftermath of some enormous explosion. If the distances between galaxies is steadily growing, it can reasonably be assumed that they were much closer together in the remote past. The farther we go back in time, therefore, the higher the average

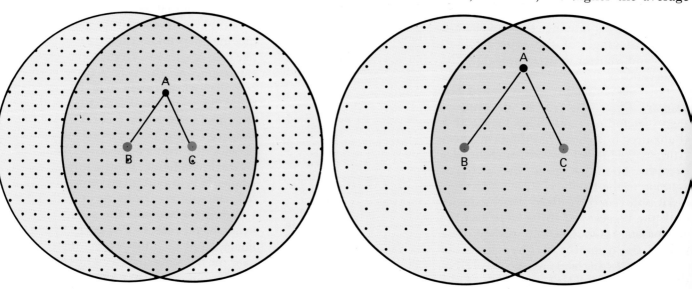

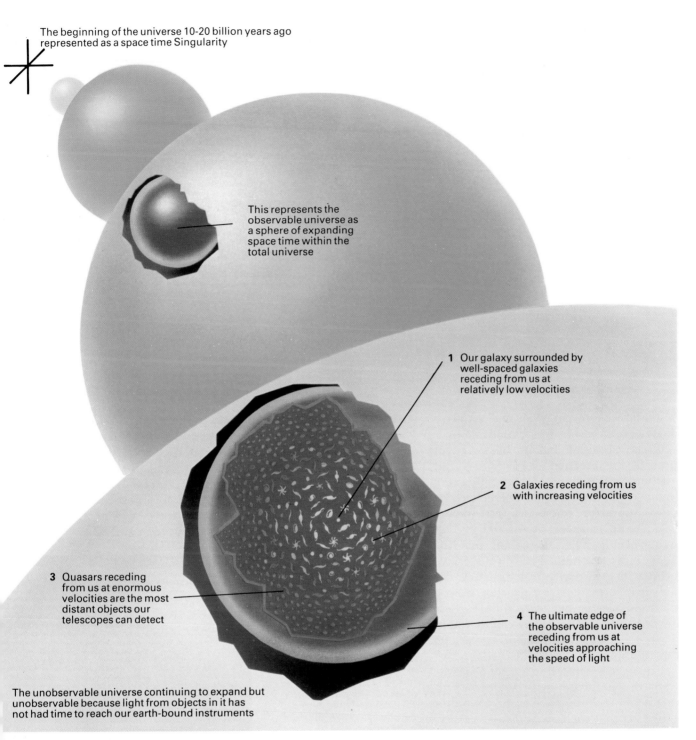

The beginning of the universe 10-20 billion years ago represented as a space time Singularity

This represents the observable universe as a sphere of expanding space time within the total universe

1 Our galaxy surrounded by well-spaced galaxies receding from us at relatively low velocities

2 Galaxies receding from us with increasing velocities

3 Quasars receding from us at enormous velocities are the most distant objects our telescopes can detect

4 The ultimate edge of the observable universe receding from us at velocities approaching the speed of light

The unobservable universe continuing to expand but unobservable because light from objects in it has not had time to reach our earth-bound instruments

density of the universe will be. At some stage the entire universe of stars, dust clouds, and galaxies must have been crushed together in a small mass — perhaps no more than a few light-years in diameter. Before this, at the very moment of creation, the density of the universe would have been infinite, with all matter squeezed to a microscopic point.

This fantastic concept is really only the start of the difficulties. The question

Above: the universe viewed historically as it expanded from a single point of space-time, 10,000–20,000 million years ago. As the millennia pass, the total universe expands, and we see within it a smaller sphere of space-time — which is all that can be observed from the Earth.

inevitably arises of what there was before this incredibly compressed universe. Did space suddenly appear spontaneously and immediately begin racing outward in all directions? If time itself "began" with space, can we meaningfully ask questions about what occurred "before"? Although traditionally this kind of problem is usually tackled only by philosophers, modern cosmology is at last applying rigorous scientific methods to examine these fundamental issues and is

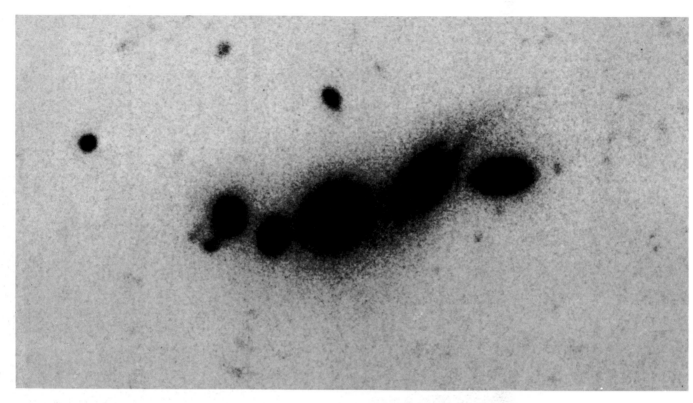

beginning to reveal what the universe must have been like in its infancy.

Today, besides looking backward in time to the creation, cosmologists and astronomers alike are probing the very limits of the observable universe. The most powerful telescopes can detect galaxies up to several thousand million light-years away. This means that the light now reaching Earth must have begun its journey not long after the universe formed. It means that we see these galaxies as they were in the earliest days of the universe. They can therefore teach scientists much about conditions during those remote epochs.

Efforts to extend the range of observations much further will be doomed to inevitable failure, however. Hubble's Law says that the farther away a galaxy, the higher its speed of recession. The most distant objects detected so far are hurtling away at over 90 percent of the speed of light. Beyond them is a horizon that will never be seen. No signal can travel faster than light – Einstein's theory of relativity showed this to be a basic law of nature. Therefore, we can only see objects still close enough for a ray of light to have had time to reach Earth since the universe was created. If the universe formed 10,000 million years ago, the horizon is at a distance of about

Above: a chain of galaxies photographed by the American astronomer Halton C Arp. This chain, VV 172, is made up of five galaxies, four of which are 600 million light-years away. The fifth galaxy is twice as distant and is thought to have been ejected from the cluster at a high velocity. There is growing evidence that small galaxies may have been ejected from larger galaxies.

Right: the Large Magellanic Cloud, which is some 40,000 light-years in diameter, is between one third and one half the size of our own galaxy. It is linked to its smaller neighbor the Small Magellanic Cloud by a tenuous common envelope of hydrogen gas. They are only about 200,000 light-years away from our galaxy.

13,000 million light-years. Moreover, if the most distant objects are actually moving at speeds faster than light measured relative to the Earth, their signals will never reach here.

Although the fundamental idea of an expanding universe is now very widely accepted, there is still some opposition

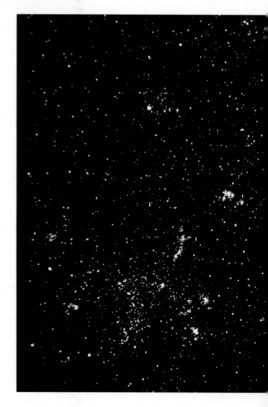

to it. The single most important piece of evidence supporting the expansion theory is the galactic red shift first discovered by Hubble. Most people today agree that the red shift is a Doppler effect caused by the galaxies speeding away from us, but a number of astronomers have suggested that the red shift has nothing to do with the motion of the galaxies. For example, the American astronomer Halton Arp of the Hale Observatory has discovered groups of galaxies, the members of which have widely varying red shifts. If the groups really are made up of neighboring galaxies, the red shift could hardly be a sign of many differing velocities in a relatively small region of space.

Other astronomers have pointed to the discovery of quasi-stellar sources, called quasars for short, as proof that red shifts must have some alternative explanation. Quasars are fairly compact luminous objects, the light from which is so strongly red shifted that they must be among the most distant objects in the observable universe. But if they really are so remote, there seems no way to account for their remarkable luminosity. To be able to see a comparatively small object – perhaps no more than a few light-years across – from a distance of several thousand million light-years suggests an unbelievable brightness. The amount of energy

generated every second would be far beyond anything the present knowledge of physics could explain. Opponents of Hubble's Law and the theory of the expanding universe say that no dilemma in fact exists. It is not our knowledge of physics that is inadequate but our interpretation of the red shift. In fact, quasars might be nearby objects of only average

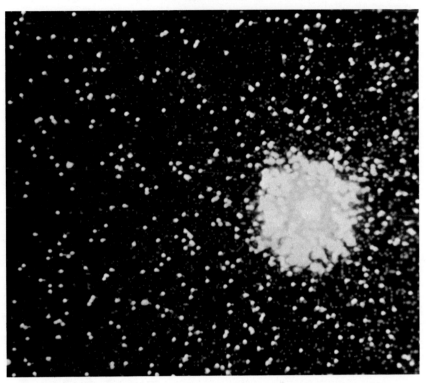

brightness but so dense that their gravitational fields "stretch" the light they emit, causing the observed red shifts in their spectra.

Despite these objections, most astronomers are now agreed that galactic red shifts are a sure sign of an expanding universe. The most convincing evidence is fairly easy to discover. Hubble's interpretation of the red shifts implies that the expansion of the universe began between 10 and 20 thousand million years ago. Therefore, if independent evidence that the universe is that old can be found, there will be strong corroboration of Hubble's ideas. In fact, studies of the way stars evolve and calculations about the amount of radioactive material to be found in the Earth's crust both tend to point to the age of this galaxy as about 15,000 million years. This fits the prediction of the galactic red shifts very well and is almost certain confirmation that the universe is expanding.

Above: the central "hot spot" of the quasar 3C 273, which is located in the central region of the Andromeda Nebula.

The primeval fireball

In 1927 the Belgian astronomer Georges Lemaître laid the foundations for what is today known as the *big bang theory*. He supposed that the universe was once condensed into a single huge mass that became unstable and exploded. He called it the "cosmic egg" and suggested that the fact that the galaxies can now be observed receding from one another was the direct result of the original explosion. The fragments of the cosmic egg, flung apart by the upheaval, have evolved into the galaxies that now exist.

The more recent version of the big bang theory updates Lemaître's idea in one very important respect. It is now known that no fragments of the primeval egg are flying from the site of an ancient explosion, but that space itself is expanding. In doing so it carried with it the stars, galaxies, and dust clouds that make up the observable universe.

It is hard to imagine space as being filled with anything, so how can it expand? Albert Einstein provided the answer with his theory of a space-time that can twist, bend, stretch, and even expand. This means that the picture of the expanding universe reveals the galaxies as stationary, moving apart only as the space between them expands.

In 1965 American radioastronomers Arno Penzias and Robert Wilson discovered that space was saturated with a faint background of microwave radiation. This was soon recognized as an echo of the big bang. The radiation is equivalent to the whole universe being at a temperature of about $-518°F$. This may not seem important in itself but working backward from this figure, cosmologists have been able to reconstruct the earliest eras showing that the young universe was extremely hot and that it has cooled to its present temperature only after a period of between 10 and 20 thousand million years.

The fascination of trying to picture the universe as it was soon after the big bang

is almost irresistible. In the last few years – using sophisticated computer techniques and the latest knowledge from the fields of atomic and nuclear physics – scientists have worked out a fairly detailed description of the events during the first 100,000 years of the universe's existence. More exciting still are recent attempts to reconstruct the first seconds after the Big Bang. These first few seconds are also of great interest to those scientists, the particle physicists, who study the subatomic world. Just as astronomers wish to discover how the universe began, so particle physicists want to know where the forces that govern the universe – gravity, electromagnetism, and the weak (radioactive decay) and strong (nuclear bonding) forces – came from, and whether they were ever all united as one all-encompassing force.

The first 100,000 years is a particularly important period because it was during this time that the basic atoms were formed from which all the natural chemical elements subsequently evolved. During that remote time, the universe was a very strange place, hardly recognizable to modern eyes. Indeed, there

Below: the mix of elements that, according to the Russian-born physicist George Gamow, were produced as the result of thermonuclear reactions during the first 30 minutes of the universe's evolution. Neutrons (green) fuse with protons, or hydrogen 1 (blue), to form deuterium, or hydrogen 2 (red), and helium 4 (yellow). Deuterium is later built up into heavier elements.

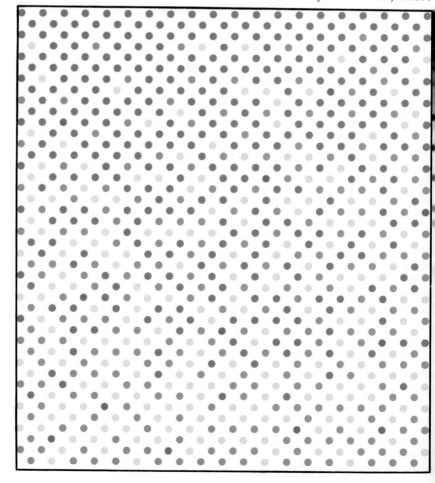

would have been nothing but a seething cauldron of radiation and electrically charged subatomic particles. Everything would have been plunged in a thick impenetrable fog. That hostile environment is called *the plasma era*. To understand it better it is necessary to know something about a plasma.

In simplest terms a plasma is just a very hot gas. It is special because at the extremely high temperatures at which it forms, the gas atoms hurtle around so violently that they literally knock pieces off each other. The countless impacts dislodge electrons from the atomic nuclei around which they were orbiting. The loss of its electrons leaves an atom with an overall positive charge. A plasma, therefore, is a turbulent cloud of electrically charged particles – negative electrons and positive atoms – at a temperature of several thousand degrees Farenheit. Plasmas are easily made in especially equipped laboratories and have been studied in considerable detail. One of their most striking properties is that they are very difficult to see through. Ordinary gases are often transparent to light but a plasma absorbs light strongly, and what it does not absorb it scatters in all directions. This is the reason why the plasma era must have been characterized by a vast luminous fog, filling the entire universe.

The plasma era was by far the longest of all the primeval epochs. It is generally thought to have begun about one second after the big bang itself. The temperature was then about 38,000 million degrees Farenheit. But as the universe expanded the temperature fell rapidly. Nonetheless, even in the first instants of the era, the basic foundations were laid for all the later chemistry of the universe. It must be remembered that in the first minutes after the big bang, events took place with incredible speed. More probably happened in the first second of creation than in all the 100,000 years of the plasma era put together.

It did not take long, therefore, to produce the building blocks of today's universe. During the first seconds of the plasma era, particles of matter were being created out of an intense primeval radiation. This kind of spontaneous creation was first predicted by Einstein who in his theory of relativity showed that matter

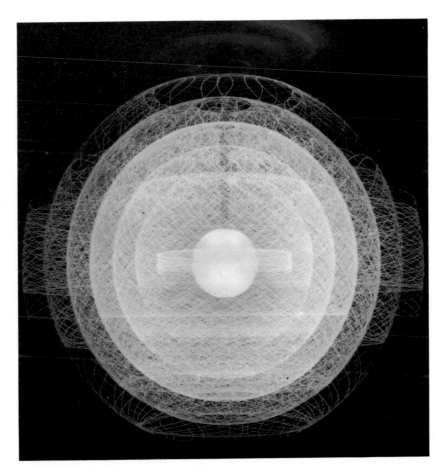

Above: uranium, the heaviest naturally occurring element, has an atom consisting of 235 or 238 protons and neutrons in the nucleus and 92 electrons orbiting it.

Below: plasma, created here in a laboratory experiment, is formed when matter is heated to such high temperatures that all electrons in its atoms are stripped away from their nuclei leaving a gas of positively charged ions. Soon after the creation of the universe, all matter existed in this state.

and energy are interchangeable. But for every particle of ordinary matter to form, an equivalent particle of antimatter must appear. In very approximate terms, antimatter is a mirror image of ordinary matter. For example, an antielectron – called a positron – is almost identical to an electron except that it carries a positive instead of a negative electrical charge. Matter and antimatter cannot exist together. They annihilate each other instantly in a blaze of radiant energy. During the first second of creation, massive amounts of matter and antimatter were forming and disintegrating, generating countless blasts of light, heat, and every other imaginable kind of radiation. Only in the opening moments of the plasma era, when the temperature fell below 1000 million degrees for the first time, did the upheavals diminish.

The aftermath of the unimaginable violence of the matter-antimatter holocaust was a huge number of protons and neutrons. As soon as the universe started to cool rapidly, the tiny particles could combine to form the first atomic nuclei without being instantly disrupted by violent impacts from other particles. Because of the relative numbers of protons and neutrons, 90 percent of the atoms turned out to be hydrogen and about 7 percent helium. Much later, the fact that hydrogen was the most abundant element made it possible for more complicated atoms to be synthesized in the cores of the first stars.

In the first violent seconds of cosmic history, events moved with blinding speed. At the very moment of creation the temperature and density of the universe was immense. A thousand millionth of a second after the big bang its temperature was about 200,000 million degrees Farenheit. A mass equivalent to the Earth would have been squeezed into a volume no greater than an average size bucket. Such extreme conditions meant that the only matter that could exist was in the form of tiny elementary particles, the building blocks of the bigger particles such as protons and neutrons. The universe at first was dominated not by matter, however, but by intense radiation ranging over all the wavelengths of the electromagnetic spectrum.

Imagine the universe in the first 100,000 million million millionth of a second after the big bang. Light will only have had time to travel the distance across an atomic nucleus. The overall cosmic density would be fantastic – the equivalent to our entire galaxy crushed into a volume less than four thousandths of an inch across.

Despite the fact that people can barely imagine conditions of this kind, scientists can still make a reasonable guess at what must have been taking place. There are indications that in the midst of tremendous turbulence the fabric of space itself would twist and buckle like the surface of a storm-lashed sea. Some particle physicists believe that it was at this point that the one, all-encompassing force that governed matter/energy began to break down into the forces as we know them today (gravity, electromagnetism, and the weak and strong forces). Out of this chaos the most basic type of subatomic particles – known as *quarks* – formed, the very process of creation helping to damp down the terrific upheavals.

Before even this flicker of time, science encounters a fundamental barrier to further knowledge. Investigators reach the so-

A Singularity time = zero temperature = infinite

B Matter/Antimatter time = $\frac{1}{1,000,000,000}$ second
temperature = 100,000 million °C
to 10,000 million °C

C Plasma
time = 1 second-100,000 years
temperature = 10,000 million °C
to 4000°C

Right: the origin of the universe according to the big bang theory. The spheres represent the universe in a fourth – time – dimension. Thus, as astronomical instruments improve and we are able to see further into the universe, we are in fact seeing further back in time. Light from sphere E, for instance, has taken billions of years to reach us, and it shows events that actually took place long ago. Similarly, we know that the rate of the expansion of the universe is slowing down so that consequently the further back we look with our telescopes, the greater the speed at which objects are moving away from us.

called *quantum era* in which the very concepts of space and time cease to have any meaning. In this brief instant, the entire universe was squeezed into a volume now occupied by the nucleus of an atom, and there was only one force governing it. There are no words to describe events under such conditions, and scientists trying to understand the ultimate mysteries of the big bang have to rely on the language of mathematics alone.

Despite every effort, no one has been able to penetrate the moment of creation itself. It remains the final challenge to cosmology and one of the most fundamental unsolved problems in science.

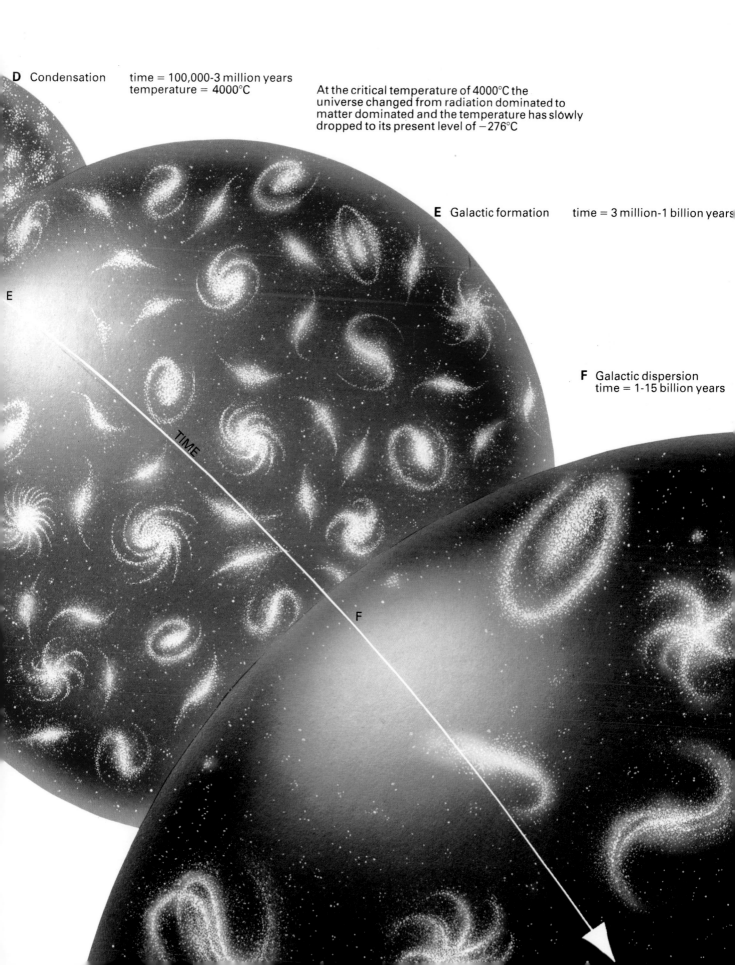

D Condensation   time = 100,000-3 million years
temperature = 4000°C

At the critical temperature of 4000°C the
universe changed from radiation dominated to
matter dominated and the temperature has slowly
dropped to its present level of −276°C

E Galactic formation   time = 3 million-1 billion years

F Galactic dispersion
time = 1-15 billion years

TIME

Echoes of the big bang

In the spring of 1964 American radio-astronomers Arno Penzias and Robert Wilson began work on measuring the intensity of radio waves emitted by our galaxy. It was a difficult task requiring extremely sensitive receiving equipment. Fortunately the Bell Telephone Laboratory had a remarkable radio antenna housed on a hill in New Jersey that was ideal for the purpose. The antenna, built a few years earlier for experiments with communication satellites, was shaped like a huge horn, its throat about 20 feet across.

The vital feature of the antenna as far as Penzias and Wilson were concerned was that it was almost totally free of radio noise, the background static caused by random movements of electrons inside most electronic instruments. This was important because the signals they hoped to detect coming from the galaxy would be so faint that even a minute amount of unwanted noise could easily drown them.

The two men took elaborate precautions to make sure that any signals they detected could only be coming from space. During the course of their checks they tuned the antenna to pick up signals at a wavelength of about 3 inches, far

Above: the American radioastronomers Arno Penzias (right) and Robert Wilson (left) who in 1965 while working at the Bell Laboratory in New Jersey, discovered and identified a faint background noise in the electromagnetic range coming from all directions of outer space. This radiation, named "cosmic background radiation," provides strong evidence to support the theory that the universe began with a high-temperature big bang.

Right: this graph of the background radiation plots weak signals coming from every part of the sky. Its wavelength corresponds to the universe being at an overall temperature 2.7° above absolute zero.

lower than the wavelength on which the galaxy would be radiating radio signals. They did not expect to pick up anything if in fact the antenna was free of noise. To their surprise, however, they discovered a weak signal coming from all directions. The signal consisted of microwaves – radiation much like radio waves but of a shorter wavelength. The signal was weak, almost too faint to be heard, but the astonishing feature about it was that its strength did not vary even if the antenna was moved to scan the entire sky.

Penzias and Wilson were sure that their receiver was faulty. They spent months overhauling it, even chasing away two pigeons roosting inside the antenna in fear that the birds might soil its delicate inner surfaces. Early in 1965 they dismantled the antenna entirely and painstaking cleaned it before reassembling and checking their results. Again they found the mysterious microwave signal apparently coming from all over the sky at a faint but steady level.

Electronic engineers often use a mathematical technique to work out an equivalent between the intensity of any background radiation and temperature. Using this idea, Penzias and Wilson calculated that the radiation they had detected was 2.7° above absolute zero on the Kelvin scale, or −518°F.

Why should all space show disturbing signs of being at a uniform temperature slightly above absolute zero? To Penzias and Wilson, their findings made no sense. Then through a chance conversation with fellow radioastronomers they were put in touch with a group of researchers at Princeton University led by the brilliant theorist P. J. E. Peebles. Peebles and his colleagues at Princeton had been working out what the universe must have been like shortly after the big bang had taken

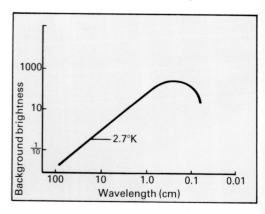

place, if indeed it had occurred at all. Their results suggested that the primeval universe was a cauldron of intense electromagnetic radiation at all conceivable wavelengths. During this remote period of the universe's evolution, its temperature would have been equivalent to billions of degrees Centigrade. Then as it expanded, the universe cooled. At the present time, Peebles calculated that the temperature would not be much more than $-518°F$.

Once Penzias and Wilson knew of the work of Peebles and other groups investigating the big bang, they realized the full significance of their own momentous discovery.

The radiation they had detected was the feeble heat glow left by the ancient explosion out of which the universe had been formed. In a sense, the weak microwave signals were a kind of echo reaching back in time to the big bang itself. Detailed studies of cosmic background radiation has subsequently enabled cosmologists to make the first systematic attempts at reconstructing the universe

Right: the radiometer at Princeton University used to observe the remaining faint evidence of the primeval fireball during which the universe was created. The antenna horns extend to left and right and are directed upward to collect radiation. A switch, microwave receiver, and an amplifier are at the center.

as it must have been soon after its creation.

Almost by chance, Penzias and Wilson had made one of the most important discoveries in the history of science. The echo they heard has become a key to the remote past revealing a glimpse of the fantastic primeval fireball that filled the universe during the first moments of its existence.

Right: diagram showing radiation wavelengths in the universe increasing and changing with time. In the first few seconds after the big bang the densities and temperatures were enormous and radiation filling the universe was mainly extremely energetic gamma and x-rays. As millennia passed, the universe "cooled," with wavelengths stretching into the visible and infrared. Today all that remains of the initial holocaust of radiation is a faint microwave signal permeating the entire universe.

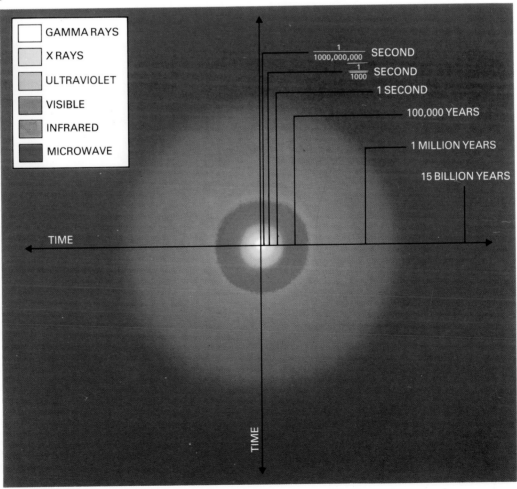

GAMMA RAYS
X RAYS
ULTRAVIOLET
VISIBLE
INFRARED
MICROWAVE

$\frac{1}{1000,000,000}$ SECOND
$\frac{1}{1000}$ SECOND
1 SECOND
100,000 YEARS
1 MILLION YEARS
15 BILLION YEARS

TIME

TIME

Steady state

The discovery that the universe is expanding and the big bang theory make it almost irresistible to see the galaxies as fragments of some huge ball of matter flung apart by an explosion of unimaginable violence. The expansion now observed is then no more than the natural evolution of a universe created by such a cataclysm.

Not everyone is convinced by this simple and graphic account. After all, why should the universe have burst forth in a single, special moment of time? And if it did, what was going on before that moment?

In 1948 three astronomers in England – Hermann Bondi, Thomas Gold, and Fred Hoyle – proposed a radical alternative to the big bang theory. They argued that, despite appearances, the universe

Below: an illustration of the "steady state" theory of the universe put forward by the British astronomers Herman Bondi, Thomas Gold, and Fred Hoyle. According to them new matter is continuously created, which means that over a long period galaxies (indicated by the dots) will be more or less evenly spaced despite an overall expansion of space. The steady state theory has many disadvantages and few cosmologists now accept it.

was unchanging in overall terms. According to this view, if people could travel either backward or forward in time, they would still see the universe looking very much as it does today.

Philosophically the new theory, known as the *steady state model*, was extremely attractive. The difficulty of having to account for the origin of the universe was solved by rejecting the idea of a beginning altogether. According to Bondi, Gold, and Hoyle, the universe always has existed and will continue to exist forever.

Despite its obvious appeal, their idea did not at first appear to fit the facts. The universe undeniably was changing – gas clouds condensing to form new stars, and stars growing to maturity and then ending their lives in total extinction. Far from being unchanging, nature seems in a constant state of flux. On the largest scale of all, the galaxies themselves are receding from each other, the gulfs between them presumably growing with the passing centuries. How could a theory based on a changeless universe account for these easily observable facts? The answer lies in the fundamental difference between a universe that is

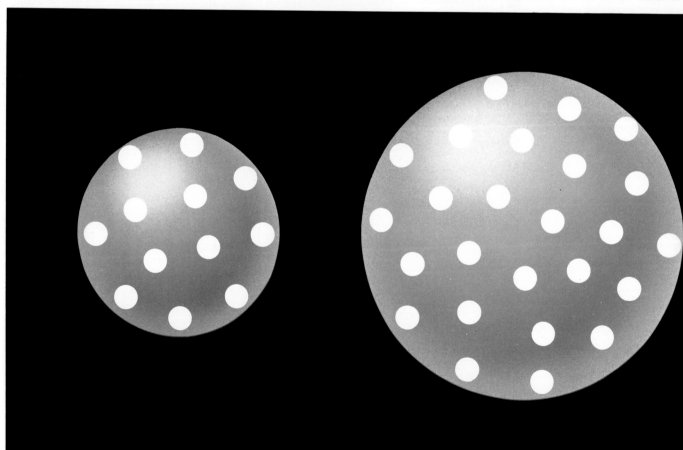

unchanging in an overall sense and one that is static and frozen in time.

Bondi, Gold, and Hoyle realized the universe was not static. They even accepted that the galaxies were receding. But they believed that the observable change is only a local phenomenon and if scientists could view the universe on a large enough scale, no overall change would be taking place. The key to their thinking was a process they called continuous creation. They argued that the universe was constantly being replenished by new matter forming spontaneously in space. It would not need to form very fast. The theory only requires one atom of hydrogen to appear out of nothing in a cubic mile of space once every 100 years. Even our most delicate instruments would be unable to detect such a minute process, so it is not surprising that no one has noticed it going on. But if anyone could observe the universe over thousands of years, the effect of continuous creation would be very obvious. As the galaxies drift apart, exactly enough fresh hydrogen is formed between them to insure that new galaxies evolve in time to fill the gaps. In this way

the universe is always expanding and small scale change takes place but the average separation of the galaxies remains the same forever.

Hoyle elaborated the theory. In particular he tried to work out a sound mathematical basis to explain the process of continuous creation. To do this he invented a new kind of phenomenon which he called the *Creation Field*. He supposed that the C-field, as it became known, contained huge amounts of negative energy. As new hydrogen atoms were created – releasing positive energy into the universe – the field gained an equivalent amount of negative energy. In this way the fundamental principle that energy is always conserved could still hold true even in the face of continuous creation. Overall, the universe and the omnipresent C-field never lose or gain energy but merely shuffle it between them.

Ultimately the steady state model was rejected in favor of the big bang theory, though not because scientists found continuous creation hard to accept. It is not, after all, any more remarkable a concept than the supposition that the universe came into being in a sudden vast explosion. Steady state failed because its central idea – that the universe looks the same from whatever moment in time it is viewed – simply is not true.

Looking across space is in fact looking backward in time. A star 10 light-years away looks as it did 10 years ago, the light from that time having only just reached Earth. The most distant galaxies, thousands of millions of light-years away, appear as they did in the remote past. The steady state model therefore requires that the distant universe must look very much like the one seen in the more immediate neighborhood. In fact, there are many marked differences. For example, looking farther away – and therefore deeper into the past – there are many more galactic radio sources than in our local region of space. There are also such mysteries as quasars, luminous objects too distant to appear as bright as they do by known physical laws.

The crushing blow to the steady state model came in 1965 when for the first time scientists detected the unmistakable echo of the big bang in the universe's background murmur of radiation.

The fate of the universe

At the heart of all astronomical study, of all the probings with telescope and dish into the depths of space and time, is one endless enigma that all astronomers seek to solve: How did the universe begin, and how will it end?

The beginning and the end are two parts of the same question; if we knew how the universe began and evolved, we would have a very good idea of how it might end. The most widely accepted theory of the

Below: an infrared thermogram of a suburban house. Such "heat scans" have a wide application and reveal the extraordinary wastage of energy that takes place on a huge scale in the world's industrialized nations. The color scale runs from white and yellow (hottest) to green and blue (coolest). As can be seen, the thin gable wall of the house and its windows are areas of greatest heat loss.

origin of the universe is that it began with a Big Bang, a sudden cataclysmic explosion and expansion of a singularity that contained all the matter/energy of the present universe in a point with no dimensions in either space or time. The theory holds that as the primeval fireball expanded and cooled, the building blocks of the universe started to take shape. First, the all-encompassing force that governed the pre-Bang universe broke down into gravity, electromagnetism and the weak and strong forces. And at this time, atoms began to form from elemental particles. Scientists agree on much of this. It's when they try to figure out how the objects we see in the sky – planets, stars, galaxies, clusters and superclusters – formed out of the early expanding soup of matter and energy that the subject gets complicated.

One thing is fairly certain – the primeval soup cannot have had a uniform density, for unless certain areas contained more mass and had stronger gravity than others, then matter wouldn't have gathered together into the lumps that formed the stars, galaxies, clusters and superclusters. But the question is: How did these fluctuations in density divide up the original soup?

Yakov Zel'dovich of the Institute of Applied Mathematics in Moscow theorized that the first fluctuations of density were the size of superclusters, and that clusters and individual galaxies formed secondarily from them. This is known as the "top down" theory. Some argue, however, that the present estimated age of the universe (10 to 20 billion years) is not enough time for galaxies to have formed from superclusters, and that galaxies were the first to form, and that they have, over time, slowly gathered into clusters and superclusters. This is the "bottom up" theory.

While many find the bottom up theory the most logical, no one really knows how galaxies were formed, nor even how they stay in one piece. Astronomers have discovered that the light that galaxies give off indicates a certain mass, but so does the way they interact gravitationally with other galaxies. The trouble is, the two values are very different. In fact, the light that a galaxy gives off indicates it has only 10 percent of the mass that we know it really has.

This is the intriguing question of *miss-*

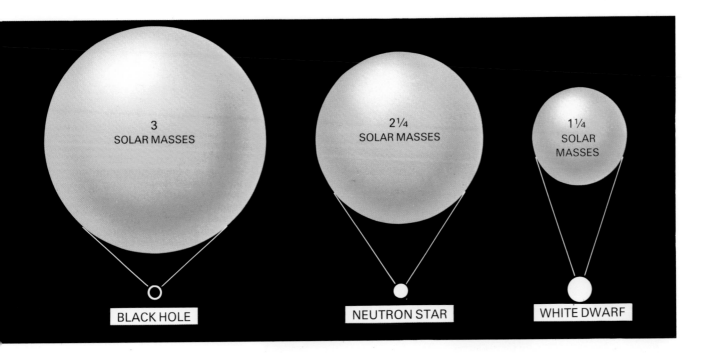

3
SOLAR MASSES

2¼
SOLAR MASSES

1¼
SOLAR
MASSES

BLACK HOLE

NEUTRON STAR

WHITE DWARF

ng mass. Astronomers want to know what makes up the 90 percent of the universe they cannot see. It is a mystery that has profound implications for the fate of the universe. If we are correct in assuming that the universe began with a Big Bang, and we believe Hubble's interpretation that the red shift of distant galaxies means the universe is still expanding, then there are three possible fates in store for the cosmos.

The universe could be undergoing "hyperbolic" expansion, in which case the expansion will continue, unabated, forever, as there is not enough matter to create enough gravity to slow it down. The universe could also be "flat," in which case there is just enough matter to slow the expansion, but not quite enough to stop it. Or, the universe could be "spherical." If spherical, there is enough matter to slow and stop the expansion, at which point the universe would start to fall back in on itself, perhaps to form another singularity, which in turn might explode, starting another universe.

If the universe is "open" – hyperbolic or flat – then it's ultimate fate will be *heat death*. Very simply, heat death is a result of the *Second Law of Thermodynamics,* which holds that hot things eventually lose their heat, for in any system, whether it be a car or the clockwork of the cosmos, some energy is lost, never to be recovered. Thus, if the universe expands forever, it will eventually cool to a point very close to

Above: the final evolutionary stages of normal stars – the formation of white dwarfs, neutron stars and black holes. In each case the critical factor in their formation is the mass of the original star. Up to one and a quarter solar masses a star is most likely to end its days as a white dwarf. Between one and a quarter and two and a quarter solar masses a more dramatic collapse to a neutron star is possible. Heavier stars, particularly those of three solar masses and above may undergo the most extreme collapse possible in nature and become black holes.

Right: our Sun, center of the solar system, is probably destined one day to become a white dwarf of approximately 24 miles radius.

absolute zero where molecular motion itself almost ceases. All stars, all galaxies, would be just cold chunks of dead drifting matter. And when all motion stopped, time too would come to an end.

Although such a fate is far beyond our reasonable concern (it would be hundreds of billions of years hence), there is still something inside us that craves the optimism of a closed universe which could collapse and begin again, even though that collapse would of course utterly obliterate all that had gone before. The dividing line between the open and closed models is the flat universe, which has just enough matter to slow the expansion, but not enough

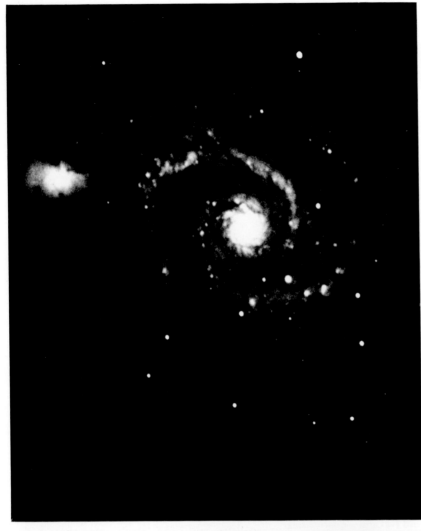

Left: How galaxies formed can tell us a lot about how the universe itself formed. "Top down" theorists believe that superclusters-size features formed first in the universe and then divided into cluster and galaxy-size objects later. "Bottom up" theorists believe that galaxies formed first and then collected into clusters and superclusters.

Below: will the universe continue to expand forever? Some scientists believe that after another 50–100 billion years, expansion will cease and the universe will fall back upon itself, becoming smaller, hotter, and more dense. Perhaps therefore in the remote future the universe will return to the state it was in at the time of the big bang itself. If this happens, some experts believe another big bang will take place and the cycle of expansion followed by contraction will begin once again.

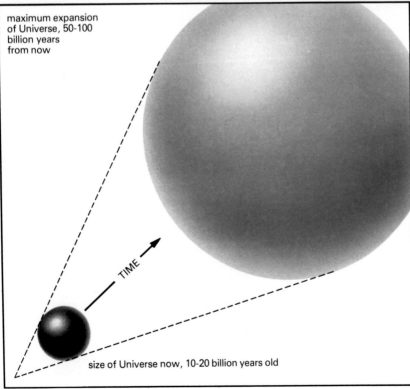

maximum expansion of Universe, 50–100 billion years from now

TIME

size of Universe now, 10–20 billion years old

to bring everything back together again. The flat model must then have a *critical density* of matter. Below the critical density point the universe is open; above, it is closed. Averaged out to the whole of the universe, this critical density has been calculated to be 3 hydrogen atoms per 1,000 liters of space.

We also know that if the universe were of the critical density it would be 13 billion years old. However, if it is older, then it's below the critical density and open; if it is younger than 13 billion years, it's closed. Now the problem is to calculate the age of the universe.

By observing the motions of galaxies and guessing their mass, we find that the universe is up to 18 billion years old and open. By another method – measuring the amount of deuterium that has survived since the Big Bang – the universe is supposedly 13 billion years old and open. However, if the deceleration of the universe's expansion as shown by the red shift of galaxies is observed directly, we find that the universe is less than 12 billion years old and is therefore closed. But all these figures can easily vary by several billion years and so there is no true answer to the question of whether or not the universe is open or closed.

This brings us back to the question of missing mass, which tells us that we can only see 10 percent of what is really out there. What makes up the other 90 percent? According to one popular theory, there is a particle in the universe, created at the time of the Big Bang, that is very plentiful, but is so fast and so light that we cannot detect it. We have only recently

Below: this map, showing the distribution and density of neutral hydrogen in the Milky Way, was made by Dutch and Australian radio astronomers. Neutral hydrogen in interstellar gas is largely undetectable by optical astronomy, but since its discovery through radioastronomy there has been much speculation that our galaxy, too, has passed through a period of violent activity.

detected the elusive neutrino, and there are 10 billion of them for every atom in the universe! If a neutrino had even one-one-hundred millionth the weight of a normal atom (and there is some evidence that these supposedly massless particles do in fact have a bit of mass) then that would account for the missing 90 percent. If such massive neutrinos, or other similar particles do exist, then they would have helped to form the galaxies. There might even be enough of them to bring the universe's density above the critical density limit, so that the cosmos will close, knowing no end, only an infinite number of beginnings.

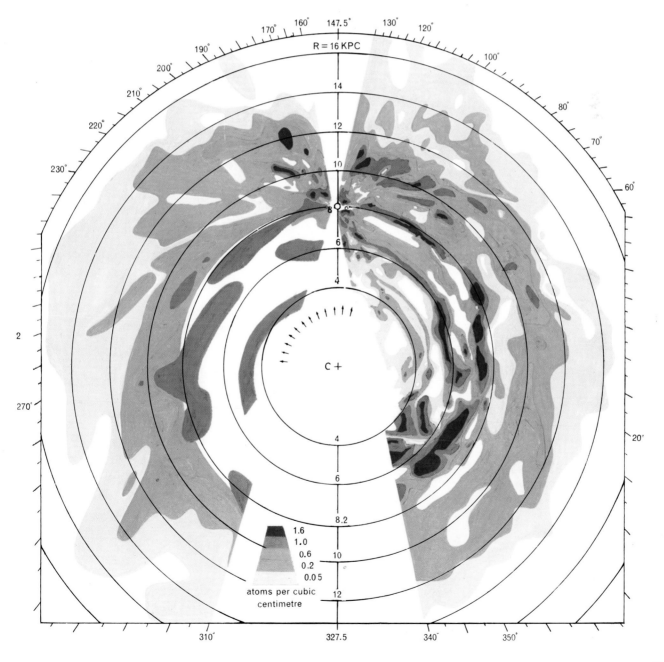

Chapter 6

Man in Space

The exploration of space is mankind's greatest adventure, but the means of reaching beyond the Earth's atmosphere demands sophisticated technology. The new skills and techniques that have emerged to meet these demands have had an uneasy birth. In the turmoil of World War II the first liquid-fuel rockets were built and flown – not for scientific research but to target explosive warheads on civilian populations.

After World War II, the mutual suspicion and distrust of the so-called cold war between the United States and the Soviet Union stimulated rocket development as part of a terrifying new arms race. Then with the first real breakthroughs in space-flight a new race began – the race to land humans on the Moon.

The early dramas of the space explorers, their first close-up sight of the surface of the Moon, and the unbearable tension of the moments before touchdown on the lunar surface, make the step into space one of the most gripping adventure stories of all time.

Opposite: United States astronaut Edwin Aldrin on the Moon, photographed by fellow-astronaut Neil Armstrong on July 21, 1969 – man's first day on the Moon. For centuries before that exciting decade of the 1960s, mankind had speculated about visiting the planets and stars. Those first tentative steps were boosted equally by a spirit of competition and scientific enquiry; but as Earth's resources become endangered, man's reaches into space are likely to be fueled as much by necessity as by scientific enquiry.

Visionaries and inventors

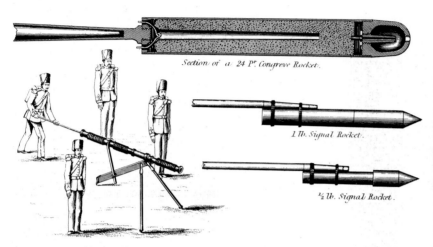

Section of a 24 P.ʳ Congreve Rocket.

1 lb. Signal Rocket.

½ lb. Signal Rocket.

The space age may be said to have begun on March 16, 1926. With an ear-splitting roar the world's first liquid-fuel rocket blasted away from a makeshift launch pad built on some Massachusetts farmland. It took 2.5 seconds to reach a height of 41 feet and to travel 184 feet down range. Its top speed was around 60 miles an hour. The brainchild of the American inventor Robert H. Goddard, the rocket – despite its modest performance – was the forerunner of giant boosters that within 50 years made possible man's first flight to the Moon.

Dreams about space flight go back for many hundreds of years before Goddard's pioneering experiments. Two thousand years earlier, the Greek philosopher Lucian of Samos wrote a remarkable story called *The True History*. Lucian told how a trader's ship, caught by a giant waterspout, had been hurled off the face of the Earth. After an eight-day journey through space the crew reached the Moon, discovering on it a race of alien creatures. Lucian's work is the first true science fiction story and it remained a classic for centuries. But more important, it is the first record of people's fascination with the idea of reaching across space and landing on the Moon.

In 1634 a new book appeared to rival Lucian's. Written by Johannes Kepler,

Above: a section of a Congreve rocket (top) and two signal rockets (below right). The soldiers (below left) are firing a 6-pound rocket. Rockets were first used as war weapons by the Chinese in the 13th century and occasionally in Europe in the innumerable wars of the 16th century. Their revival as weapons in Europe was brought about by William Congreve in the early 1800s.

Above right: a gigantic cannon blasts a rocket and its crew on a journey to the Moon. The illustration is from the 19th-century French author Jules Verne's *From the Earth to the Moon*. He foresaw many of the realities of modern space flight. Remarkably he sited the cannon some 150 miles from what is today the Kennedy Space Center – and his astronauts took only 50 minutes longer than the crew of Apollo 10, who circled the Moon in 1969.

Right: a drawing by the Russian pioneer rocket researcher Konstantin Tsiolkovsky. It shows how he attempted to grapple with the problem of the enormous thrust required of a rocket to escape from the atmosphere and gravitational pull of the Earth.

one of the architects of modern astronomy, the story was a strange mixture of supernatural and scientific ideas. Kepler called the book *Somnium* meaning *The Dream*. He imagined demons somehow projecting a sleeping man through space on a lunar journey.

Despite the elements of occult fantasy,

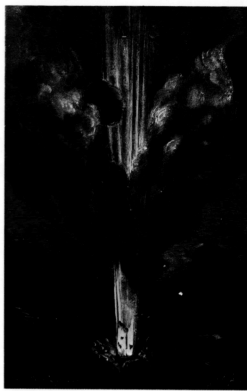

Somnium also reveals glimpses of Kepler's remarkable scientific insight. Kepler describes the air above the Earth's atmosphere as becoming increasingly rarefied and even suggests that the Earth exerts some kind of pull on anyone trying to leave it. These are the first recorded speculations about the vacuum of space and the universal force of gravity. Once on the Moon, Kepler's hero discusses the length of the lunar day and the shape and structure of the Moon's surface.

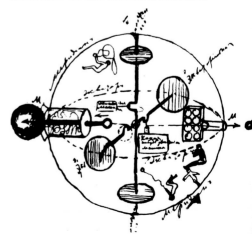

A few years after *Somnium* was published, the English cleric Francis Godwin was inspired to write about another fictional lunar expedition. Like his two famous predecessors he used a bizarre means of propulsion to send his hero into space. He conjured up a flight of wild geese, born on the Moon and compelled by their migratory urges to make regular trips between their lunar home and the Earth. Godwin's hero simply hitched a cart to the geese and was carried through space by the powerful beat of their wings. The story is important because it contains the very first account of weightlessness, an inconvenience of space flight well-known to modern astronauts.

In later years, writers invariably relied upon utterly fantastic methods of sending their characters into space, although they became more sophisticated. The great French visionary Jules Verne used the concept of a 900-foot cannon, its barrel buried vertically in the earth, to blast a crew on a voyage to the Moon. Some years later, the British writer H. G. Wells used the idea of antigravity as a means of propulsion. Both these men had considerable scientific knowledge and powerful imaginations. Yet neither realized that the motive power of space flight lay in a simple device that had been in

Above: Robert Goddard, the American pioneer scientist, with part of his team at work on a rocket in their workshop at Roswell, New Mexico. In 1932 he flew the first rocket stabilized by gyroscopically controlled vanes, and in 1935 he launched a rocket to the record altitude of 7500 feet.

Below: Robert Hutchings Goddard (1882–1945), known as "the father of American rocketry," was the first scientist to construct and launch a liquid-fueled rocket. NASA's Maryland center, the Goddard Space Flight center, was named in his honor.

use for about 1000 years – the rocket.

Rockets were probably invented by the Chinese. As far back as 1232, they used them as a weapon against the invading Mongols in the battle of Pien-Ching. We also know that the Chinese used rockets and other fireworks as a means of celebrating festivals and holy days. For centuries war rockets were a standard part of European arsenals and their use only declined when artillery became more accurate.

In the early 1800s the British engineer Sir William Congreve suggested the use of rockets as a battle weapon alongside artillery. Through his efforts they were used with devastating effect against the French in the early part of the century and against American troops during the War of 1812. "The rockets' red glare," a familiar phrase in the United States anthem, actually refers to a massive bombardment of Fort McHenry in Baltimore. The last use of Congreve's war rockets was against the Russians during the Crimean War.

At about that time the young Russian scientist Konstantin Tsiolkovsky was beginning to grasp the wider significance of rockets. Within a few years he produced a series of remarkable papers explaining for the first time how rockets

could be used not to kill, but to carry humans into space. Tsiolkovsky, the first of the modern visionaries, is now widely regarded as the father of space flight.

Born near Moscow in 1857, Tsiolkovsky suffered from scarlet fever when he was 10 years old and the illness left him totally deaf. Cut off from the world by his handicap, Tsiolkovsky turned increasingly to scientific research. In 1876 he became a teacher of physics and mathematics in a secondary school. At first he was fascinated by airship design but gradually his work on the science of flight kindled the interest that dominated his life – the concept of space travel.

Unlike many other space prophets, Tsiolkovsky realized that humans could only survive in space in a sealed capsule containing life-support systems to provide oxygen, temperature control, food and water, and other necessities. Most important of all, Tsiolkovsky knew that the propulsion system would have to be a combustion engine that carried its own oxygen for burning with it, an engine that developed a forward thrust by venting gas at high speed in a backward direction. Tsiolkovsky recognized the

Above: a V1 flying bomb German missile photographed immediately after launching during World War II. The V1 was powered by a pulsejet engine and could travel about 150 miles at a speed of 360 miles an hour. It was regulated by both a magnetic compass and a clock mechanism. The missile was launched toward England and after a preset time the clock locked the missile's controls and crash-dived it to the ground. The research for the V1 and the later V2 rocket was carried out at a secret establishment near the northern German village of Peenemünde on the Baltic coast. It was thus at Peenemünde that the prototypes of the first modern rockets were built.

rocket as the ideal device. He was well ahead of his time when he concluded that a liquid-fuel rocket, perhaps propelled by burning kerosine, would provide the answer for the necessary power and control.

Throughout his life Tsiolkovsky was interested in space travel, not just rocket propulsion. He wrote several books in which he discussed the use of artificial satellites, details of space medicine, the mechanics of long-term space flight, and even the idea of people colonizing the universe. The epitaph on his grave at Kaluga bears the statement of his certain belief in the future of space flight: "Man will not always stay on the Earth."

Tsiolkovsky was little known outside Russia so that Goddard was inspired more by writers such as Jules Verne. In the early 1900s Goddard devoted himself to making Verne's dream of space flight a practical reality. Single-handedly he worked out remarkable designs for both solid and liquid fuel rockets.

Goddard knew that he needed backing to develop his ideas. In *A Method of Reaching Extreme Altitudes,* his historic paper published in 1919, he emphasized the practical uses of rockets for investigating the upper atmosphere. At the end of the paper, apparently as an afterthought, he mentioned that a rocket might actually be used to carry people to the Moon. His ploy worked and, despite his growing reputation as an eccentric, he attracted funds to continue his research. His painstaking work culminated in 1926 in the first liquid-fuel rocket flight in history.

For three years Goddard used a cow pasture near Auburn, Massachusetts as his launching site. During that period he was apt to utilize rudimentary scientific instruments as payloads for his rockets. By 1929 his reputation in the area as a kind of mad professor was so widespread that the authorities were forced to take action. Claiming that his work was endangering life, the Massachusetts fire department prohibited more experiments.

Goddard moved to a remote part of New Mexico and continued to develop increasingly sophisticated rockets until 1941 when the United States entered World War II. By then he had already

Below right: the German V2 guided missile was the first modern rocket. It stood about 46 feet high and weighed some 12 tons – of which 8.5 tons were fuel and oxidizer. Among its instruments were gyroscopes that stabilized high-altitude flight by controlling the movement of vanes introduced into its exhaust blast. The rocket's external fins helped to stabilize it in the upper atmosphere, just as an arrow's feathers do. The fins became useless in the thinner air of the upper atmosphere, however, and the internal vanes that redirected the rocket blast were crucial in controlling the flight of the V2 and other high-altitude rockets.

Below: a captured V2 rocket being raised on a wagon. Many of the Peenemünde scientists made their way to the British and American-held territory after World War II, where they were given every facility to continue their research and development.

Militarists had received a dramatic demonstration of this potential in the closing months of 1945 when German V-2 missiles were launched against Britain. These terrifying liquid-fuel rockets carried a warhead of 1000 pounds of TNT, reached altitudes of about 60 miles, and had a range of 200 miles. The origins of this technological breakthrough goes back to the enthusiasm of some amateur rocket engineers who in 1927 formed the German Society for Space Travel. Its leading figures included Hermann Oberth and Wernher von Braun. When war loomed, German military leaders invited these brilliant

built 11-foot rockets that could travel at over 500 miles an hour. He had also developed ideas on gyroscopic control systems, multistage rockets, and variable thrust engines – all of which today play a vital role in space flight.

Ironically, Goddard was virtually ignored during his lifetime and his work recognized only after the war when American military experts grasped the destructive potential of rocket-powered missiles.

innovators to work at a special research establishment on the island of Usedom off the Baltic coast, near a small town called Peenemünde.

During the War years, German rocket engineers made tremendous advances in rocket design. The techniques and systems evolved during their design and testing were the cornerstone of all post-war rocket development. The dreams of the great visionaries of the past had become reality.

The principle of the rocket

In the mid-1600s Isaac Newton formulated a fundamental law of physics which became known as his Third Law of Motion. Three centuries later, the same law has carried astronauts to the Moon and boosted hundreds of satellites into orbit around the Earth. Newton's Third Law is today the motive power of space flight.

In straightforward terms, the law says that for every mechanical action in nature, there is an equal and opposite reaction. To understand how this makes a rocket work we can reformulate the law using a simple idea called momentum. As soon as an object moves, it takes on some momentum. The faster it moves, the more momentum it acquires. Exactly how much momentum a body has can be worked out by multiplying its mass by its velocity. A high momentum can be produced by a big mass moving slowly or a small mass moving very fast.

Left: the two basic types of rocket. Outer diagram shows the solid-fuel rocket – like the sky-rockets used in fireworks displays – which consist of a combustion chamber where the propellant burns explosively, generating gases that are expelled from a vent in a jet causing thrust. Inner diagram shows a liquid-fuel rocket that has three separate chambers, one containing the oxidizer (blue), another containing the fuel (green), and a combustion chamber. The fuel and the oxidizer are mixed and ignited in the combustion chamber. The gases produced create the rocket's thrust.

Restated on the basis of the idea of momentum, Newton's law of motion says that on producing a momentum in one direction, an equal amount of momentum must appear in the opposite direction. For example, if a revolver is fired, the bullet leaves the barrel at high speed carrying away momentum equal to the bullet's mass multiplied by the gun's muzzle velocity. Equal and opposite momentum appears in the much more massive gun by moving it backward at an appropriate velocity. The result is a sharp recoil.

A rocket is like a continuously firing gun, the unnumerable recoils being smoothed out as a steady thrust from the rocket engine. The simplest rockets burn a chemical fuel such as gunpowder in a combustion chamber, a sealed container with only one small outlet – the rocket nozzle. The fuel burns rapidly and produces huge volumes of gas. The pressure in the chamber builds up, relieved only by gas vented at high velocity through the engine's nozzle. The mass of gas expelled together with its high exhaust speed creates a large backward momentum. The reaction to this is a powerful continuous thrust in the opposite – or forward – direction.

The importance of the rocket engine to space flight is that unlike other kinds of propulsion it will work equally well in the Earth's atmosphere and in the vacuum of space. While a jet engine needs to suck air into its body to provide the oxygen needed to burn its fuel, rockets carry the oxygen in the fuels themselves.

In designing a rocket, engineers try to get as much mass expelled from the engine at the highest possible exhaust velocity. In this way the engine generates maximum thrust. To lift off its launch pad, a rocket's thrust must be greater than the combined weight of its engines, fuel, and payload. To actually get a payload into orbit, the thrust has to be powerful enough to accelerate the vehicle to around 18,000 miles per hour.

There is no reason to haul more load than is absolutely necessary. For example, as fuel is burned up, the tanks holding it become no more than dead weight. To avoid wasting thrust, modern rockets use the principle of staging. A staged booster is basically a series of rockets stacked on top of each other. The first rocket needs to be the most powerful because it has to lift the entire vehicle off its pad. Once its fuel is exhausted, it separates and falls away leaving subsequent stages to continue boosting the vehicle to the required velocity. Saturn V, mainstay of the United States manned

space program, is a three-stage booster. Its first stage engines develop a colossal 7.6 million pounds of thrust – equivalent to the power of 500 Hoover Dams. By contrast, the second and third stage engines produce 1.2 million and 230,000 pounds thrust respectively.

The principle of the rocket works whether or not the engine is burning chemical fuel. The basic requirement is for mass in one form or another to be expelled to create a forward momentum. For high thrust over a short period,

Above: the enormously powerful boosters of the Soviet Union's Vostok rocket are capable of delivering 1,340,000 pounds of thrust. They were used for the successful Soviet Earth-orbit program between 1961–1963.

boosters using either solid or liquid chemicals are ideal. Steady thrust over the extended periods needed to build up velocities that will make long-range flights practical requires more exotic techniques such as ion-drives. Emitting high-velocity streams of electrically charged particles, such engines can only provide thrusts of a few pounds. But over months or even years, the gentle power of ion propulsion can build up velocities high enough to bring the nearest stars within reach.

The first space boosters

After World War II, rocket development was influenced as much by political considerations as by advances in technology. The closing months of the war had given a grim demonstration of the destructive potential of rocket-powered missiles. When American and Soviet

Above: the first vehicle to be launched from Cape Kennedy was Bumper 8 on July 24, 1950. It was a two-stage rocket incorporating a V2 as its first stage. Bumper was the first successful multistage space booster in the United States.

troops closed on Berlin, rocket engineers from Peenemünde were surrendering themselves to one army or the other. In their new homes, they later spearheaded a new arms race between the United States and the Soviet Union, the two emerging world superpowers.

The Soviets, aware of the United States' powerful long-range bomber force, launched a major program to develop massive liquid-fuel rockets – the first Intercontinental Ballistic Missiles (ICBMs). Guided by German engineers, the Soviets fired their first V-2 rocket on October 30, 1947 and by late 1949 had developed an improved missile which they called the T-1. By 1954 they were building multistage vehicles, the first of a generation of long-range missiles that could hurl atomic warheads at enemy bases thousands of miles away.

American experts also used the V-2 as the foundation of a new military technology. Between 1946 and 1951, 66 V-2's were fired from White Sands in New Mexico. But, secure in the strength of their long-range bombing force, the Americans were less concerned about developing large missiles and concentrated on smaller tactical ones.

In 1947, however, work was started on plans for an ICBM in case it was necessary to match the Soviets in this new technology. Three different designs emerged. The first was called Teetotaler because it did not use alcohol in its fuel; the second was called Old Fashioned, because it was based on the V-2; and the third was dubbed Manhattan, because it would carry an atomic bomb, the end-product of the so-called Manhattan project. ICBM development eventually led to the famous Atlas missile, but in the course of the design and testing programs a number of other boosters appeared.

The first was a simple modification of the V-2 called Bumper, made up from a V-2 first stage and a second stage consisting of a rocket called the Wac Corporal. The vehicle failed repeatedly and was soon abandoned. But American engineers had learned invaluable lessons in many of the essential techniques of rocketry.

Soon after the Bumper program the Army's missile team built the first operational missile. The team was led by the former German engineer Wernher von Braun who later became an American citizen, based on the V-2, the rocket was nearly 70 feet high and was named the Redstone. It made its first flight in August 1953.

Following the successful launch, American military experts were faced with a crisis. They had suddenly realized that their relatively low-priority approach to rocket development had allowed the Soviet Union to gain a lead in developing military missiles. In particular, it was clear that the Soviet capability of launching a nuclear attack by ICBM could not be easily countered by ordinary jet bombers. To overcome the gap between Soviet and American missile potential, the development of the Atlas became a crash program. Compared with the Redstone the new booster was to be a giant. Atlas would generate a thrust of 367,000 pounds and have a range of 9000 miles compared with the Redstone's

75,000 pound thrust and 200 mile range. In late 1955 and early 1956 work was also begun on two intermediate range missiles, Thor and Jupiter. Together with Redstone these vehicles formed the mainstay of the United States missile armory. Within a few years these devices also provided the foundations of the family of boosters that carried the first American astronauts into space.

The year 1955 was crucial for space flight. The international scientific community decided to make it a year of concentrated effort to learn more about the Earth, calling it the International Geophysical Year. Before long it had become a propaganda battlefield. Both the United States and the Soviet Union announced their intention to launch artificial satellites. In neither case was this merely a bold scientific venture. The two nations were demonstrating to each other their powers in missile technology. But what-

Above: a Thor guided missile, here modified as the Thor Delta, can also be used for peaceful purposes, and has been widely used as a launch vehicle for many scientific and commercial satellites. This one is being used to place Tiros I, the first weather satellite, into orbit.

Left: the United States' most powerful rocket in 1952 and 1954 was the Redstone rocket. Developed by a team working under the famous German-born rocket scientist Wernher von Braun, the Redstone had a range of 200 miles and 75,000 pounds of thrust.

ever the motives involved, space flight was firmly in the political arena. The space race was on.

The Soviet Union planned to use its big ICBMs as space boosters. Having no missiles of comparable power fully developed, the United States pinned its hopes on Project Vanguard. The idea was to use an existing liquid-fuel rocket, the Viking, as the first stage with small solid-fuel rockets as second and third stage boosters.

But the United States was trying to move too quickly. Vanguard was a disaster, and failure followed failure with vehicles either crashing back to the launch pad or not even lifting off. Although Vanguard boosters eventually managed to blast satellites into orbit, their early failures made American experts realize that only wholehearted commitment to space technology could ensure success.

The Vanguard fiasco was made more obvious by dramatic news from the Soviet Union. On October 4, 1957 the 184-pound satellite Sputnik 1 was launched into orbit, circling the Earth every 96 minutes. The Soviet booster was essentially a perfected ICBM. A month later they launched the 1200-pound Sputnik 2 and its passenger, a dog named Laika.

The United States was galvanized into action. Ignoring the unreliable Vanguard, a new rocket was developed. Under the leadership of von Braun, a select team of engineers built Jupiter – an enlarged version of the Redstone topped by a second stage cluster of solid-fuel rockets. On January 31, 1958 – just 84 days after von Braun had been given the go-ahead – the first Jupiter boosted a 31-pound, pencil-shaped satellite called Explorer I into orbit. The United States was at last moving forward in the space race.

In the years that followed, the original Soviet lead in big boosters was whittled down by the American teams. In one sense, the late start by the United States turned into an advantage. Because their boosters did not have an ounce of thrust to spare, all onboard systems had to be as light and efficient as possible. This tough constraint forced American engineers to develop highly sophisticated mechanical and electronic systems long before the Soviet Union. Soviet boosters

were so powerful that there was little need for engineers to concentrate on improving this aspect of their technology. The United States took a lead that eventually insured their success in putting the first astronauts on the Moon.

In April 1961 the Soviets used a Vostok booster to send the first man, Yuri Gagarin, into orbit. The Vostok, the Russian word for "swallow," later dominated the Soviet space program. A direct development of their intercontinental missile, Vostok featured the first use of strap-on boosters bolted onto the main body of the vehicle. Later versions of the booster used up to 20 first-stage engines developing over one million pounds of thrust at lift-off.

American booster technology relied on several distinct families of vehicles.

The successful Jupiter built by von Braun was quickly developed into the Juno. This consisted of an elongated Jupiter first stage generating 150,000 pounds of thrust. The second stage was a cluster of 11 solid-propellant engines giving a total thrust of nearly 20,000 pounds. The third stage was a similar cluster but with only three such engines. A final fourth stage consisted of a single solid-propellant rocket attached to the payload itself.

Two other important booster families known as Atlas and Titan, grew out of the ICBMs. It was on top of an Atlas-D in 1962 that astronaut John Glenn became the first American in orbit. The main Atlas first stage used a five-engine cluster developing over 350,000 pounds of thrust. The Titan boosters were successfully used to launch all the two-man spacecraft of the Gemini program. The basic vehicle had two stages. The first stage produced over 400,000 pounds of thrust; the second, about 100,000 pounds. A remarkable aspect of Titan was the "fire in the hole" staging technique by which the second-stage engines ignited before the first stage had separated to give efficient staging and additional thrust as the second stage took over from the first.

The most important family of American boosters has been the giant Saturn launch vehicles. As early as 1957 von Braun argued the need for an ultra-powerful booster developing over a million pounds of thrust. In 1959 the Saturn project had come into being. The name was chosen because Saturn is the next planet after Jupiter, the name that had been chosen for von Braun's successful previous booster.

The first step was the development of Saturn I. Over 200 feet high, the massive booster produced about 1.5 million pounds of thrust from its first stage engines. As the first Saturn 1s were being flown, Saturn V was already being developed. The ultimate development in booster technology, Saturn V became the mainstay of the Apollo program of lunar flights. Over 360 feet high, Saturn V's first stage engines delivered a fantastic 7,610,000 pounds of thrust.

Saturn was the pinnacle of von Braun's achievements and its perfect success record was a fitting testimony to his genius. But just as developing the right boosters was vital to space exploration, the spacecraft and the people to fly them were equally important. With the new technologies developed to a high degree, humans could at last take their first steps into space and so begin one of mankind's greatest adventures.

Above: the Vostok rocket. So powerful were the Soviet boosters that they had little need to reduce the weight of their onboard equipment and it was in this field – of highly sophisticated and lightweight equipment – that the United States overtook its Soviet rivals.

Left: Explorer 1, the American satellite launched in 1958, carried scientific experiments including one that discovered the Van Allen Radiation Belt.

Right: the Saturn V rocket being carried to its launching pad on a crawler transporter. The Saturn rockets – which would eventually take people to the Moon – were gigantic in comparison to any that had been designed before.

Satellites

Artificial satellites are almost common-place today. Thousands have been launched since the first Sputnik orbited the Earth in 1957, and hundreds are still in space. On a clear night many are easily visible with the naked eye as tiny fast-moving points of light.

Today satellites perform a wide range of tasks. Best known of these are the world wide communication networks made possible by complex telecommunication satellites that can relay radio and television signals between any points on the Earth. The United States' Intelsats are giant cylindrical satellites while the Soviet Molniyas are characterized by large windmill panels of solar cells. Both were launched in 1965.

Satellites such as Tiros (since 1960) and Nimbus (since 1964) are used to study weather patterns and ocean currents. Earth Resources Satellites – now called Landsats – provide a wealth of information about the Earth's surface while others make a wide range of specialized astronomical, geophysical, and solar observations, radioing results from orbit to earthbound scientists.

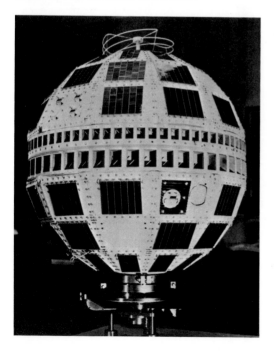

Military satellites are mainly intelligence gathering devices. High-resolution cameras aboard such American devices as SAMOS (since 1961), Big Bird (since 1971) and the KH-11 (since 1976) can distinguish individual missiles, tanks and artillery weapons from altitudes of over 100 miles. The highly classified pictures from these spy satellites are returned either in a special reentry capsule, in the case of a Big Bird, or in the form of electronic telemetry, in the case of a KH-11. As well as spying, these satellites now form an integral part of the war contingency plans of both the United States and the Soviet Union.

There is no such thing as a typical satellite. Just as their jobs vary widely, so their shapes and sizes range from metal spheres six inches in diameter to huge cylindrical devices weighing over 500 pounds and standing eight feet or more in height. But every satellite is made up of several important subsystems and, although they may vary in size and composition, they always perform the same basic function.

The power supply subsystem provides the electricity needed to run all the satellite's on-board systems. The most common technique for generating power in space today is to use solar cells either in winglike panels or covering the surface of the satellite itself, to convert the energy of

Left: Telstar II, the second low-altitude communications satellite to be launched. Around the center can be seen two rows of microwave antennas that receive and retransmit ground signals.

Above: an Intelsat II satellite, which was used both for commercial communication over the Atlantic and Pacific oceans, and for linking the Apollo Space Program stations.

sunlight into electricity. A satellite's communications subsystem is the satellite's link with the Earth. Through it, the satellite can relate information (its state of health, position, what it has recorded, etc.) as well as receive instructions (adjust its position or turn on equipment, etc.).

Guidance and control subsystems keep the satellite on its planned course while the attitude control subsystem makes sure it is correctly oriented. An environmental control is essential to maintain the right temperature and atmosphere pressure within the body of the satellite. It also protects the other subsystems from excessive radiation or mechanical dangers such as vibration and acceleration forces during launching or course corrections in orbit.

Many satellites have some kind of onboard propulsion subsystem. Usually a low-thrust rocket engine, it is most commonly used for adjusting an orbit or changing it altogether.

Perhaps most fundamental of all is the structural subsystem, the basic skeleton of the satellite. This provides support and space for all the other subsystems. Usually constructed from aluminum and magnesium, the framework is a delicate balance between structural strength and lightness.

Just as the shape and configuration of satellites vary with their jobs, so do their orbits. Communications satellites orbit at the geostationary, or geosynchronous, altitude of 22,300 miles. At that height, the time it takes for a satellite moving at the orbital velocity of 18,000 miles an hour to encircle the Earth is the same time it takes for the Earth to rotate – 24 hours. Therefore, at that height, if the satellite is positioned over the equator, it will appear to hover over one spot, making it ideal for communications relay. Weather satellites generally orbit at from 300 to 800 miles, while spy satellites will fly overhead as close as 80 miles.

The advent of artificial satellites has brought a wide range of information to scientists, technologists, military strategists, and business people. These satellites are not just American and Soviet built; there are Canadian, French, British, Chinese, Indian and Indonesian satellites in space as well. With increasingly sophisticated designs, satellite capabilities have continued to grow, and consortia now construct and launch them as they prove themselves profitable. While the drama of the manned spaceflight continues to hold the spotlight of public attention, the steady development of satellite technology is working a quiet revolution that is already bringing benefits – and some changes – into everyday life.

First steps in space

Early in April 1961 foreign correspondents in Moscow reported rumors sweeping the city that the Soviets had put a person into orbit. Then, on April 12, the official news agency Tass electrified the world with the following dramatic announcement:

Below: *The Vostok in Orbit*, painted by the Soviet cosmonaut Alexey Leonov. The scene was remembered from his own flight in Voskhod 2.

"The world's first spaceship Vostok, with a man on board, has been launched from the Soviet Union in a round-the-earth orbit. The first space navigator is Soviet citizen pilot Major Yuri Alexeyvich Gagarin."

Mankind had at last taken its first step into space. Launched from the Soviet spaceport at Tyuratam, Vostok 1 completed a single orbit reaching a maximum altitude of 203 miles. After 108 minutes, Gagarin made a safe landing inside the spherical command module of his space-

Right: Soviet pilot Major Yuri Gagarin was the first man to fly in space, orbiting the Earth once in his spacecraft Vostok 1 at 17,000 miles per hour and reaching an altitude of 203 miles. After 108 minutes, Gagarin fired retro-rockets and parachutes brought the pilot and craft safely down near Saratov, in European Russia.

ship. Despite its short duration, the flight was a historic achievement and a major challenge to American space experts.

The world did not have to wait long for an answer to Vostok from the United States. Only 23 days later on May 5, 1961, astronaut Alan B. Shepard waited in his cramped spacecraft Freedom 7 to become the first United States astronaut. His launch vehicle was a Redstone booster. In sharp contrast to the secrecy surrounding the Soviet space efforts, Shepard's flight began in a blaze of publicity. At 9.34 am millions of Americans watched the television transmission direct from Cape Canaveral to see the Redstone lift off from its pad on a tongue of flame. The pilot's voice could even be heard crackling over the radio link to the ground controllers, saying: "Roger. Lift-off and the clock is started. This is Freedom 7."

Shepard's flight lasted only 15 minutes 22 seconds. It was a simple ballistic shot with no possibility of achieving an orbit. The trajectory took Shepard to an altitude of 116.5 miles at a top speed of 5180 miles an hour. He was weightless for about five minutes.

The successful flight of Freedom 7 signaled the beginning of Project Mer-

cury, a pioneering program of manned flights designed to test systems and train pilots for more ambitious missions planned for the future. Subsequent orbital missions relied on Atlas boosters but the spacecraft itself remained essentially unchanged. Bell-shaped, the Mer-

cury capsule was only just large enough for an astronaut and essential equipment. The main conical section contained the pilot, life-support and electrical power systems, and flight controls and displays. The smaller cylindrical section held the main and reserve parachutes. The heat shield covering the conical base of the spacecraft was made of a mixture of glass fibers and resins. During reentry into the Earth's atmosphere, the friction of the air heated the shield to over 3000°F. As it glowed red-hot, the resin mixture vaporized, boiling off and conducting the worst of the heat energy away from the body of the spacecraft.

Project Mercury consisted of six flights by astronauts. Most famous of these was the first orbital mission flown in Friendship 7 by John Glenn. After numerous frustrating delays and postponements, Glenn went into orbit on February 20, 1962, 11 months after Gagarin. The mission lasted just under five hours and Glenn covered over 83,000 miles, reaching a maximum speed of 17,549 miles an hour. Just before reentry Glenn noticed that one of his on-board monitoring systems showed that his heat shield was no longer locked in position. Flight controllers decided to let Glenn keep the special retrorocket pack, normally jettisoned before reentry, on the heat shield in hopes this would keep it securely in place. As Friendship 7 plunged back into the atmosphere, Glenn could be heard calmly reporting the break-up of retropack with the blazing fragments hurtling past his viewports. "There's a real fireball out there," Glenn commented. Later he admitted that he thought the fire was due to his heat shield disintegrating. Fortunately the shield held

Above right: Vostok 1 after it had returned to Earth with Gagarin. The spherical command module was brought to Earth safely using braking rockets and parachutes.

Above left: Alan Shepard, America's first man in space, being lifted to the recovery helicopter after his 15-minute suborbital flight. The Mercury capsule, which landed in the Atlantic Ocean, is below him.

throughout the descent and Glenn received a hero's welcome from the waiting recovery forces.

Unlike the Soviets, American space experts had decided to land their astronauts in the sea. The splashdown was carefully controlled by parachute systems and immediately after hitting the sea the pilot climbed out of the spacecraft and waited in a rubber dingy for a recovery helicopter. Soviet pilots either made a controlled parachute landing on land

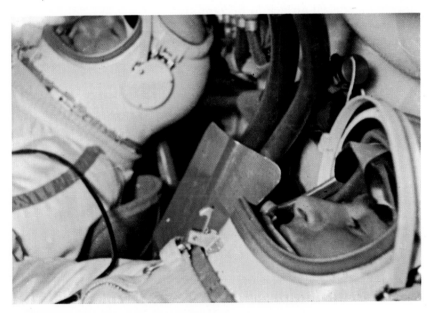

Above: Konstantin Feoktistov, a scientist, and Vladimir Komarov, the pilot, photographed by the third member of the Voskhod 1 crew, space doctor Boris Egerov. The three cosmonauts orbited the Earth 15 times in just over 24 hours. It was the first time a spacecraft had been manned by more than a single spaceman.

or used an ejector seat to leave the spaceship at an altitude of about 2500 feet.

As the United States Mercury program continued, more Vostoks were launched in the Soviet Union. In August 1962 Vostok 3 and 4 made the first ever group flight, the pilots maneuvering their vehicles to within three miles of each other. A second group flight in 1963 aroused worldwide interest because one of the pilots was the first woman in space

– 26 year-old Valentina Tereshkova.

Because the Soviets had far more powerful boosters than the Americans, their spacecraft were more spacious than the cramped capsules of Project Mercury. The Vostoks were solid steel spheres about 8 feet in diameter, each weighing 10,400 pounds – about two and a half times as much as a Mercury spacecraft.

On October 24, 1964 the new Soviet spacecraft Voskhod was launched. Many non-Soviet experts now believe that Voskhod differed only slightly from Vostok, apart from the fact that it carried a crew of three. Their opinion was partly based on the flight's timing, which came before the proposed two-man space flights of Project Gemini by the United States. Such scientists believed that three cosmonauts were simply crammed into a stripped down Vostok and blasted into orbit atop an uprated booster.

In the end, Voskhod 2's flight on March 18, 1965 took only two cosmonauts into orbit. The reason for this became obvious when one of them donned a bulky spacesuit and crawled through a special airlock to become the first man to walk in space. Tethered to the spacecraft by umbilical and safety

Right: the ghostly figure moving outside the spacecraft Voskhod 2 is Aleksey Leonov, the first man to walk in space. His pilot, Pavel Belyayev, controlled the craft and watched his companion's 10-minute walk on television.

Below: Gemini 7, photographed by astronauts in Gemini 6. Both craft, launched within two weeks of each other in December 1965, rendezvoused in Earth orbit and flew in tight formation for six hours.

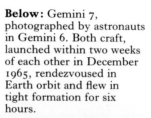

lines, Aleksey A. Leonov floated weightless in space for about 10 minutes.

Leonov's spectacular space walk was an important milestone in space flight. A human had at last directly entered the hostile environment of space and returned safely. It was a sign of a new confidence in the technologies developed to keep people alive in space.

The Gemini program began in March 1965, two years after the last Mercury flight. But once the program was underway, its 10 astronaut missions were flown at the rate of five each year – a sure sign of confidence in both spacecraft and boosters. The Gemini spacecraft drew on proven technology, but it also represented a major advance. Nearly twice as heavy as a Mercury capsule, Gemini was also more complex and versatile. Its crew of two worked in small but comparatively comfortable quarters. The missions themselves were also much more ambitious. They included rendezvous and docking tests and several long space walks.

The aerodynamics of Gemini spacecraft actually allowed them to develop some lift like an airplane, and the vehicle could be flown in a very limited way. It was the first spaceship that could be maneuvered to change the plane of its orbit. The 7000-pound craft was laun-

quently developed into the mainstay of the Soviet cosmonaut space program. Soyuz spacecrafts are still used regularly today and despite some setbacks have proved extremely versatile in a wide range of missions.

Mankind's first steps in space had shown that despite the hazards, space flight could be made into a practical reality. Vostok and Voshkod, Mercury and Gemini were acid tests of both people and machines. The wealth of new knowledge learned by Soviet and American space agencies enabled increasingly ambitious plans to be developed. The Soviets concentrated on complex Earth orbital missions while the United States chose sending the first human being to the Moon as its goal.

Against the background of success and optimism, the fatal accident of Komarov was a grim warning against complacency. The price of any serious failure in space is almost certainly death.

Below: American astronaut Edward White, pilot of Gemini 4 space mission, floats in space on June 2, 1965, secured to his spacecraft in the foreground by a 25-foot umbilical line and a tether line, both covered with gold-foil tape to protect them from the Sun's unfiltered rays.

ched by modified Titan boosters, a direct development of an intercontinental missile.

Soon after the successful conclusion of Project Gemini by the Americans, Soviet space experts began a major new program of flights using a highly advanced spacecraft called Soyuz – the Russian word for "Union." The vehicles were a logical development of the Vostoks. Over 30 feet long and about six feet in diameter in their spherical center region, the vehicles use 30-foot wings of solar panels to provide electric power for on-board systems.

The first flight of Soyuz was a disaster. Reports suggest that soon after Soyuz 1 reached orbit, a systems failure caused it to begin tumbling in space. Hour after hour, pilot Vladimir Komarov fought to control his spacecraft. Then in a desperate bid to return to Earth, Komarov attempted reentry under almost hopeless conditions. Just after a radio blackout, Komarov was heard saying quietly, "The parachute is wrong." It was his last transmission. The parachute lines had been tangled by the spacecraft's wild gyrations following the uncontrolled reentry. Nothing further could be done and Soyuz 1 hit the plains of Central Russia with a terrible impact.

Despite the tragedy, Soyuz subse-

Probes to the Moon

but the trajectory was wrong and the craft missed its target by well over 4000 miles. A few months later, however, Luna 2 made up for the earlier disappointment. On September 12, i crashed to destruction on the lunar surface. Three weeks later the Soviet achieved an even more spectacular feat

Before people could contemplate visiting the Moon they had to learn as much as possible about it. Direct observations by telescopes showed a moonscape of rugged highlands and ridges of high craggy mountains. Between these were mysterious dark plains that early astronomers had mistaken for seas. Accordingly, they had been called maria, the plural of the latin word *mare* meaning "sea."

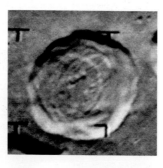

Many theories had been proposed to explain the Moon's structure. One in particular had grim implications for any planned lunar landings. Some experts had suggested that the maria were not seas at all but vast fields of fine dust accumulated over millions of years. Any spacecraft attempting to touch down would, they warned, sink in the dust and be swallowed up as if it had landed in quicksand.

To check this frightening theory and to build up the background of firm knowledge essential to a mission involving people, the Moon and the space around it had to be thoroughly scouted. This was done by probes without a human crew.

On January 2, 1959, the Soviets launched their 3245-pound probe Luna 1 toward the Moon. It was a bold effort

Above: the Moon's Ritter Crater photographed from an altitude of 150 miles by one of the Ranger series of probes, Ranger 8.

Above right: the six cameras mounted on the Ranger probes were fitted with different lenses for high- and low-resolution pictures.

Left: Luna 1, the first artificial object to escape the Earth's gravitational field. It was launched on January 2, 1959 by the USSR. It was instrumented to detect the Moon's magnetic field, if any, but it was unable to collect any information. It missed the Moon by 4660 miles.

Luna 3 was successfully placed into lunar orbit and radioed back close-up photographs of the Moon's surface. At last space experts could have a detailed look at all the surface features. Luna 3 was even able to send back the first sight of the side of the Moon forever turned away from the Earth.

Despite the remarkable quality of the photographs, American space engineers were more impressed by the obvious mastery of the Soviet engineers in building sophisticated guidance, control, and communication systems within their spacecraft. In the United States, early efforts to reach the Moon were plagued by repeated failures. In the first series of American moon probes, called Pioneer, the first four launches were almost total failures. The closest any of the probes came to the Moon was about 37,000 miles. The spectacular Soviet successes made the failures a particularly bitter blow for the newly formed National Aeronautics and Space Administration, nor did it help when NASA switched its lunar program to a different series of probes called Ranger.

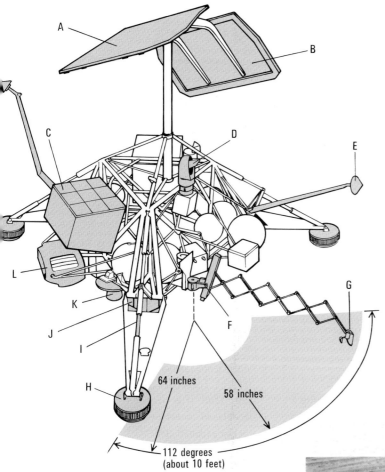

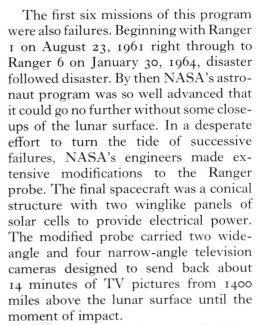

The first six missions of this program were also failures. Beginning with Ranger 1 on August 23, 1961 right through to Ranger 6 on January 30, 1964, disaster followed disaster. By then NASA's astronaut program was so well advanced that it could go no further without some close-ups of the lunar surface. In a desperate effort to turn the tide of successive failures, NASA's engineers made extensive modifications to the Ranger probe. The final spacecraft was a conical structure with two winglike panels of solar cells to provide electrical power. The modified probe carried two wide-angle and four narrow-angle television cameras designed to send back about 14 minutes of TV pictures from 1400 miles above the lunar surface until the moment of impact.

Rangers 7, 8, and 9 were brilliant successes. Ranger 7, launched on July 28, 1964, made a flawless 68-hour flight and radioed back 4308 superb pictures including some of the closest ever views of the lunar surface. Surface features only one-foot across could be clearly

Above: this schematic diagram shows the main features of the American Surveyor lunar probe. The band of color indicates the area – some 24 square feet – in which the surface sampler could operate.

Right: the Surveyor during preflight tests deployed in the same position as the diagram above.

Below: photograph taken by Surveyor on the lunar surface, showing the soil sampler in action.

A Antenna
B Solar panel
C Thermal compartment
D Television camera
E Antenna
F Elevation pivot axis
G Surface sampler
H Footpad
I Shock absorber
J Crushable block
K Vernier engine
L Landing radar

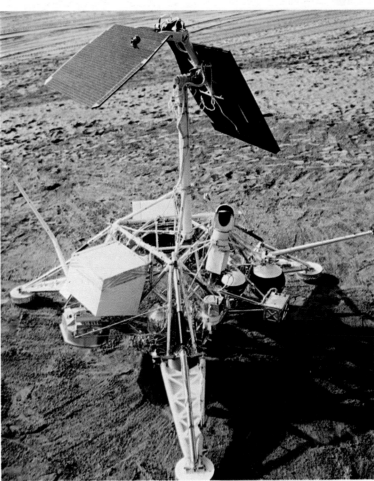

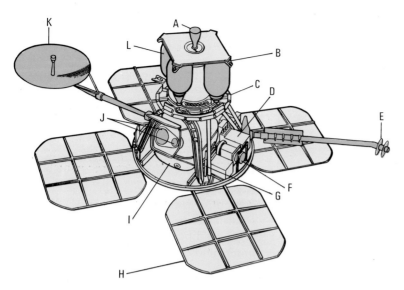

A Rocket engine
B Attitude control jets
C Micrometeoroid detector
D Programmer
E Omnidirectional antenna
F Star tracker (locked on Canopus)
G Radiation detector
H Solar panel
I Photographic subsystem
J Lenses
K Directional antenna
L Propellant tank

distinguished. Rangers 8 and 9 sent back a total of 13,000 pictures, revealing the lunar landscape over areas of both highland and plain.

After the Ranger series, the United States concentrated its efforts on two more probes known as Surveyor and Orbiter. While the hardware for these two programs was still being developed, the Soviet Union made the first attempt to soft-land an instrument package on the Moon. The idea was for a 220-pound package of cameras and power and communication systems to be ejected from a larger probe in the mountains before it crashed into the lunar surface.

The first five attempts failed but on February 3, 1966, Luna 9 safely delivered its spherical capsule. When the package came to rest, petal-like protective panels fell away and the camera switched itself on. Before its power was exhausted, the camera sent back 27 pictures of the Moon's surface.

By spring 1966 United States experts were also ready to attempt a soft landing. Their spacecraft, Surveyor, was a more sophisticated device than the Soviet packages. The flight plan for the 2200-pound vehicle included a carefully controlled landing using a complex series of retrorockets. The craft would be slowed to rest at an altitude of about 14 feet and then would fall the rest of the way to the

Above: diagram shows a Lunar Orbiter as it looked in flight. During the launch, the solar panels and antennas were folded closely to the body of the craft, and opened only after leaving the atmosphere.

Right: the first probe to soft-land on the Moon's surface, the Soviet Union's Luna 9 capsule. It touched down on February 3, 1969 in the Ocean of Storms. When the spherical capsule landed, its petal-like protective panels opened and the camera switched itself on. For three days it sent back panoramic views of its surroundings.

surface. Under lunar gravity, the impact would be fairly minor and Surveyor's three spiderlike legs had built-in shock absorbers to make sure that none of the spacecraft's systems were endangered.

Between 1966 and 1968 NASA had five successes out of seven with Surveyor probes. Four of these touched down in maria that were potential sites for the first astronaut landings. Each probe was equipped with television cameras that could be swivelled about to provide a variety of views. Their high-resolution lenses could make out particles as small as one tenth of an inch across. Other instruments included a remote control soil sampler on the end of an extendable arm and devices to measure surface radiation, soil structure, and local magnetic fields.

From the data accumulated during the Surveyor program, American space scientists could at least be fairly sure that it was feasible to land astronauts. The fear that the maria might be dangerous bowls of dust seemed groundless. Surveyor made possible the first definite ideas about lunar structure. It was clear that the maria were covered in a fine light-colored powder but that the under-

lying dark soil was firm enough to support a fairly heavy spacecraft. In fact the average bearing strength of the soil was similar to that of soil found on Earth.

The final program of probes before NASA's first attempt to land people on the Moon was project Orbiter. The series began on August 10, 1966 and ended a year later. The missions were successful.

The Orbiters were characterized by four large panels of solar cells extending like paddles from the central body of the spacecraft. The flight plan was for the Orbiter to be placed first in a circular orbit around the Moon at an altitude of about 500 miles. This orbit was then to be altered into an elliptical one, carrying the spacecraft as far as 1150 miles out into space on the far side of the Moon and as low as 22 miles on the near side.

The first three Orbiters circled the Moon's equator photographing the most promising landing sites for astronaut expeditions. They were so successful that the final two Orbiters were sent into Polar orbits, enabling them to record about 99 percent of the lunar surface as they passed over it.

In addition to photographic work, the Orbiters also reported on radiation levels in the space around the Moon. They discovered no belts of trapped radiation comparable to the Van Allen belts surrounding the Earth. They also found that the rain of micrometeoroids falling onto the Moon was not sufficiently dense to trouble any future astronauts.

One important discovery was that the Moon was dotted with concentrations of heavy material, called mascons, lying between 15 and 75 miles beneath the surface. These caused local variations in the Moon's gravitational field which made spacecraft in orbit wobble as they passed over them.

While NASA's lunar probes paved the way for the astronauts and their successful Apollo expeditions, the Soviet Union was concentrating its efforts on perfecting their mastery over remote control systems aboard probes. It was clear by the mid 1960s that the Soviets no longer had plans to send people to the Moon but, as if to demonstrate that they had no need to rely on cosmonauts, achieved spectacular success in landing probes, collecting lunar samples, and returning them safely to Earth.

Their most remarkable mission utilized a highly complex probe called Lunokhod, the first remote control "Moonwalker." Once a safe touchdown had been achieved, a roughly circular vehicle with eight wire mesh wheels rolled down a special ramp. Guided from Earth, the vehicle could rove across the lunar surface analyzing soil and taking photographs. Lunokhod's success surpassed all Soviet expectations.

Left: Lunokhod 1, the remote-controlled self-propelled Moon vehicle landed on the lunar surface by the Soviet probe Luna 17. It was about the size of a small automobile made largely of magnesium alloy and had four spoked wheels on each side. The Lunokhod's movements were controlled from Earth by a crew that consisted of a commander, driver, engineer, navigator, and radio operator. Altogether the robot Mooncar traveled over six miles and mapped 95,000 square yards of the lunar surface, made texts on the Moon's surface, and took thousands of photographs.

The Apollo adventure

Projects Mercury and Gemini in the early 1960s were the starting points of American ambitions in space. With the data and experience collected from the projects space experts turned their attention firmly to the Moon. The complex preparations for this exciting objective were handled by the vast civilian space agency NASA – the National Aeronautics and Space Administration set up

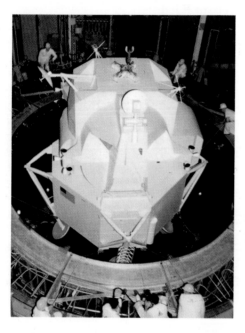

Left: the environmental chamber at the Manned Spacecraft Center at Houston, Texas. It was used to test a complete Apollo vehicle and its crew under the near-vacuum conditions that are found 200 miles above the Earth.

Below: also at the Manned Spacecraft Center, Houston, is this centrifuge used to train astronauts to withstand the forces of acceleration. The centrifuge also tested vital components and systems of the spacecraft to insure they too could withstand the forces.

in 1958. There was no illusions about the complexity of the task. New spacecraft and launch vehicles were needed together with huge ground facilities to build, test, and assemble them. Project Apollo was to be the greatest challenge ever to man's technological ingenuity. More than this, it was to be a test of human determination and courage. The astronauts, in particular – hand-picked and trained to perfection – would be facing danger during every moment of their complex missions.

The center of activity was the Kennedy Space Center of Cape Canaveral. Here the huge boosters were assembled and the Apollo launches themselves took place. But in addition to the new test rigs, assembly buildings, and launch pads, enormous worldwide investments were

made in extensive telecommunications facilities.

Questions like, "Where are they now?" "What is happening on board the spacecraft?" had to have answers that could be instantly available to ground control teams to ensure the safety and success of the Apollo missions. Four completely instrumented ships were especially built to augment the existing land-based tracking and communication network. But more ground systems were also built across the world stretching from the Pacific island of Guam to Madrid, Spain. The complex network of radio, radar, and optical systems called the Manned Space Flight Network had two distinct nerve centers – the Goddard Center in Greenbelt, Maryland and Apollo Mission Control at Houston, Texas. Linked to the myriad of other tracking and communications facilities worldwide by satellites, undersea cables, landlines, and microwave transmitters, Greenbelt and Houston were the crucial establishments where vast amounts of complex data had to be handled fast and efficiently throughout each Apollo mission.

The Apollo mooncraft had a family resemblance to both Mercury and Gemini vehicles. In particular the Apollo crew module shared the basic conical shape, slightly rounded base and heat shield to protect returning astronauts during their 25,000 mph reentry into the Earth's atmosphere. But although the Apollo spacecraft was based on the development experience of the earlier

manned programs, it was a vastly more complex vehicle.

The mooncraft was made of three modules. The 5-ton command module was nearly 12 feet high and 13 feet in diameter at its base. Packed into its interior were the flight controls – including a small onboard computer – and communications equipment, life support systems, and recovery parachutes. Except for the excursion to the lunar surface, the crew remained in the command module throughout the flight. It was the only part of the mooncraft designed to return to Earth.

Directly behind the command module was the 27-ton service module, a 25-foot-long cylindrical structure housing oxygen supplies for the crew, fuel cells for electricity, communications equipment, and a system of small thrusters for maneuvering. From the astronauts' point of view the most important part of the service module was the large rocket engine used to enter and leave orbit around the Moon. The service module was not separated from the command module until only moments before the crew re-entered the Earth's atmosphere. When they plunged back toward the safety of the recovery forces in the splashdown area, the service module burned up like a meteor.

The third component of the mooncraft was the delicate spiderlike lunar module (LM). During lift-off the LM was stowed behind the service module on top of the booster. Once the mooncraft left Earth orbit on its translunar voyage, the third stage of the booster with the LM attached separated from the command and service modules. The crew then turned their spacecraft around and relinked nose-to-nose with the LM. The maneuver completed, the spent third stage was jettisoned.

The nose-to-nose link between command module and LM contained a tunnel through which the astronauts could crawl. Once in orbit around the Moon, two of the crew – the mission commander and the lunar module pilot – transfered to the LM, separated from the command module and descended to the lunar surface. By carefully varying the thrust of the LM's descent engine the astronauts were able to control the rate of descent and make a soft landing on a chosen site.

Above: a lunar landing device at the Langley Research Center, Hampton, Virginia. It was used to train astronauts on the Moon missions to maneuver their lunar landing vehicle under the gravitational conditions found on the Moon.

The lower portion of the LM had four spiderlike legs to support the vehicle on the lunar surface. It also acted as a launching platform from which the upper part of the LM blasted off to return to orbit above the Moon. Rendezvous and docking with the combined command and service modules was fol-

lowed by a crew transfer. The LM was then jettisoned and the crew prepared for their return journey to Earth.

Work on the command and service modules began in early 1962 and on the LM in 1963. In 1969 their combined cost had reached a staggering 8000 million dollars, exceeding even the cost of the giant booster used to launch them.

The development of the booster itself really began long before Project Apollo was put on the planning boards. The original concept of an ultrapowerful booster came from the German-born rocket expert, Werner von Braun. He argued that if America had a future in space, that future must involve putting increasingly bigger payloads into orbit. His idea was to bind together a cluster of powerful rocket engines to produce a superrocket capable of boosting over 120 tons of payload into orbit.

By 1961, Saturn – the ultimate in rocket technology – was a reality. Saturn 1 the first member of the family, used a cluster of eight engines for its first stage, generating a thrust of 1.5 million pounds of lift off. But Project Apollo needed a still more powerful. launch vehicle and within six years the first Saturn V was ready for flight testing.

Everything about Saturn V was gigantic. Topped with the Apollo moonship, the vehicle stood 363 feet high and weighed over 3000 tons. Its first stage consisted of five engines each generating as much thrust as the entire Saturn 1 – a fantastic total of 7.5 million pounds. The second stage consisted of five engines developing

Above: the burned-out interior of the Apollo 4 command module. During a full-dress rehearsal for the Apollo 4 space flight in January 1967, fire broke out destroying the capsule and killing the three astronauts – the commander, Virgil "Gus" Grissom, veteran of Mercury and Gemini missions; Edward White, the first American to walk in space; and Roger Chaffee, who was going up for the first time. An investigation blamed an electrical spark from faultily installed equipment, which easily ignited in the 100 percent oxygen atmosphere inside the capsule. After the shock of the accident new, more stringent safety measures were introduced into the space program.

over a million pounds thrust. The third stage used a second stage Saturn 1 engine, delivering 200,000 pounds thrust. At lift off, the first stage engines burned 700 tons of fuel every minute and the sound alone could cause a minor earthquake in the vicinity of the launch pad. In all, more than 125,000 people spent 7000 million dollars to design, build, and test the giant rocket. But the immense efforts were well-rewarded. The 13 Saturn Vs used during the Apollo flights worked perfectly and even showed an unexpected degree of versatility, one even survived being struck by lightning during a launch.

Saturn boosters were assembled at the Kennedy Space Center in a huge building – the largest in the world. From here they were carried on the back of a gigantic transporter called the "crawler" to the launch pad three miles away. The slow trip took almost eight hours.

The launch pad itself was a masterpiece of engineering – a huge truncated pyramid of concrete and steel topped with gantries and masts to carry fuel and electricity umbilicals for the booster and moonship. In the center of the pad was a 42-foot-deep flame trench to deflect the tremendous blast of the first stage engines. And deep below ground, a special "blast room" was readied where astronauts and ground crews could take cover in case of a fire on the pad. Happily it never had to be used, but engineers were confident that the blast room would survive even the holocaust of the entire Saturn V blowing up directly overhead.

The first manned flight test of the moonship was Apollo 4. But disaster struck on January 27, 1967, during one of the final dress rehearsals some months before the schedules launch. The crew was commanded by astronaut Virgil "Gus" Grissom, veteran of Mercury and Gemini missions. With him aboard the moonship were Edward H. White – the first American to walk in space – and a new astronaut Roger B. Chaffee. It was 6 pm. The astronauts, who had been strapped into their couches since noon, were going through the final stages of countdown just as though it were the real thing. Even the cabin had been pressurized with the 100 percent oxygen required by American spaceship design.

Suddenly, at exactly 6.31 pm a frantic

voice came over the radio link – "Fire – I smell fire!" Immediately ground controllers saw their instruments showing a rapid rise in temperature within the command module. The astronauts struggled to free themselves but no quick escape system had been built into their ship. There were sounds of unintelligible shouting and a last despairing cry from Chaffee – "We're on fire – get us out of here!"

The tragic deaths of the three astronauts was a blow to the Apollo program. Months of investigation followed and eventually, with massive new safety systems designed and tested, Project Apollo was revived. On January 22, 1968, the unmanned Apollo 5 was blasted into orbit by the same Saturn on which Grissom, White, and Chaffee had died a year earlier.

Subsequent manned tests went without a hitch and by December 1968, mission controllers were ready for the first big test of the Apollo concept. In probably the most dramatic space flight of all time, Apollo 8 blasted out of Earth orbit on the first journey to the Moon. The flight plan did not include a landing, merely a number of lunar orbits before returning to Earth. But this was the first time that anyone had left the vicinity of the Earth and ventured into deep space. Many newspapers carried headlines accusing NASA of rushing ahead too fast. The mission was the most dangerous ever attempted. It was only two years after the Apollo 4 tragedy – were the spaceship's systems ready for the test?

History has shown that NASA had judged its capabilities perfectly. Astronauts Frank Borman, James Lovell, and William Anders flew a trouble-free mission. And for the first time, television viewers on Earth shared the sense of awe and wonder of the astronauts, skimming over the forbidding lunar surface, just 70 miles below. On Christmas Day, from lunar orbit, the crew transmitted a message of goodwill to the people of the world. They compared the gray lifeless moonscape below them to the beautiful living colors of what they called "the good Earth." And as their cameras recorded the bleak and cratered scene below, Borman slowly read the opening Old Testament verses of Genesis.

No subsequent crews were ever so

Below: the reliability of the giant Saturn V rocket was proved when, in 1968, Apollo 7 with a three-man crew was launched to test its power and capabilities. The photograph shows stage B of the Saturn falling away after boosting Apollo 7 into space. After 163 orbits on a very nearly perfect mission, Apollo 7 splashed down in the Atlantic.

moved by their experience as the men of Apollo 8. In a sense, their's was the greatest adventure of all – man's first real step into deep space.

In the two missions that followed Apollo 8, all spacecraft systems were thoroughly tested. NASA was at last ready for the culmination of all their efforts – the lunar landing mission itself.

Launch day was July 16, 1969. The crew of Apollo 11 was commanded by Neil Armstrong. He would be accompanied to the lunar surface by lunar module pilot Edwin "Buzz" Aldrin. The man with the thankless task of staying in Moon orbit with the command module during the historic lunar landing was Michael Collins.

As the countdown entered its final

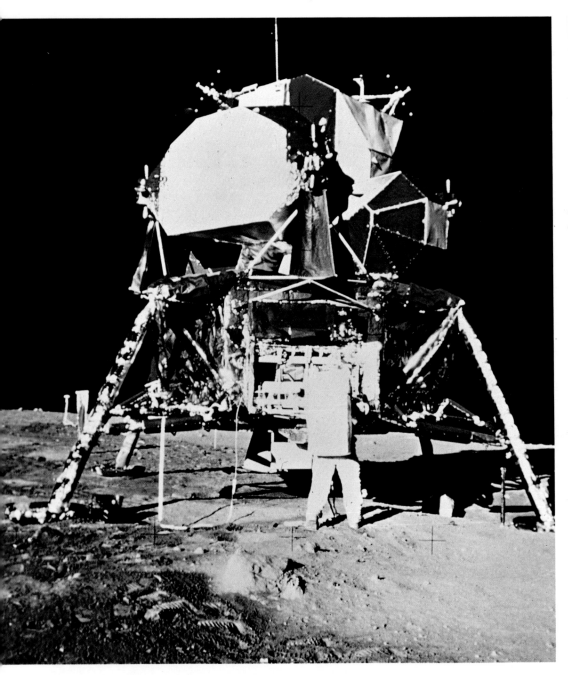

Above: Neil Armstrong's historic footprint on the Moon's surface.

Left: Armstrong checks the *Eagle* to see if any damage had occurred during landing.

Below: Edwin Aldrin on a Moonwalk. The astronauts set up three scientific experiments on the lunar surface: a seismometer to measure the earthquakelike shocks on the Moon and a two-foot-square reflector pointing toward the Earth so that scientists could bounce a laser beam from Earth onto the reflector and back again and give scientists a precise measurement of the distance between Earth and the Moon. The third experiment involved trapping some of the particles coming from the Sun in the solar wind.

seconds, a million people around Cape Canaveral strained their eyes for a glimpse of the launch. Around the world hundreds of millions watched the great moment on their television screens.

Four days later the mission was approaching its breathtaking climax. Armstrong and Aldrin crawled through the connection tunnel and entered their lunar module codenamed *Eagle*. Within a few hours they were drifting away from the command module to begin their descent toward a landing sight in the area known as Mare Tranquillitatis (Sea of Tranquillity).

In the last moments before touchdown Aldrin and Armstrong could be heard calmly calling out velocity and altitude figures. Then at 4.17 pm Eastern Daylight time came the report from Armstrong that the whole world had been awaiting, "Houston," Armstrong radioed, "Tranquillity base here. The *Eagle* has landed."

Some hours later the men left the lunar module and stepped on to the Moon's surface. As Armstrong's left boot touched the dusty ground, he radioed the now famous words "That's one small step for a man, one giant leap for mankind."

Although subsequent Apollo missions

brought back much more scientific data, the Apollo 11 flight will remain forever the symbol of NASA's success. At last the supreme test had been faced and men and machines had met the challenge.

Ironically the blaze of publicity surrounding Apollo 11's magnificent achievement created a serious public relations problem for NASA. The first moon landing had been a peak of excitement and despite every effort of NASA's publicity machine, all that followed seemed to the public to be anticlimax.

Interest revived momentarily during the ill-fated Apollo 13 flight. An explosion in the service module soon after leaving Earth orbit left the spacecraft dangerously underpowered. The chances of crew men James Lovell, Jack Swigert,

Right: an important part of the scientific work done by the two astronauts involved the collection of soil and rock samples using coring tubes.

Below: the Apollo landing module *Eagle* rises from the surface of the Moon to rendezvous with its command module, *Columbia*. When the two craft had docked the two astronauts made their way into the command module and the *Eagle* was jettisoned into lunar orbit. On July 24, 1969, just eight days after they blasted off from Cape Kennedy (Canaveral), Apollo 11 splashed down in the Atlantic. Humans had landed on the Moon safely and returned.

and Fred Haise getting back alive looked extremely slim.

But in a remarkable demonstration of flexibility the vast manned spaceflight network began to examine all possible ways of saving the men. It was NASA's capacity to play a totally unforeseen situation by ear, to throw away the flight plans and extemporize, that ultimately returned the men safely to Earth.

Apollo 13 was a grim reminder of the dangers of spaceflight. But the later flights followed one another with clockwork precision and despite the wealth of scientific data returned by these highly complex missions, NASA's funding was cut. Apollo 17 had to be the last of the moonflights. Its crew brought back a record 240 pounds of moonrock after spending well over seven hours on the lunar surface.

Although Apollo 17 was the last moonflight, it was not the end of Apollo. The finale was a mission long dreamed of by those who envisioned space as a frontier that all mankind would share in peace. On July 17, 1975, at 2:17 p.m. Houston time, 10:17 p.m. Moscow time, an American Apollo spacecraft docked in orbit with a Soviet Soyuz vehicle. During this Apollo-Soyuz Test Project, the two craft remained together for 42 hours as the crews performed several transfer operations and joint solar eclipse experiments. Now, as the superpowers begin to arm the heavens, it seems that the opportunity for a repeat of this cooperation has vanished.

Right: Apollo 14 on the Moon's surface. The landing module Antares landed in Fra Mauro, less than 60 yards from the designated spot. Alan Shepard and Edgar Mitchell completed two EVA's (extra-vehicular activity).

Far Right: an infrared photograph taken in space from Apollo 9. It shows the rich vegetation of the Imperial Valley in southern California (right center edge) surrounded by desert. The camera system used here was a forerunner of one that was later used in the Earth Resources Technology Satellite, which was launched in 1971.

Below: the space missions central control room at Houston, Texas. It is here that every aspect of the flight is checked and double-checked, from the temperature inside the spacecraft, to the astronauts' heartbeats. These are the people who decide whether a mission is "go" or "abort."

Below right: astronaut James Irwin of Apollo 15 gives the salute to the American flag on the mission that first used a self-propelled lunar roving vehicle, seen here at the far right of the photograph.

Interplanetary probes

communications subsystem has to be powerful enough to receive and send signals over immense distances.

Power is usually provided by panels of solar cells, converting the Sun's rays into electricity. The appearance of most probes therefore is dominated by these winglike structures. Maneuvering rockets are also an important feature of all probes, enabling corrections to be made to flight trajectories to ensure that at the end of their long journeys, the probes arrive at their intended destinations.

The most obvious targets beyond the Moon were the two nearest planets to Earth – Venus and Mars. Of these Venus in particular had for long intrigued astronomers. Sometimes its orbit around the Sun brought it as close as 24 million miles to Earth, and it had for many years been considered as Earth's sister planet. Their sizes are almost identical. But despite its relative nearness to Earth, Venus was a total mystery. Enshrouded

The first probes – Soviet Lunas and American Pioneers – were sent into space to find out what was there. They were relatively simple craft carrying perhaps a few measuring devices and radio. Since the first pioneering launches, the USSR and the USA have worked through numerous programs of probes ranging from the exploration of Earth's nearest planetary neighbors, Mars and Venus, to the awesome voyage of Pioneer 10, the first probe to be sent into interstellar space.

Just as the objectives of their flights have become increasingly ambitious, so the probes themselves have grown in sophistication and complexity. Basically there are no fundamental differences between a probe and an artificial satellite. The main distinction is one of mission rather than structure. Free of the need to fly in planetary atmospheres, probes and satellites alike have functional shapes with no aerodynamic grace about them. The subsystems inside a probe are essentially the same as those in a satellite. But the designers of probes have to take account of the fact that their vehicles will generally spend much longer operational periods in space and may travel many millions of miles from the Earth. This means that the probe has to be protected against prolonged attack by micrometeors and radiation. Additionally, the

Above: an artist's concept of the slingshot technique that was used to speed Pioneer 10 to the outer planets. The spacecraft passed Jupiter in 1973 and is now headed for the outer reaches of the solar system. Pioneer 10 will eventually become the first man-made object to pass beyond the solar system when it crosses the orbit of Pluto in late 1986.

Opposite top right: the Venus probe Venera 3, the Soviet Union's third attempt to reach the surface of the planet Venus, successfully reached the surface in March 1966 after a 106-day journey that took the space probe some 24 million miles. It failed to eject the instruments it was carrying, however.

Right: Mariner 10, the NASA spacecraft that used the planet Venus' gravity to swing it on to Mercury, the closest planet to the Sun. Mariner 10 sent back pictures of very high quality during its close fly-by mission. Because of its closeness to the Sun, Mercury has always been difficult to observe, and little was known about it until the Mariner 10 mission.

n thick, light-reflecting clouds, not even the most powerful telescopes could reveal anything about the planet's surface. Some astronomers even believed that beneath the clouds was a world similar to the Earth as it was millions of years ago when life was first evolving.

The first successful probe to Venus was the mission of Mariner 2 launched by the American space agency NASA on August 27, 1962. After a 109-day journey, the probe flew past the planet at a distance of 21,600 miles. The fly-by provided scientists with 35 minutes of instrument scanning during which the idea of life on Venus was given a crushing blow. Surface temperatures appeared to be hot enough to melt lead.

The next Mariner probe to fly-by Venus was Mariner 10. The mission was a complex one involving the spacecraft using the gravitational field of Venus to help it on its way toward the Sun and the first fly-by of the innermost planet,

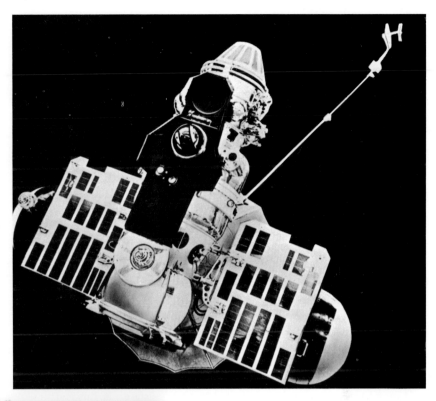

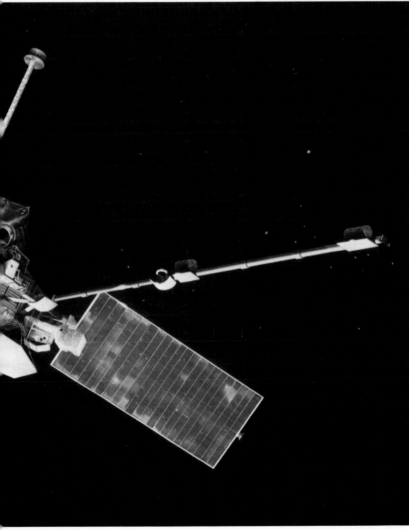

Mercury. On February 7, 1973, the probe passed within 3500 miles of Venus radioing back a wealth of data on the planet's cloudy atmosphere. Then on March 29 Mariner 10 swept within 200 miles of the surface of Mercury. During an 11-day period around the point of closest approach, ground controllers received a total of 2300 superb TV pictures. The quality was remarkable. Man's first glimpse of Mercury showed a cratered, Moonlike surface. Many of the craters overlapped and had tall central peaks – much like those seen on the Moon. The first surface feature of Mercury to be given a name – a large crater with rays of material surrounding it – was called Crater Kuiper. It was named for Gerard Kuiper, pioneer designer of some of the early lunar probes and until his death in 1973 a key member of the Mariner 10 team.

While the Americans relied on Mariner for their early investigations of Venus, Soviet space scientists developed a remarkable probe named Venera for the planet it was designed to explore. The flight of Venera 1 launched on February 12, 1961 was the Soviet Union's first attempt at interplanetary flight. Disappointingly, ground controllers lost touch with the spacecraft only 5000 miles away from Earth and although it probably passed within 64,000 miles of Venus, no data was radioed back. Later flights made up for early failures and the Venera program eventually provided a wealth of detailed information about Venus.

The Soviet concept of interplanetary probe design differed in one fundamental way from the American approach. While NASA was content with close fly-by missions, Soviet scientists wanted to gain data more directly by dropping landing capsules from the main probe as it by-passed the planet.

The first demonstration of this approach in the Soviet investigation of Venus came in October 1967. Venera 4 succeeded in dropping a 400-pound instrument package into the Venusian atmosphere. During its 94-minute parachute descent, the lander ended its transmissions, presumably crushed by the tremendous atmospheric pressure near the planet's surface.

The Russians learned a great deal from their first experiment and later landing capsules were substantially stronger. Transmissions from the surface of Venus were achieved. Then on October 22, 1975 Venera 9's capsule made a parachute landing. Some 15 minutes after touchdown centuries of speculation were ended with the transmission of a single historic photograph of the planet's surface. It showed a barren, rocky terrain with no sign of life.

Exploration of Mars has been spearheaded by American Mariners and Russian Mars spacecraft. The first attempt to send a probe to Mars took place in November 1962 when the Soviet Mars 1 was launched. Unfortunately radio contact was lost nearly five months later as the probe was nearing its objective.

In November 1964 Mariner 4 began its long journey to Mars and on July 14, 1965 flew past the planet at a distance of just over 6000 miles. The first pictures radioed back crushed any hopes of advanced life forms. The surface was heavily cratered and dry with no trace of surface water. In 1969 Mariners 6 and 7 sent back sensationally detailed photographs from a distance of about 2000 miles. One of the most spectacular pictures was taken by Mariner 6 – a gigantic volcano more than 15 miles high and 40 miles wide at its base.

Above: the Viking spacecraft. Two of these highly sophisticated craft soft-landed on the Martian surface – in July and September 1976 – and sent back spectacular photographs of the red planet. The landings were automatically controlled by a built-in computer brain that selected a suitably smooth and safe site for touch down.

In 1971 Mariner 9 spent 90 days in orbit around Mars and photographically mapped its entire surface. It also gave close-up shots of a tremendous dust storm that for a time enveloped most of the planet.

The first successful soft-landing on Mars was achieved by Soviet probe Mars 3 on December 2, 1971. Using a complex parachute brake coupled with a retro-jet system, a landing capsule equipped with a television camera made a successful touchdown. The camera operated for only 20 seconds before a systems fault put an end to the transmissions. The main probe from which the lander had been dropped remained in orbit providing data for a further three months.

Russia's most ambitious onslaught on Mars followed in 1973 when two pairs of spacecraft were launched within three weeks of each other. Mars 4, 5, 6, and 7 through soft-landings and reconnaisance from orbit provided scientists with an unprecedented amount of information about the planet.

Two American landers in the sophisticated Viking series touched down on July 20 and September 3, 1976. Superbly detailed panoramic views of the Martian surface were radioed back to Earth while complex remote-controlled experiments were performed to sample the surrounding soil and test for signs of life. The results were tantalizingly inconclusive and although some scientists believe the

Viking results do prove the existence of primitive Martian organisms, most prefer to await the results of later missions.

Some of the most exciting missions so far flown by probes have been flights to the more remote regions of the solar system. The American Pioneers 10 and 11, for example, were the first probes designed for extremely long journeys, able to function for over seven years at up to 1500 million miles from the Sun.

Pioneer 10, launched on March 3, 1972 was given an initial speed of 32,400 miles an hour – faster than any other man-made object had ever traveled. It swept past the Moon just 11 hours after leaving Earth orbit. In November it began threading its way through the orbits of the moons of Jupiter passing within 815,000 miles of the giant planet. Pictures and sensor readings provided scientists with new data of overwhelming

Below: Voyager 1, the probe that passed close to the planet Jupiter in March 1979 and radioed back amazing photographs of the planet's cloud formation.

Bottom: Jupiter's faint ring system photographed by Voyager 2 in July 1979 from a distance of 900,000 miles. The picture was taken in Jupiter's shadow – which cuts short the lower ring image – and reveals the rings as two light orange lines.

has flung the probe straight up and out of our solar system. Voyager 2 however, continued on through the Saturnian system and is expected to accomplish a flyby of Uranus in 1986 and of Neptune in 1989. Most recently, Soviet, Japanese and European Space Agency probes have been launched to rendezvous with Halley's comet in 1985–86. In years to come there will be NASA's ambitious Galileo Project, to launch May 1986, which will explore Jupiter's moons for months, and will even drop a probe into Jupiter's atmosphere. Also set to go is a mission to map Venus with radar, scheduled for launch in April, 1988. Other suggested missions include flybys of the asteroids, putting a probe in orbit around Mercury to examine the sun, dropping a balloon probe into the Venusian atmosphere, and sending a small, remotely piloted plane over the surface of Mars.

importance.

Pioneer 10 then passed on toward the outer solar system. It is expected to pass into interstellar space in 1987 and will probably reach the vicinity of the star Alderbaran in 8 million years' time!

Pioneer 11 passed even closer to Jupiter than its sister probe and then continued on a similar trajectory toward Saturn.

In August and September 1977 Voyagers 1 and 2 were launched by NASA to follow the Pioneers. Both Voyagers completed successful flybys of Jupiter and Saturn, yielding spectacular photographs and reams of data that is still being studied. Voyager 1 was sent close to Saturn's intriguing moon Titan, and Titan's gravity

The first space stations

With the success of the Apollo lunar expeditions behind them, America's space agency NASA began serious development of the first manned space station to be placed in Earth orbit.

Skylab – launched on May 14, 1973 and subsequently visited by three three-man crews – was a remarkable demonstration of NASA's ability to snatch success from the jaws of disaster.

Originally called the Apollo Applications Program, the concept of Skylab was at first an effort to find further use for the hardware and technical expertize amassed during Project Apollo itself. It turned out to be a major step forward in mastering and exploiting the space environment, comparable in scientific importance to the moon landings themselves.

The Skylab Workshop consisted of a modified Saturn S-4B third stage – the type used in the Saturn V and Saturn IB launch vehicles. Coupled to the modified rocket stage was a multiple docking

Below: Skylab, America's first experimental space station. It carried eight telescopes and various sensors, which are invaluable aids to astronomers. The helicopter-vanelike apparatus is the telescope mount, with a four-panel solar array to generate the Skylab's electrical power. Behind the octagonal "nose" is the orbital workshop – a forward compartment containing food and equipment storage, and a crew compartment. Barely a minute after lift-off in May 1973 the micrometeoroid shield was torn away from the exterior of the cylindrical workshop, carrying with it one of Skylab's solar wings.

adaptor incorporating an airlock module and a complex solar observatory called the Apollo Telescope Mount. The rocket stage itself was converted into an orbital workshop by using an aluminum grid floor to divide it into a two-story cabin. Outside the Workshop, an aluminum shield protected the vehicle from micrometeors and the intense heat of the Sun. The rear section of the interior was divided into a sleeping area, a diningroom-lounge area, workroom, and a waste-management area. The rest of the vehicle was equipped as a spacious laboratory.

The Apollo Telescope Mount was deployed at right angles to the adaptor where Apollo spacecraft dock to bring up or return Skylab crews. Powered by four large panels of solar cells, it was basically a solar observatory. Visual images of the Sun could be relayed to the crew by television cameras while other instruments examined the different regions of the solar spectrum. The astronauts could maneuver the telescope using on-board control system while Earth-bound astronomers instructed the crew by radio on what they wanted to see.

The Skylab mission, launched by Saturn V, began to go badly wrong almost immediately after lift off. Although the space station had been placed in the correct orbit just over 250 miles above the Earth, the vehicle had obviously been damaged during the launch. Vibrations had torn away the combined micrometeor and radiation-shield, which in turn had ripped one of the Workshop's main solar cell panels. To make the situation worse, the other panel had only partially deployed, debris apparently jamming the opening mechanism.

With no protection against the Sun's rays and little on-board electrical power, temperatures inside Skylab soared to over 190°F. NASA's plans seemed to be in ruins.

The launch of the first Skylab crew was delayed while experts analyzed the options, searching for some way to retrieve the desperate situation. Eventually they decided to attempt some in-flight repairs. The first crewmen – Charles "Pete" Conrad, Joseph Kerwin and Paul Weitz – were launched in their Apollo spacecraft on May 25, 10 days later than originally planned. Seven hours later Conrad was to give a firsthand

description to a hushed Mission Control of how one of the great solar panels of Skylab had been torn off while the other was jammed by the badly buckled micrometeor and radiation-shield.

In a dramatic bid to free the jammed panel Conrad maneuvered his spacecraft to within a few feet of Skylab. Weitz, his head and shoulders out of the spacecraft's hatch tried vainly to release the panel using toggle cutters on the end of a long pole.

The astronauts then deployed a large orange shroud over the area of Skylab turned directly toward the Sun. At last the men were able to board the Workshop and within a few days – though still uncomfortably warm – Skylab was at least habitable. The most serious problem facing the Project was lack of electrical power. Somehow the jammed solar panel had to be freed. After three days of discussions and rehearsals, the situation deteriorated rapidly, and Conrad and Kerwin suited up for a dangerous "space-walk" in a last attempt to deploy the crucial panel.

Conrad lay on the outer surface of the Workshop within a few feet of the damaged panel guiding Kerwin as he tried to free it with the long pole that had been used a few days earlier. Then Conrad fastened a rope to the panel and hauled on it. Suddenly it swung out and clicked into place. Instantly power began flooding into the Workshop's eight main batteries and the two jubilant astronauts returned inside to wild congratulations from Mission Control.

By the end of their 28-day mission the crew had performed about 80 percent of their planned experiments. In particular, they made important observations of a huge solar flare and conducted numerous tests to see how well men adapted to long periods of weightlessness.

The second Skylab crew – astronauts Alan Bean and Jack Lousma and a civilian engineer named Owen Garriott – were launched on July 28. They took aboard Skylab with them six mice, two spiders named Anita and Arabella and a number of minnows.

Their 59-day mission included a record-breaking 6½-hour space-walk by Garriott and Lousma. But the main purpose of the flight was to see how well men stood up to very long periods of

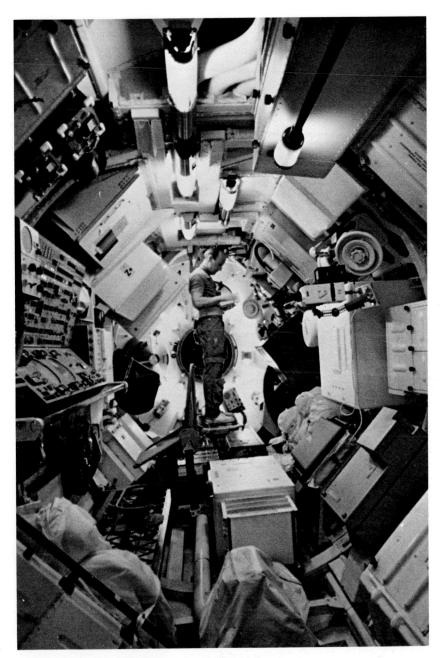

Above: the interior of the Skylab. Several experiments were carried out from Skylab, such as using the Apollo telescope mount to obtain solar spectra and to photograph the Earth's resources and the passage of the comet Kohoutek. Other experiments involved the effects of weightlessness on men and animals. The crew also carried out repairs to a damaged solar wing and rigged a solar shield over the orbital workshop area.

weightlessness. Despite some loss of strength in muscles and bone, the crew returned to Earth in good physical condition. The final Skylab mission was flown by Gerald Carr, Edward Gibson, and William Pogue. It lasted 84 days during which the crew traveled an astonishing 34 million miles. They brought home with them about 75,000 pictures of the Sun and 17,000 of the Earth – enough raw data to keep scientists busy for years.

The Soviet space station program began in 1971 with the launching of Salyut 1. During the six years that followed five further stations were launched and a number of manned missions were flown in them. But it was not until 1975 that Western experts had their first close look at the Salyut vehicle itself. The opportunity came during the Paris Air Show where visitors were allowed to inspect a full-size mock up and even climb aboard to sample the space stations accommodation.

Salyut differs in many ways to Skylab. The Soyuz spacecraft used to ferry crews between the space station and Earth are an integral part of the vehicle. The station is only complete therefore when docked with a Soyuz. The overall length of the Salyut-Soyuz vehicle is about 60 feet and its weight is almost 25 tons.

The main living quarters are divided into sections used for stowing equipment, working, recreation and sleep and waste management. Most of the station's power is provided by externally mounted panels of solar cells. Compared to Skylab, the controls and equipment are fairly rudimentary and unsophisticated. But the continuing success of the Salyut program suggests that the underlying spacecraft systems are highly efficient and versatile.

The latest successes in the Salyut missions were accomplished at considerable cost. Almost as soon as Salyut 1 was launched, Soviet experts were facing a major disaster. The second crew to board the station – Dobrovolsky, Volkov, and Patsayev of Soyuz 11 – spent 28 days in orbit. All went well and it seemed that the success of the mission was assured. Then during reentry into the Earth's atmosphere their Soyuz spacecraft suddenly depressurized. The three cosmonauts died instantly.

The tragedy was a major blow to Soviet space plans. It was two years before another manned flight was risked and even following this more problems were in store. On April 3, 1973 Salyut 2 was launched. But before any crews were sent aboard a massive systems failure wrecked the station. Western observers reported that the entire vehicle had apparently broken up in orbit.

A year later Salyut 3 provided a much-needed success. The two-man crew of Soyuz 14 spent 15 days on board and returned safely to Earth. Salyut 3 was an improved version of the earlier vehicles. Its solar cells were more extensive and automatically rotated to keep facing the

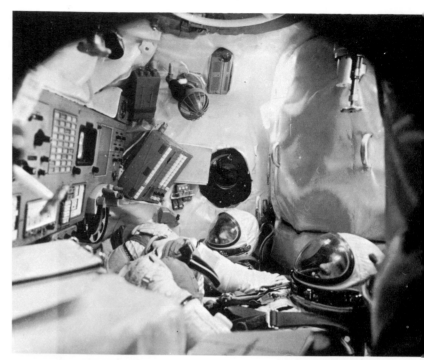

Left: cosmonauts Aleksey Gubarev and Georgi Grechko in training for their Soyuz 17 space flight. The two cosmonauts were launched in January 1975 on a wide-ranging scientific mission that involved observation of star fields for bursts of radiation that would identify and fix the position of pulsars and neutron stars; detection of x-ray emission from the Crab nebula; and the obtaining of spectra from the Sun in the ultraviolet region.

direction of the Sun. In addition, the living quarters featured an interesting psychological trick. A "floor" and "ceiling" were created by painting one surface of the cabin black leaving the illusion of a lighter colored ceiling above it. Salyut 3's flight ended in January 1975 when ground controllers fired its retro-rockets and the vehicle plunged back into the Earth's atmosphere, breaking up harmlessly over the western Pacific. It had functioned for nearly seven months, twice as long as had been originally planned.

Salyut 3 was still in orbit when in December 1974 its successor was launched. The crew of Soyuz 17 – Gubarev, and Grechko – spent over 28 days on board in the course of an almost perfect mission. The next team scheduled to board Salyut 4 had a dramatic escape. Nine minutes after launch their rocket

booster began to veer off course. Automatically the Soyuz capsule was blasted clear. Crewmen Lazarev and Makarov suffered an uncomfortable emergency landing in the foothills of the Altai Mountains of Siberia.

Later advances in Salyut technology has included multiple dockings. For example in June 1978 the Soyuz 30 crew joined the crew of Soyuz 29 on board Salyut 6. At last Soviet space stations could be reprovisioned. Latest developments include automated supply ships sent by remote control to dock with Salyut. Soviet experts expect these techniques to extend the useful lifetimes of their space stations to five years or more. And, despite a series of near catastrophes (astronauts stranded in orbit, the failure of a robot supply ship to dock with a Salyut), the Salyut program has generally been regarded a success.

Left: the Soviet spaceships Soyuz 4 and Soyuz 5 in the position they took up as an orbiting space station in January 1969. The Soviet news agency Tass hailed the feat as "the world's first experimental cosmic station," and involved two of Soyuz 5's cosmonauts spacewalking across to Soyuz 4. They were the first men to transfer from one craft to another in space.

Chapter 7

Mysteries
of
Space

Astronomy is a fast-moving science. In the last 50 years we have learned more than in all the preceding 5000 years of its history put together. But these dramatic advances have often made the limitations of our knowledge all too obvious. New discoveries have often revealed fresh mysteries to perplex and baffle astronomers. For example, the emergence of radio-astronomy as a real force in observational astronomy has immediately plunged us into the enigma of quasi-stellar sources – immensely bright objects located on the very edge of the observable universe. The first use of an x-ray telescope revealed mystifying x-ray activity in the centers of distant galaxies. Every new step in probing the universe presents us with more questions to answer. Even in the rapidly developing world of modern astronomy every discovery is a reminder that the mysteries of space remain a continuing challenge for future generations.

Opposite: a line of dish aerials forming a radiotelescope complex at Cambridge, England. Complex instruments such as these are today probing the furthermost reaches of the universe and revealing new mysteries to perplex and intrigue modern astronomers and cosmologists.

Time and space

In the early 1900s the German mathematician and physicist Albert Einstein worked out a theory that revolutionized our ideas about the universe. For nearly 300 years scientists had regarded space and time as absolute and unchanging. Einstein's theory of relativity, published in 1905, showed that every measurement we make – whether a measure of mass, length, or time – depends on how we are moving in relation to whatever we are measuring.

Einstein's theory of relativity is not easy to understand, and at times even seems contrary to common sense. But this is because in our everyday lives the effects of relativity are so slight that they are not noticeable. Because our ideas of what we consider commonsense are derived from our daily experiences, it is not surprising that the effects of relativity – only significant at very high velocities – seem to defy logic.

Einstein's work on relativity is in fact supremely logical. His starting point was the simple assumption that the speed of light (186,000 miles per second) is absolute, and that nothing can travel faster than it. In other words, no matter how an observer was moving in relation to a beam of light, it would always appear to be traveling at 186,000 miles each second. This in itself appears to contradict one of our most basic ideas of relative motion. After all, if an astronaut in a starship was traveling towards an on-coming light beam at half the velocity of light, he ought to meet the beam at one and a half times the velocity of light. This is exactly what we expect on a highway when two automobiles approach at 50 mph. The speed at which they close on each other is the sum of their individual speeds, 100 mph. But Einstein assumed that light is a special case. No matter how fast an object travels, or in what direction, the velocity of light is always the same. Having made this assumption, Einstein went on to see what it implied in nature.

His findings, formulated as The Special Theory of Relativity, were remarkable. For example, according to Einstein, astronauts in a spaceship traveling at

Below: Einstein's special theory of relativity tells us that the measurement of such basic things as mass, length, and time depends on the relative motion of the person making the measurement. The schematic diagram (below left) shows a starship at rest on its launch pad. While there is no relative motion between the crew on board and scientists at the launch site, both agree on the passing of an hour and that the spacecraft is 328 feet (100 meters) high and weighs 220,200 pounds (100,000 kg). Once in flight and approaching the speed of light (below right) the situation changes dramatically. Although the crew can detect no changes, earthbound observers see the ship's mass increase to 15,435 pounds (7000 kg) while its length appears to shrink to about 46 feet (14 meters). Even more bizarre – the crew seem to age more slowly. One hour on Earth becomes equivalent to about seven minutes of ship time.

99 percent of the speed of light would age seven times more slowly than people on Earth. The length of their spaceship – if it could be measured by an Earthbound scientist – would have shrunk to a seventh of its original size while its mass would have increased seven times. To the astronauts themselves nothing would have changed. Only when they returned home would they discover that the people on Earth had been ageing more rapidly. In fact, if they traveled fast enough during their space voyage, they might return to Earth to find themselves younger than their own grandchildren!

These fantastic concepts are really no more than logical implications of Einstein's initial assumption about the speed of light. Imagine the astronaut traveling at half the speed of light closing on an in-coming light beam. Despite his own speed, when he measures the speed of the light beam as it reaches his ship, he finds it exactly 186,000 miles per second. The only explanation for this is that his measuring instruments have changed in some way because of the high velocity of his ship. In a sense, this is precisely what has happened. If he climbs onto the outer hull of his ship and measures how quickly the light ray traverses it, he will inevitably find the time taken implies that the beam is traveling at the speed of light. He can only conclude, therefore, that the ship has actually shrunk so that the beam is in fact traveling a shorter distance than it would when the ship was motionless, and that his clocks are running more slowly measuring the time the ray takes to traverse the hull. But there is no way he can confirm these ideas. Every-

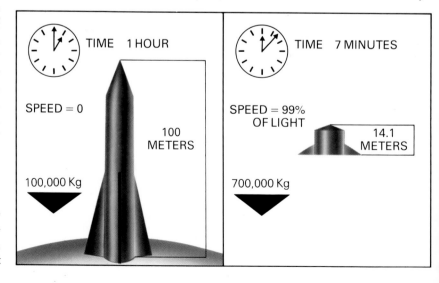

thing looks and feels normal to him. But if people on Earth could somehow observe him, perhaps with an ultra-powerful telescope, they would see a very strange scene. Instead of a normal looking astronaut and a normal looking spaceship, they will see a pencil thin man standing on the hull of a strangely squashed up ship. If they could also measure the mass of the ship they would find it slightly heavier than when it left Earth.

This is the central idea of the theory of relativity. All the wierd effects predicted – the shrinking of length, increase in mass, and the slowing of clocks – is a

Right: The graph shows the variation of mass, length, and time as a spaceship approaches the speed of light. The changes are only observed by someone who is more or less stationary compared to the spacecraft. The crew on board, for example, would detect no changes. The "decrease" in time shown by the downward curving line implies that as the speed of light is approached, clocks on board the spacecraft run more slowly than clocks on Earth. At 99.9 percent of the speed of light the spacecraft's mass soars toward infinity while its length shrinks almost to zero. Clocks on board seem to have almost stopped while time on Earth races by. Compared to people on Earth the crew will hardly age at all.

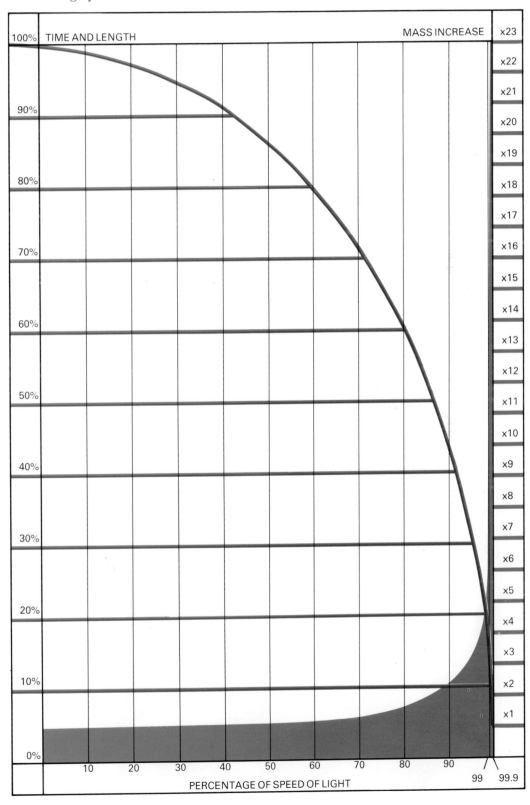

relative effect. They do not occur in the absolute sense but only in so far as a particular observer sees them happening. The deciding factor determining the extent of the changes is the relative velocity between the observer and the object he observes.

Einstein's work was only finally accepted when experiments had been performed to confirm his predictions. Even then, many scientists were sceptical. Since the days of Isaac Newton space and time had been thought of as absolute and unvarying. No matter how you were moving in Newton's universe, the fundamental quantities of mass, length, and time were always the same. But Einstein was to deal an even greater blow to classical ideas. Extending his early work, he formulated the General Theory of Relativity. This completely demolished Newton's ideas about gravity.

To do this Einstein developed the concept of a four-dimensional universe made up of something called space-time. The number of dimensions in our universe is determined by how many we need to specify the position of any object in it. Roughly speaking, an object can be a certain distance in front of us, above us or to one side of us. This defines the three dimensions length, depth, and breadth. But to completely specify the position of an object we must also say *when* it is. For example, a brick may be six inches long, four inches wide and three inches thick. But as each second passes, the brick changes. It dries out,

Below: a flat rubber sheet representing four-dimensional space time. The effect of the ball on the sheet illustrates the way in which the mass of a large body such as a star, can distort the fabric of space itself. The behavior of other smaller bodies – such as planets orbiting the star – is governed by the shape of the space in which they move. As Einstein is said to have put it: "matter tells space how to behave and space tells matter how to behave."

Right: the dotted line represents the Earth's elliptical orbit, the shape of its path around the Sun. But as well as moving through space, the Earth – and everything else in the universe – is also moving through time. If we represent this movement as a vertical displacement in the diagram, the orbit is no longer a simple ellipse. Instead it becomes a spiral, the shape of the Earth's orbit through four-dimensional space time.

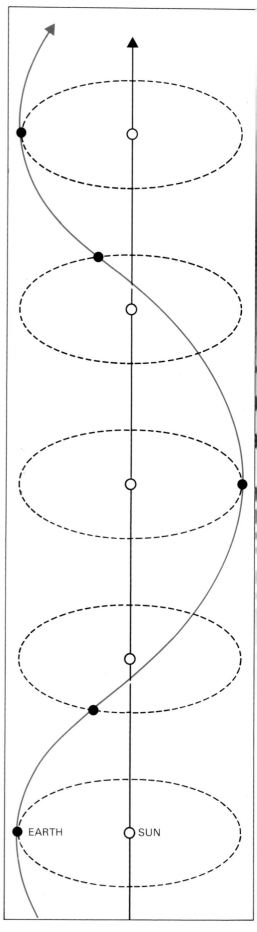

EARTH   SUN

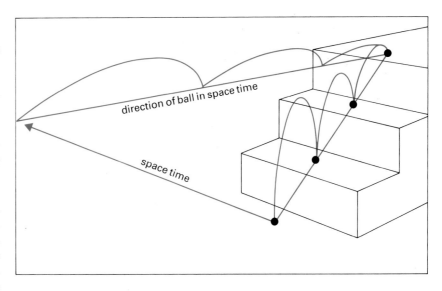

direction of ball in space time

space time

it is worn away by the wind, it undergoes chemical changes and eventually disintegrates. To be exact in specifying the brick we have to say not only "where" it is in space but "when" it is in time. Space and time therefore are inextricably linked, and the idea of space-time is simply a recognition of that fact.

Newton believed that gravity was the fundamental force pushing and pulling the universe in its ceaseless motions. Einstein agreed that gravity was fundamental to nature but he rejected the idea of it being a force somehow transmitted between all objects. Instead he argued that it was a basic property of space-time itself. Einstein's idea was that all of space-time was gently curved and that the presence of anything possessing mass caused a localized distortion in the overall curvature. For example, the huge mass of the Sun creates a distortion in the space-time around it, forming a kind of elliptical corridor through which the Earth travels in its orbit. All gravitational fields are made up of infinite numbers of such corridors filling the universe. Gravity is therefore a result of the geometry of space-time itself, and not a force at all.

The distortion of space and time by gravitating bodies was even more difficult to accept than Einstein's earlier ideas, but General Relativity was also born out by experiment. In 1919, for example, during a total eclipse of the Sun, British astronomer Arthur Eddington discovered that the light from stars bends as it passes close to the surface of the Sun. The bending is caused by light traveling through a region of space-time distorted by the Sun's mass.

Another implication of this concept of gravity is that particularly violent events should cause ripples in the fabric of space-time. In 1969, American physicist Joseph Webber, in an elaborate experiment, discovered evidence of these so-called "gravity waves" and determined their origin to be the center of our galaxy. Unfortunately, successive experiments have been unable to confirm Webber's findings.

The full implications of General Relativity are still being discussed. But in the 1960s, American physicist Professor John Wheeler – one of the inventors of the hydrogen bomb – suggested an exciting concept that may one day make possible instantaneous travel to any part of the

Above: the diagram contrasts the way a ball can be shown bouncing down a series of steps in three-dimensional space, to its trajectory in space time. The displacement of the ball in time is represented as a movement perpendicular to its path down the stairs.

universe. Wheeler worked out that there might exist beyond our ordinary space a region called Superspace where many of our natural laws do not apply. He believes that the fabric of normal space-time is made up of invisible particles of pure gravitational energy called *geons*. But the structure is not smooth and regular. It is more like a sponge, filled with innumerable flaws and holes. Wheeler thinks that the holes in the geon walls may be entrances to Superspace.

If we could ever get into Superspace we would find a very strange environment indeed. Every event in Superspace takes place simultaneously. If we could find our way through a nearby hole in normal space-time and navigate through Superspace to another hole, we might emerge at the opposite end of the galaxy. But because of the timelessness of Superspace, the awesome journey would have taken place instantaneously.

Since Wheeler first proposed his theories some experimental evidence has emerged which may support them. The gravity waves that Webber discovered emanating from the center of the galaxy, may actually be the aftermath of massive amounts of energy slipping between our normal universe and Superspace. If this is the case and the existence of Superspace is confirmed, it may only be a matter of time before we can travel to the remotest parts of the universe with the ease of stepping through a door. Two problems that may delay such a breakthrough are first finding the doors into Superspace and then making sure that having found an entrance, we can also find a way out.

The vanishing neutrinos

Evidence recently came to light which made astronomers take a renewed interest in the internal workings of the Sun. For the first time signs were discovered suggesting that the Sun's core is undergoing some kind of crisis and that the nuclear reactions on which its stability depends have either slowed down or stopped altogether.

The key to the mystery is the elementary particle the neutrino. Among the many elementary particles now known, the neutrino has some extremely unusual properties. For example, it has no mass and spends its entire lifetime traveling at the speed of light. But the most astonishing feature of neutrinos is that they very rarely react with any other particles. This makes them difficult to detect. To a neutrino even a block of lead a thousand miles thick is no real barrier.

The Sun is a prolific source of neutrinos. Vast quantities are generated as a by-product of the nuclear fusion reactions that keep the Sun burning. In fact, about three percent of the total energy output is in the form of neutrinos, the rest being released as various kinds of electromagnetic radiation. Most of this radiation remains within the dense solar core for thousands of years before eventually working its way to the surface. Neutrinos, however, are hardly affected by the great bulk of the Sun, and within seconds of being formed they are hurtling away into space. After a journey of about eight minutes a small proportion of them hit the Earth, passing through it in a fraction of a second to continue their virtually unstoppable journey into interstellar space. It has been estimated that about 200,000 million neutrinos cross every square inch of the Earth each second.

This vast flux of neutrinos could provide solar researchers with a unique means of studying the innermost processes of the Sun almost as they happen. The big problem is to stop enough of the elusive particles to examine them. In 1968 American physicist Raymond Davis Jnr built a remarkable neutrino detector

Below: Neutrinos can only be detected by their effect on other subatomic particles. This bubble-chamber photograph shows what happens when a neutrino, which has no mass and leaves no track, enters the chamber top right. It has collided with a proton causing a reaction that can be seen as a dark three-pronged track going from right to left across the photograph.

deep down in a South Dakota coal mine. Using a 100,000-gallon tank of dry cleaning fluid and shielded by the surrounding Earth from stray radiations, Davis was able to count individually the few solar neutrinos stopped by the atoms of the cleaning fluid.

It took months of patient waiting for reliable results. But Davis' hard work was rewarded by a startling discovery. Instead of confirming the intense stream of neutrinos bathing the Earth, Davis found only a fraction of the expected particles. The world's solar scientists were deeply worried by the findings. Either their theories about the processes in the Sun's interior are completely wrong or the processes are simply not taking place at present. This alarming prospect is not as far-fetched as it sounds. The fact that the Sun still rises each morning as bright as ever

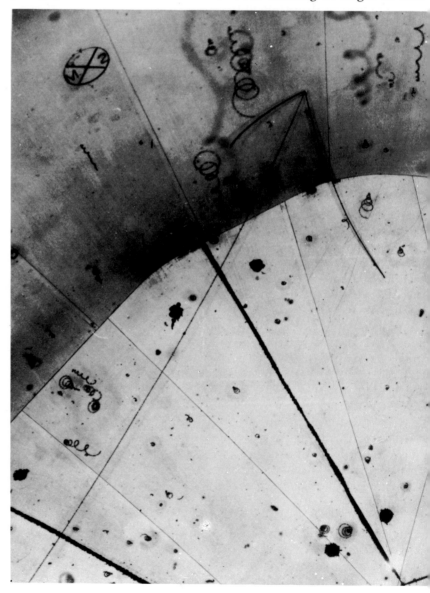

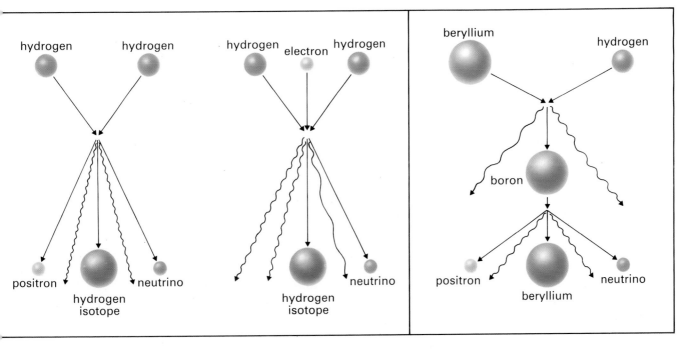

does not mean that fundamental changes are not taking place at its core. The Sun takes many thousands of years to adjust to internal changes and the radiation being emitted at present really reflects the Sun's inner condition in the remote past. Only the neutrinos can give an up to date picture of the situation inside.

It is very unlikely that our ideas of nuclear physics and stellar evolution are badly wrong. Therefore, the missing neutrinos seem to tell us that, at least since 1968, the life-giving inner processes of the Sun have not been taking place. Several theories have been suggested to account for the predicament. Perhaps the most comforting is that the Sun's furnace has only switched off temporarily and that the effect is not an uncommon interlude in a star's evolution. If this is so, the vital nuclear reactions should start up again before too much damage is done. But the cooling now taking place in the Sun's core may eventually show itself at the surface. This could be an explanation of past Ice Ages on Earth.

There is another possibility however. Some scientists theorize that while nothing is wrong with the Sun, there is something wrong with the neutrino. Theoretically, the neutrino may have cousins: the μ and the τ neutrino. It is possible, in the strange world of subatomic particles, for a neutrino to change into one of its cousins, then back into itself again. Our present neutrino counters do not detect μ and τ neutrinos, and so, if neutrinos do change

Above: parts of the Proton-Proton Chain, one of the processes by which the Sun converts hydrogen into the heavier isotopes such as boron and beryllium emitting in the process gamma radiation (wavy lines), positrons and neutrinos. On the left is shown two early stages of the sequence that results in low-energy neutrino emission, and on the right a later stage showing the decay of boron and the emission of high energy neutrinos. It is the neutrinos from this stage that were sought by Raymond Davis Jnr. and his team.

Right: a vast tank of common cleaning fluid, tetrachloroethylene, located deep underground in the Homestake Mine, South Dakota. The tank was set up to measure the strength of the flux of neutrinos from the Sun.

identities, spending only a third of the time as regular neutrinos, then that would account for the detectors being only one-third the number they should. Another implication of the theory of the neutrino's changeable identity is that if they do oscillate back and forth then they may also have mass, and if they actually have a little mass, then that would go a long way toward explaining another great enigma of the day – the mystery of missing mass.

Cosmic rays

Scientists have known about cosmic rays since the early 1900s. Observations showed that an unknown kind of radiation was knocking electrons out of the ordinary atoms of the atmosphere, turning them into positively charged ions. The puzzle lay in working out where the radiation came from. At first, scientists assumed that the radiation was coming from the soil where naturally radioactive substances could sometimes be found. In 1911, to settle the matter, Austrian physicist Victor Hess made a series of balloon ascents to see whether he could detect a decrease in the radiation high above the Earth. If it came from the soil as most people suspected, the atmosphere between the ground and the balloon should absorb decreasing amounts of radiation as the balloon gained height. In fact, Hess found the opposite was true. As he went higher, the radiation actually increased.

In the years following Hess's pioneering experiments, scientists realized that the radiation was striking the Earth everywhere and could only originate somewhere beyond the atmosphere in space itself. In 1925, American physicist Robert Millikan coined the phrase "cos-mic rays" to describe them, and the name has been used ever since.

But even though modern scientists know a great deal about the nature of cosmic rays, it is difficult to pinpoint where they are coming from as the magnetic fields of the solar system deflect them so much. Recent research indicates that the primary

Above: checking a cylinder of 48 geiger counters deep in a disused salt mine. The counters and other scientific apparatus were part of an experiment that took place in 1949 to study cosmic-ray particles.

Below: an x-ray picture of the Crab Nebula. It is believed that nebulae such as this one may be one of the sources of the cosmic radiation that seems to fill space.

source of cosmic rays is the nuclear furnace at the core of our galaxy and that the rest come from other galaxies. In the near-perfect vacuum of interstellar space, the sub-atomic particles of which cosmic rays are largely composed could easily travel hundreds of light-years without being slowed down or absorbed by collisions with matter. The rays contain all the particles that might be expected from powerful nuclear reactions.

Studying cosmic rays has helped physicists work out many of their theories about atoms and the elementary particles of which they are made. Examples of most of the known particles can be found in cosmic ray showers together with a good deal of energy in the form of gamma radiation. One of the most important features of cosmic rays is the wide range of energies possessed by the constituent particles. These vary from modest levels to energies well beyond the reach of even the most powerful existing particle accelerating machines.

These high-energy particles are especially significant to physicists because they are most likely to hold the key to future discoveries about the nature of matter. Scientists of all nations are struggling to build bigger and more powerful machines with which to fire particles at other particles at great speed.

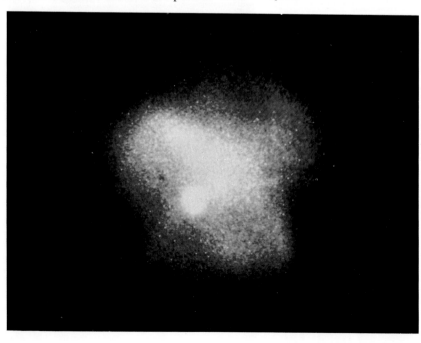

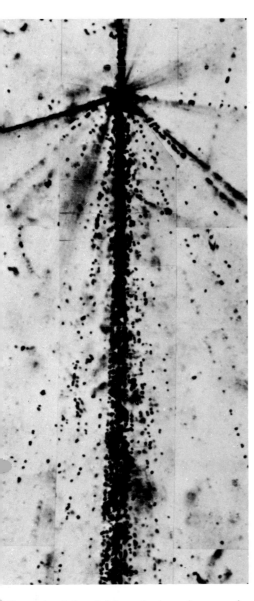

Below: a photograph showing a highly magnified (× 800) view of the effect of cosmic ray impacts on the surface of a space helmet.

Tachyons were predicted theoretically some years ago when physicists noticed a curious loophole in Einstein's theory of relativity. Although the terms of the theory forbids faster than light travel under normal circumstances, particles that traveled faster than light could exist so long as they never traveled more slowly than light! Imprisoned in their faster than light world, tachyons would be extremely difficult to detect. Fortunately, however, the theory also predicts that, if they exist at all, tachyons would produce a kind of optical "shock wave," similar to the sonic boom created by an aircraft flying faster than sound. It is just such a shock wave that a team of Australian researchers think they have spotted while studying cosmic ray showers. If they are right we may be on the threshold of verifying one of the most fantastic concepts in science. If tachyons do exist, they will actually be traveling backward in time. This means that if we

The principle of this technique is to study high energy collisions and to deduce new findings from their results. The giant accelerating machines needed are enormously costly and new breakthroughs can only come with even more powerful and expensive devices. The irony of the constant battle for funds to build them is that cosmic rays provide a limitless supply of high energy particles free of charge. Indeed, many particles have energies much greater than our present technology could produce no matter how much money was made available.

Physicists are only slowly beginning to appreciate the rich potential of cosmic ray studies. But one team of researchers has already uncovered evidence suggesting the existence of *tachyons*, the most remarkable elementary particles in the universe.

can progress from merely detecting tachyons to using them as a means of signaling we may eventually be able to communicate with the past. Some scientists believe that rather than reaching back into our own past, we may be able to contact one of the countless time-streams they believe exist, linked to universes parallel with, but different to our own.

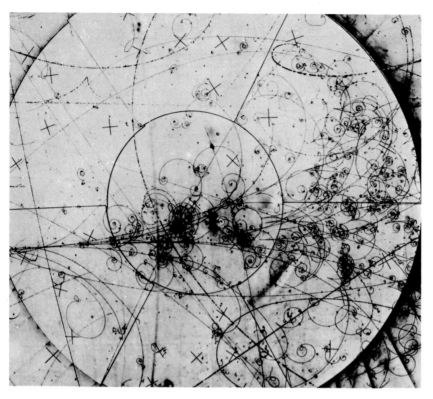

Antimatter universe

In 1930 British mathematician and physicist Paul Dirac predicted the existence of *antiparticles*. In Dirac's day matter was thought to be made up of atoms consisting simply of protons, neutrons, and electrons. No other elementary particles were then known. On purely theoretical grounds Dirac believed that antiparticles should also exist, and that they should be similar to the known particles but with their key properties reversed. This meant, for example, that an antiproton would have a negative electric charge while an antielectron would be positively charged. Neutrons, although electrically neutral, have a magnetic field associated with them which has a characteristic orientation. Therefore, an antineutron would have a field identical in strength but in the opposite direction.

The immediate objection to Dirac's idea was that at that time no antiparticles had ever been observed. Why should they be so difficult to detect?

Above: a photograph of a bubble-chamber experiment showing the tracks of fast-moving subatomic particles. The large spirals are electrons and positrons, turning in opposite directions.

There was a simple answer. Dirac' calculations showed that when a particl meets an antiparticle, they completel annihilate one another, releasing a brie flash of energy in the form of gamm radiation. In a universe dominated b ordinary matter therefore, antiparticle would have an extremely short lifetime probably less than a millionth of a second

Most scientists regarded Dirac's worl as little more than a deft exercise in highe mathematics having no relationship t reality. But in 1932 a chance observatior showed that Dirac had in fact made fundamental breakthrough in physics While studying cosmic ray showers American researcher Carl Andersor noticed a curious anomaly among th traces of particles produced when cosmi rays struck a target made of lead. Ther were particles present which appearec to have all the expected properties o electrons. The only difference was tha in a magnetic field the particles curved ir a direction opposite to the one an electror would take.

This could only mean that this elec tron-like particle had a positive electrica charge. After a careful check of hi results, Anderson announced the dis covery of the first of Dirac's antiparticles the antielectron – now called the positron The discovery of antiprotons and anti neutrons did not happen until 1956 Since then the counterparts of many of th less common elementary particles hav also been detected.

Antiparticles have all the properties o particles. They can do anything particle can do. So, if protons, neutrons, anc electrons can combine to form atoms o ordinary matter, there is no reason why the corresponding antiparticles shoulc not combine to produce atoms of anti matter. In fact, in 1965 an Americar team managed for a fleeting instant tc produce a nucleus of antideuterium, the antimatter version of an isotope of hydro gen. Like any antimatter in a world made entirely of matter, the unique anti nucleus quickly vanished, annihilatec along with a fragment of ordinary matte in a tiny pulse of gamma rays.

Today physicists know that just a matter and antimatter can disappea producing gamma rays, particularly in tense gamma rays sometimes spon taneously turn into a particle-antiparticle pair. This perfect illustration of Einstein's prediction that mass and energy are interchangeable is nowadays frequently observed in high-energy cosmic rays. More interesting still is the suggestion that some of the gamma radiation in cosmic rays may well have themselves originated in large-scale matter-antimatter annihilations in parts of the universe remote from Earth.

There seems no good reason why nature should have a preference for the ordinary matter of which our cosmic neighborhood is made. On grounds of symmetry alone, scientists believe that there must be equal amounts of antimatter somewhere in the universe. It seems possible therefore that among the stars and galaxies observed from Earth, at least some are composed of antimatter.

A collision between an antimatter galaxy and one composed of ordinary matter would be one of the most awesome events imaginable. The resulting explosion would be visible throughout the known universe. The fact that such an event has never been observed does not mean that a collision of this kind has never happened or never will happen. The period during which people have had even the simplest instruments with which to study the heavens is infinitesimally

Right: when a particle and its antimatter counterpart meet, they annihilate one another in a burst of energy sometimes accompanied by the creation of totally new particles. Top diagram shows the collision of a proton and an antiproton resulting in their transformation into mesons. These decay rapidly and as they do so emit neutrinos (blue and yellow) and gamma radiation (wavy lines) leaving only electrons (red) and positrons (green). When electrons and positrons collide, lower diagram, the result is mutual annihilation with the release of energy as gamma radiation.

Below: the galaxy NGC 5236, photographed in 1959 (left). In 1972 (right) a star could be seen exploding in a spectacular supernova. Many astronomers and physicists believe that exploding stars are one of the sources of cosmic rays which themselves give rise to particle-antiparticle pairs – notably electrons and positrons.

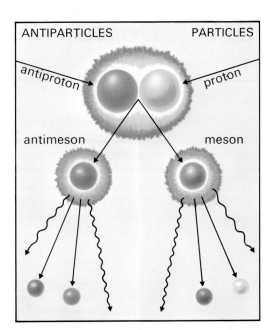

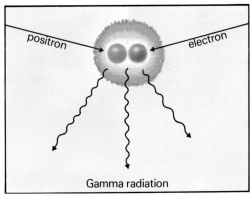

Gamma radiation

short compared with the time scale of galactic evolution. Before the few centuries of modern history we may well have missed seeing several such cataclysms. Even so, the theory that matter and antimatter on a scale as huge as entire galaxies would never come into complete contact is probably correct. As their outermost regions brushed, the initial blast of energy might well hurl the galaxies away from each other like a force of mutual repulsion.

Galactic encounters of this kind would certainly produce powerful gamma ray bursts, detectable even over vast distances. Within the last few years the first satellite-born gamma ray detectors have recorded bursts of the kind that might be expected. As yet we do not have the necessary technology to pinpoint the sources of the radiation. When this can be done we may at last have the means to verify whether our own universe of ordinary matter is mirrored by a co-existing universe of antimatter.

JUNE, 1959   MAY, 1972

Black holes

The more that is discovered about stars and galaxies, the more violent and terrible the universe seems. But of all the cosmic dramas, none is as final and complete as the total gravitational collapse of a star. It is now believed that such an event represents not just the end of a star, but the end of matter itself. The collapse obliterates a star from the observable universe, leaving in its place one of the greatest mysteries of space – a black hole.

A black hole is a region of space from which the escape velocity is greater than the speed of light. In this age of spaceflight, the idea of escape velocity is familiar. Every planet has its own. The strength of the Earth's gravity, for example, means that any object thrown upward from the surface with a velocity of less than 25,000 mph must eventually be pulled back to Earth. But once the escape velocity is reached, the object breaks free of Earth's pull and travels on into space. The Moon's gravity is much weaker than the Earth's, so the escape velocity is much lower – only about 5400 mph.

To work out the velocity needed to escape from a planet or a star, both mass and density must be calculated. Mass is what produces the gravitational field, so the more there is, the stronger the field is likely to be. Density is important because the more dense an object is, the closer its surface is to its center from where gravity always seems to operate. The Earth weighs about 6000 million million tons and has an average diameter of 8000 miles. If it shrank to half this size, the same mass would be crammed into a much smaller volume. On its surface a person would be only about 2000 miles from its center and gravity would be four times stronger than normal. If the Earth continued shrinking, its gravity would eventually become so strong that the velocity needed to escape its surface would exceed the speed of light. Calculations show that if the Earth's diameter fell below one inch, not even light could escape its surface gravity.

In 1916 mathematician Karl Schwarzschild applied Einstein's General Theory of Relativity to the concept of escape velocity. One of the most important things the theory tells us is that gravity is not a force acting between objects but a property of space itself. Einstein said that the universe exists in a four-dimensional amalgam of space and time. Wherever a mass exists, in the form of a star or a planet, the space-time around it is distorted. In the simplest case it is

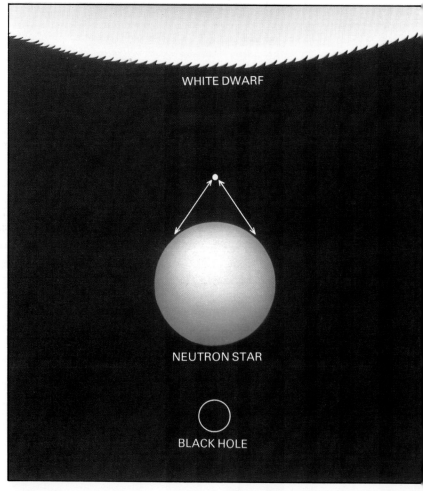

Above: stars that have around the same mass as the Sun are likely eventually to collapse to a size similar to the Earth's, becoming white dwarfs (top). Stars with more than 1.4 times the Sun's mass may undergo a more extreme kind of collapse forming a fantastically dense neutron star, perhaps less than 10 miles in diameter. Stars of much greater mass may continue to collapse beyond the neutron star stage until they become black holes – objects so dense that not even light can escape their powerful gravitational fields.

curved. The effect of this localized curvature is exactly the same as if the mass causing it was exerting a gravitational pull on objects nearby. In other words, gravity is actually a curvature in space itself.

Schwarzschild showed that an escape velocity greater than the speed of light is equivalent to a region of space curved so much that it closes in on itself. According to the theory of relativity, nothing can travel faster than light so nothing could ever escape from such a region. It would effectively be cut off from the universe. Schwarzschild worked out that there should be for any mass some critical density which will produce this closed region of space. If a mass is squeezed within a certain radius – now called the *Schwarzschild radius* after its discoverer – ts escape velocity will exceed the speed of light. Any object which emits no radiation is said to be perfectly black. An object compressed to less than its Schwarzschild radius would therefore behave just like a black hole in space.

Few paid much attention to Schwarzschild's work at the time. No one – not even Schwarzschild himself – believed that black holes could actually form. But n the 1960s astronomers made discoveries about the life cycles of stars suggesting that black holes might exist after all. Among the debris of enormous supernova explosions, superdense remnants were discovered. These neutron stars were solid balls of neutrons formed by the gravitational collapse of a once normal star. Caught in the grip of its own gravitational field, the material of which the star is made is pulled closer and closer to its center. Eventually the nuclei of the star's atoms are themselves crushed together and form a neutron star no bigger than a few miles across.

On average such a star is 1000 billion times more dense than water. A matchbox full of its substance would weigh 1,000,000 tons or more. More remarkable still, scientists found that gravitational collapse could led to still greater densities. If a star has a mass of 1.4 solar masses or less (the Chandrasekhar limit), it will collapse into a white dwarf; above that, but below 2.4 solar masses, it can collapse to form a neutron star. But what about a bigger star, with 3 solar masses or more? When a star this size collapses, it will go past the white dwarf stage, past the neutron star stage, and beyond, to a point where it will be a single fragment of nuclear material, with a radius so small that gravity will fold space in on itself, leaving a black hole.

With this discovery, interest in Schwarzschild's work suddenly revived. Faced with the likely existence of black holes, scientists wanted to know as much as possible about them. This seemed a hopeless enquiry at first. How could anything be learned about an object which never emitted even the smallest piece of information about itself? The Schwarzschild radius is a one way barrier. Nothing inside it – not even a ray of light – can get out. Matter or energy passing into it

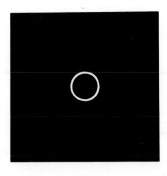

Above: the actual calculated size of a black hole of the same mass as the Earth. You would have to be about 3700 miles away from it for the gravitational force to feel the same as that of the Earth. At two feet from it you would experience a force 100 million million times stronger.

Right: a space-time diagram of the possible gravitational collapse of a star with a mass of over three times that of the Sun. After the star has collapsed beyond a critical radius – the *Schwarzschild radius* – its gravitational field would be so intense that not even light could escape from within it. Inside the Schwarzschild radius the black hole star continues its collapse until it reaches a geometric point in space, infinitely smaller than a pinpoint – a *singularity*.

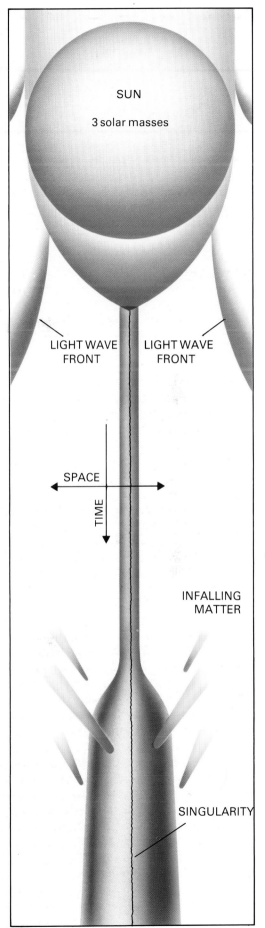

SUN

3 solar masses

LIGHT WAVE FRONT

LIGHT WAVE FRONT

SPACE

TIME

INFALLING MATTER

SINGULARITY

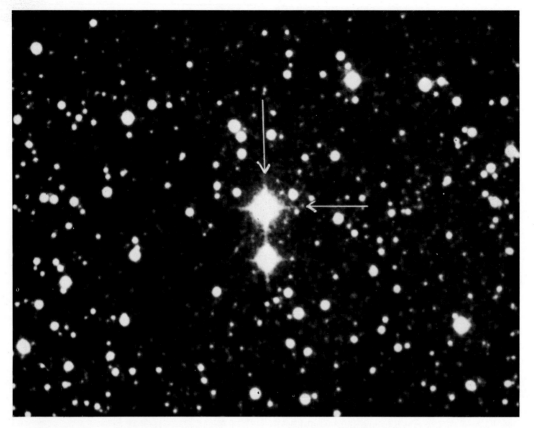

Left: The mysterious Cygnus X-1 is apparently a double-star system. One of its components, the star HDE 226868, is arrowed here. Observations of the tremendous x-ray, radio, and light emissions from this star system have led some scientists to suggest that Cygnus X-1 contains a black hole in orbit around the visible star.

Right: a model for the suspected black hole in the x-ray source known as Cygnus X-1, the dark companion to a giant primary star in the constellation Cygnus.
A shows gas being drawn off the giant primary star on the left by the gravitational attraction of the black hole. The hole is in orbit around the primary star and its movement causes some of the gas to miss the hole. The gas captured by the hole spins around in a flattened disk (blue).
B shows the flattened disk of gas around the black hole. Pressures in the swirling gas cause a central bulge.
C shows the vortex of swollen gas caused by heat from the x-rays (red) given off near the black hole.
D represents the core of the disk where the gravity is so strong that the bulge of hot gas is flattened.
E highlights the central 120 miles of the core, where the pressure and heat cause the gas to give off the x-rays that can be observed on Earth. At around 50 miles from the center, the disk becomes violently turbulent and immensely hot. The disk terminates near the black hole where the gas, no longer able to move in orbit, is sucked into the hole.

is trapped forever. The boundary around a black hole is often referred to as an "event horizon," because it is impossible to discover anything about events taking place beyond it.

Black holes seem therefore to be utterly beyond our reach. In physical terms they probably are. But despite this it is possible to explore them by using mathematics rather than the usual tools of astronomy such as telescopes. Applying Einstein's theory of relativity, it is possible to deduce not only what a black hole is like seen from a safe distance, we can reconstruct a descent into its interior as well.

Suppose that a black hole has been located and that a team of research scientists has been despatched by spacecraft to investigate it. What would they find? The only visual impression they could hope for is an area of darkness silhouetted against a background of stars. But if they came close enough to the black hole to see this they would already have been captured by its powerful gravitational pull, and would be spiraling into it with no hope of escape. Aware of the dangers the team would keep their spaceship at a safe distance, locating the hole by using instruments to plot its

gravitational field.

Having assumed a safe orbit around the black hole, one of the scientists volunteers for a suicide mission. He agrees to leave the spaceship and allow himself to be drawn into the black hole. As part of the experiment he plans to keep flashing a light signal to his colleagues at regular intervals.

As he drifts away from the spacecraft he will soon feel the gravitational pull of the black hole and begin a spiraling plunge toward it, accelerating all the time. At first his colleagues safe aboard their spacecraft would simply see the man falling away from them, his light pulsing regularly. But as he approaches the black hole's event horizon, bizarre relativistic effects begin to show themselves. Because space and time close to the horizon are vastly distorted, the rate at which time passes for the falling scientist accelerates. As far as he is concerned his light flash continues to signal at the same fixed intervals. But his colleagues will notice the intervals between one flash and the next becoming successively longer. Eventually the interval will become longer than the lifetimes of the observers, and even those of their descendants.

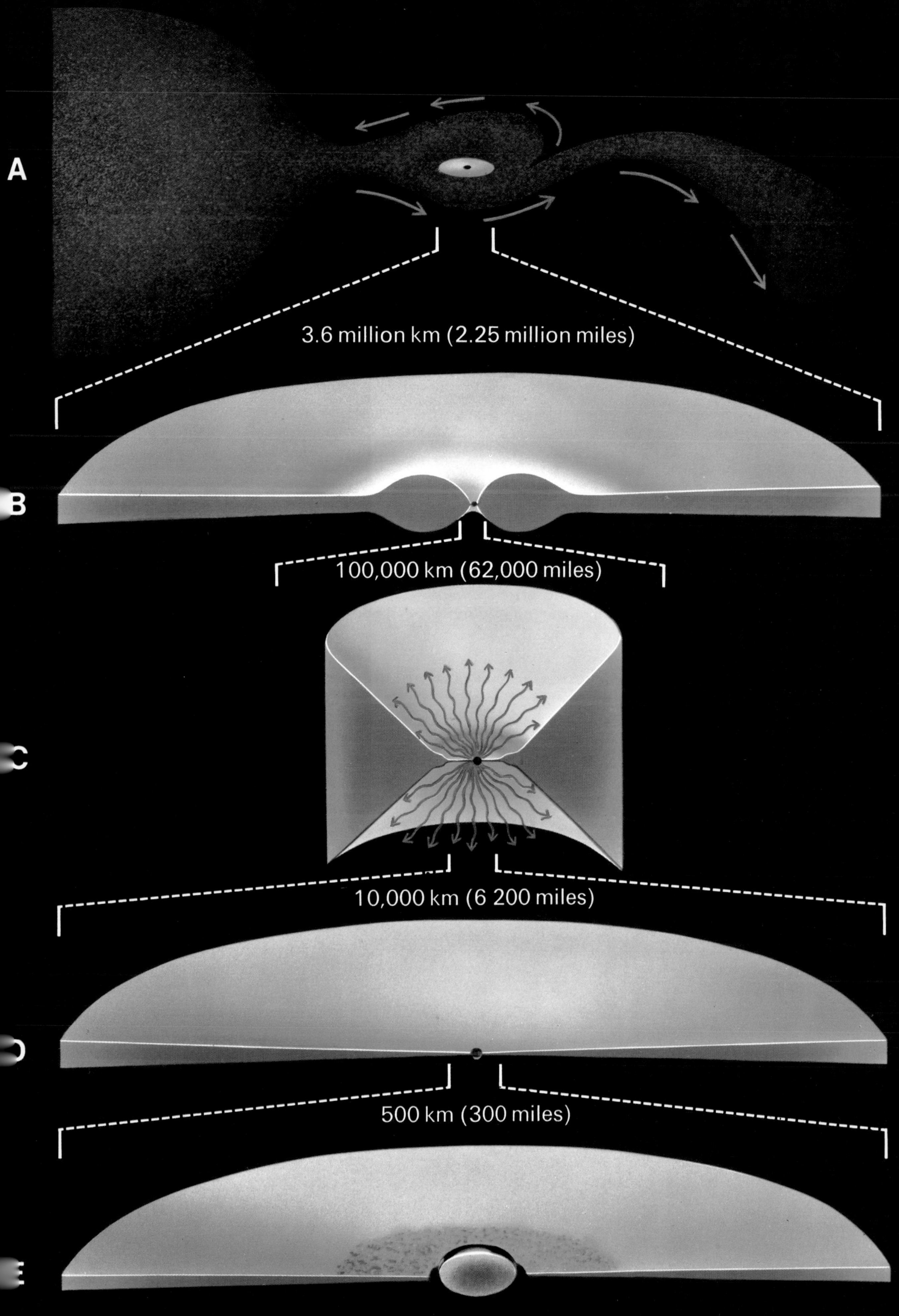

A

3.6 million km (2.25 million miles)

B

100,000 km (62,000 miles)

C

10,000 km (6 200 miles)

D

500 km (300 miles)

E

Left: Some theorists believe that not only do black holes represent the end of some stars' life, but that they also are involved with the birth of galaxies – that they were the cores around which galaxies began to form in the early days of the universe.

The appearance of the brave scientist will be even stranger. The observers will suddenly see his fall slowing up as he approaches the event horizon. Ultimately, he will appear to have ceased falling altogether and will hang frozen in space just above the event horizon. As far as the doomed man is concerned, time will seem to be passing normally and he will pass through the boundary of the black hole in a fraction of a second. But even a second in his time may now be equivalent to the passing of millions of years in the universe he has left behind.

What will he find beyond the event horizon? Even if he survives to observe his surroundings he will never be able to communicate his findings to the universe outside. He will probably be killed even before he reaches the event horizon. As he approaches, the pull of gravity soars. If he is falling headfirst, the pull on his head will be much greater than on his feet, even though they are only slightly further away from the center of the gravitational field. This will result in a

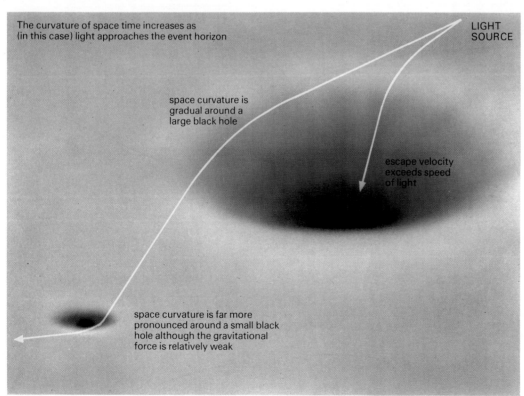

The curvature of space time increases as (in this case) light approaches the event horizon

LIGHT SOURCE

space curvature is gradual around a large black hole

escape velocity exceeds speed of light

space curvature is far more pronounced around a small black hole although the gravitational force is relatively weak

Left: the behavior of light in the vicinity of black holes of different mass.

Above: a black hole probe. The picture depicts a rotating black hole of more than 30 solar masses making up a binary system with a star 20 times the mass of our Sun. The star has evolved into a red giant and has swollen to 100 times the diameter of the Sun. The black hole's gravitational field is drawing material from the outer edges of the star into an orbit around the

hole as a disk of hot
luminous gas. The light
and x-rays emitted by the
disk has revealed the
presence of the hole to a
starship and scientists on
board have sent a probe
into the hole (top right).
Packed with instruments
designed to give readings
until it vanishes beyond the
event horizon, the probe
also contains very tough
inner compartments that,
it is hoped, will survive
passage through the hole
possibly into another
universe.

gruesome death. The man will be stretched into a long, thin thread in the instants before reaching his objective.
Scientists call this powerful gravitational
turbulence around the event horizon,
"tidal forces" after the far gentler but
essentially identical forces that control
the tides of the seas on Earth.

If he somehow survived these devastating tidal forces, the scientist would
probably find himself in an environment
too strange to comprehend. It is not
known for certain what he would find
but many of our most basic ideas of space

and time would no longer apply. Some
scientists have suggested that inside the
event horizon time might actually be
running backward or may even cease to
have any meaning at all. Perhaps no
human concepts exist to describe the
reality of a black hole and we must simply
accept that to enter one is to step into
another kind of universe, one completely
alien to mankind.

Scientists often enjoy wild ideas as much
as science fiction writers, hence it was not
surprising when several of them began to
investigate the possibility of traveling into

Above: the British satellite, Ariel 6. It is designed to carry experiments that will tell scientists more about the death of stars, particularly supernovae and the final stages of massive stars — neutron stars and black holes.

black holes. They knew that any person or ship would be torn apart. Even if a black hole was massive (a thousand solar masses or more) or supermassive (a million or more solar masses), and the gravitational curvature around the hole was less abrupt, the result would be the same. If a spaceman was going headfirst into the hole, his head would be closer to the center of gravity than his feet and he would soon be stretched thinner than a piece of piano wire.

However, it has been theorized that a black hole might have properties that could mitigate the devastating gravitational effects. For example, what if a black hole carried an electrical charge? Let's say the black hole was negative, and you dropped in it your positively charged spaceship. Although the natural electromagnetic repulsion might counter the gravitation crush enough for you to descend slowly through the hole, it appears that the electrical charges at that scale would annihilate a spaceship nevertheless.

More promising are rotating black holes. If a hole was rotating fast enough, the centrifugal force would counterbalance the gravitational force. This would create an *inner-event horizon* where the two forces would form a wall at the point where they balanced. This wall would be impenetrable, but it would form a ring, rather than a sphere around the center of the black hole. This means that someone could avoid the wall of the inner-event horizon by aiming right for the center of the hole.

But why on Earth would anyone want to travel into a black hole? Theory suggests that the depression a black hole would cause in space-time would go so deep that the black hole would eventually punch out into another universe, or a different part of our own. A black hole would then emerge as a *white hole,* spewing out matter rather than drawing it in. The connection between a black hole and a white hole is known as an *Einstein-Rossen bridge,* or *wormhole.*

It has also been theorized that black holes can come in a variety of sizes. Many believe that in the beginning of the universe, as galaxies formed, they built themselves around massive black holes, and that it is the tremendous amount of energy thrown off by the charged accretion disks of matter surrounding black holes of this size that accounts for the very distant, very old and very bright quasars which we see. Indeed, there may be a black hole at the heart of our galaxy. At the other end of the scale, brilliant British mathematician Stephen Hawking theorized that the Big Bang may have created a number of primeval *mini-black holes,* as small as atoms. A curious thing about these mini-black holes is that over time they would get lighter as they sucked in negative energy, and eventually they would even explode.

There's really not much to fear from black holes. Space travelers of the future will be able to detect their gravitational field long before they come upon them or pass anywhere near an event horizon. The idea that black holes are cosmic sink holes that suck anything and everything up is mistaken. When a star collapses, its gravity doesn't change, just its size. If the Sun were to collapse right now into a black hole (it can't, of course, being too small a star), the Earth would continue on its orbit, oblivious except for the loss of sunlight. In fact, the only time a black hole would be truly dangerous would be on the rare occasion that a rotating black hole spins so fast that its centrifugal force would surpass its gravitation. In that case its inner-event and outer-event horizons would merge or even exchange places. Then it would be possible for light, indeed anything, to escape from the hole. This would be a *naked singularity* and it would be very dangerous. Scientists do not know what happens at the heart of a black hole, in the singularity. Indeed, some think the laws of space-time would not apply – that *anything* could happen! A naked singularity could possibly disrupt the entire universe, changing fundamental laws at random, reversing time, and, in general, wreaking havoc in the cosmos. So catastrophic could the existence of such a thing be that a number of scientists believe there has to be a law prohibiting their formation.

One big question about black holes remains: Do they in fact exist? As they are invisible, they are impossible to observe directly. But an accretion disk of highly charged matter would form around the event horizon as matter was drawn into the hole, and this disk would release a tremendous number of x-rays. There was

Below: an infrared view of the galactic center. In the color-coded picture the radiation intensity increases from blue to red. The center of the galaxy (which lies obscured by dust clouds in the southern Milky Way) is marked with a cross. The galactic equator is indicated by the thin white line. Infrared emissions are strongest along the equator, as shown. Some scientists have suggested that a black hole is situated in the galaxy's center. Evidence to support the idea has been provided by American physicist Joseph Weber who claims to have detected a powerful source of gravitational waves in the heart of the galaxy, possibly a black hole steadily devouring planets and stars and creating vast outpourings of energy as it does so.

considerable excitement a few years ago when high-intensity x-ray emissions were detected from the star system Cygnus X-1 in the constellation of the Swan. Observation showed the system consisted of a normal visible star and a small dark companion. The companion might be a neutron star, but its calculated mass is 10 solar masses, far above the Chandrasekhar limit for neutron stars. So, the companion may well be a black hole. Recently it has been argued that Cygnus X-1 may be a triple star system and that there is no black hole. But that would not account for the intense x-ray emissions and so the black hole hypothesis still remains viable.

There is one other remarkable notion about black holes. While it may turn out that travel through them to other universes or distant parts of our own may prove to be unfeasible, black holes could be used for another benefit. If a civilization were to set up a strong magnetic field around a black hole the resulting interaction would create an enormous amount of energy. Enough energy, perhaps, to power that civilization until the end of time.

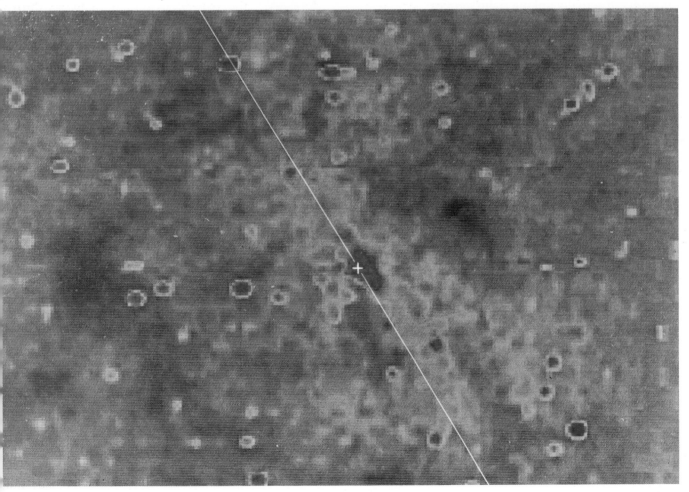

Quasars

On March 16, 1963 radioastronomy came of age. For the first time a radio-telescope revealed a totally new kind of cosmic phenomenon. The discovery challenged traditional ideas of the way stars generate their energy. Radioastronomers had discovered compact, starlike bodies, each of which appeared to be producing as much light and other radiation as several large galaxies put together. How could such relatively small bodies produce such fantastic amounts of energy? The mysterious objects – called *quasistellar sources* or "quasars" for short – were soon the subject of intense scientific interest and controversy.

For some years before the discovery of the first quasar, radioastronomers had been picking up signals from sources assumed to be stars somewhere in our own galaxy. The processes by which stars could emit strong radio signals was an interesting theoretical puzzle, but the difficulty of actually linking radio sources to visible stars discouraged most astronomers from investigating them in detail.

Despite the problem of precisely locating them, a British team led by Martin Ryle compiled a comprehensive list of known sources – called the Third Cambridge Catalog of Radio Stars. Each star was given the prefix 3C, after the catalog's name. To help astronomers find the stars, a general idea of their location in the sky was provided.

Using the catalog, an Australian group of radioastronomers led by Cyril Hazard decided on an all-out effort to become one of the first to pinpoint a radio star and link it to a visually observable source. They used a simple but effective technique. They knew that the Moon was soon to pass between the Earth and the general area of sky in which radio star 3C-273 was located. As the Moon swept through space there would come a moment when the curved leading edge of the lunar surface eclipsed the radio star, blocking its radio signals from the Earth. Shortly afterward, the radio star would emerge from behind the trailing edge of the Moon and the transmissions would be received once again. Fixing these two moments exactly, the astronomers could

check to see where the Moon should be at the two times (the lunar orbit is easily predicted by consulting astronomical tables). The technique would therefore give an accurate pinpoint location of the radio star based on well-known data about the Moon's path in the sky.

Hazard's team went to remarkable lengths to make certain the technique would work. To ensure they could maneuver their instruments to the unusually low angle of elevation required, they sawed several tons of metal off their radiotelescope. In the hours before the eclipse, local radio stations broadcast appeals for all non-essential radio transmissions to be shut down so as not to interfere with the delicate observations. Patrols were even sent around the observatory's grounds to stop motor vehicles approaching.

The moments of the eclipse finally came. Two duplicate sets of measurements were made and, using separate aircraft for safety, two members of the team rushed the data to a group of optical astronomers. Their efforts were not

Below: Martin Ryle, the British Astronomer-Royal, pictured with one of the giant radio-telescope aerials at Cambridge, England. Ryle led the earliest investigations of so-called radio stars and paved the way for the discovery of the first quasar.

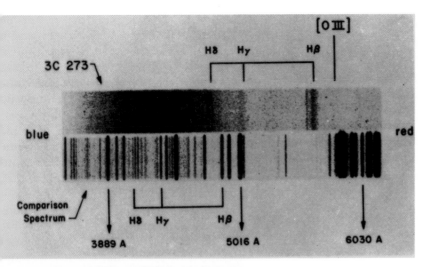

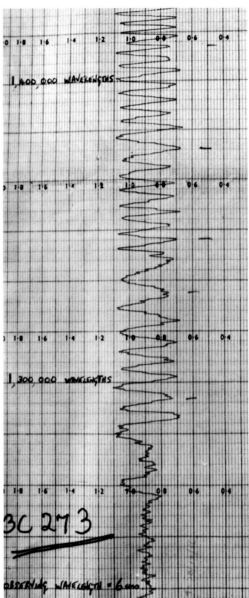

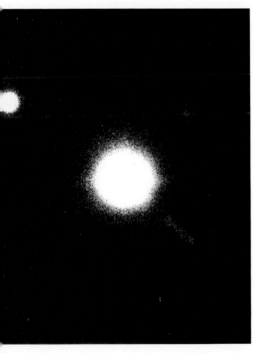

Above: the spectrum (top row) produced by light from quasar 3C 273 (photograph left), compared with a laboratory spectrum. The hydrogen emission lines Hδ, Hγ, and Hβ, which are marked on both spectra show the enormous red shift of the quasar. Using the spectrum, astronomers have worked out that the quasar lies well outside our galaxy.

Right: radio emission from quasar 3C 273, recorded with the 250-foot radio telescope at Jodrell Bank, in Britain. The intensity of the emission, together with its apparent enormous distance from Earth, seems to indicate that 3C 273 is as bright as 200 galaxies combined.

wasted. News soon came that radio star 3C-273 had been correlated with an ordinary star apparently located in our galaxy. Exciting as this news was, no-one was prepared for the discovery that followed. More detailed examination of the light coming from 3C-273 showed that whatever else the source was, it was not an ordinary star. The first hint that it was a totally unknown kind of astronomical body came when astronomers examined the wavelengths of the light it emitted. Their results showed a pronounced shift in the spectral lines toward the red end of the spectrum. Astronomers are familiar with this effect when studying distant galaxies. It tells them that the galaxies are receding at very high speed. Indeed, the concept of the red shift is central to our idea of an ever-expanding universe. Finding a receding star was not unprecedented. What astonished astronomers was the size of the red shift and the velocity of recession which it implied. If the observations were to be believed, 3C-273 was receding at over one sixth of the speed of light (about 31,000 miles per second). But using established principles of cosmology, speed of recession is linked to the distance of a body from the Earth. Calculations showed that 3C-273, apparently an ordinary star, was well outside our galaxy.

How could a starlike object be luminous enough to be visible even over such vast distances? It would have to be emitting as much energy as millions of normal stars. But 3C-273 was only the first of

variations of this magnitude to occur, the entire radiating body must change, and it can't change any faster than the time it takes for the change to be transmitted throughout the body. For example, if a body changed every day, it could be no bigger than a light-day across. In the case of quasars, they would be no more than a light-week or light-month in size. Some may be no bigger than our solar system. And yet, they are 100 times brighter than our entire galaxy.

The immense outputs of energy needed to account for this level of luminosity are very hard to credit. Many early theories attempted to prove that the mystery did not exist in the first place because quasars were not in fact distant objects at all. One idea was that quasars were extremely massive, with gravitational fields so strong that the light they emitted was somehow "pulled" toward the red end of the spectrum creating the appearance of a normal red shift. For a few years this theory seemed a workable answer to the dilemma. But before long, fundamental theoretical objections emerged ruling out the idea that gravity caused the red shift. Others, still insisting that quasars cannot be as distant as their red shifts indicate, have theorized that they are hot, luminous objects ejected from nearby galaxies.

many quasars. In the years that followed, many were located. Now, over 1,500 quasars have been discovered.

The more recently discovered quasars began to make 3C-273 seem a near neighbor. Many higher red shifts were found. In 1967 quasar PKS/0237.23 was discovered to have a huge red shift indicating a velocity of more than 80 percent of the speed of light (about 149,000 miles per second). Translated into distance, this quasar is nearly 9000 million light years from Earth. Recently, quasar PKS/2000-330 was found to have a red shift indicating a velocity of 90 percent of the speed of light (167,400 miles per second), making it, at 13 billion miles, the most distant visible object in the universe.

The mystery deepened when astronomers were able to make accurate assessments of the sizes of quasars. The key to determining their sizes came when it was found that quasars often vary in intensity over a period of weeks and months. This variability limits their size; in order for

Above: a chart of the radio sources from a small part of the sky above Cambridge, England. The sets of peaks on the chart represent a radio galaxy or a quasar. The large number of other small peaks – indicating faint signals – are caused by cosmic background radiation.

Right: the radio-quiet quasar BSO 1. Quasars have been categorized into two types – those with, and those without radio emission sources. According to the American astronomer Allan Sandage the radio quiet quasars are run-down, or dying, quasars. Because of their bluish color he called them Blue Stellar Objects (BSOs).

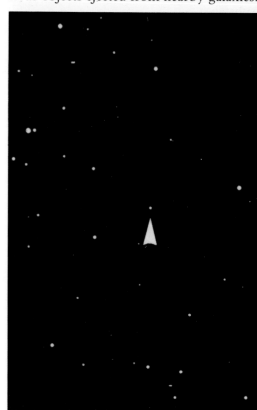

If this is the case, however, out of 1,500 quasars some would surely have been ejected toward us and would therefore exhibit blue shifts. No blue shifted quasars have been yet found.

Forced to accept that quasars were extremely remote and fantastically luminous, astronomers began devising some novel ideas to account for them. One ingenious suggestion was that a quasar was actually two ordinary galaxies lined up in such a way as to emit great luminosity to the Earth-bound observer. If the two are directly in line, one behind the other, when viewed from the Earth, light from the more distant galaxy might be bent by complex gravitational effects around the nearer one. Like a kind of gravitational lens, the bending might focus the light making it seem to the astronomer on Earth like the light from a single extremely bright body. However, for this to occur with 1,500 quasars is highly unlikely.

Another proposal is that quasars are the debris of some kind of galactic catastrophe. Some experts believe that huge explosions can occur affecting entire galaxies, similar to supernova among stars but far more violent. The idea is that instabilities sometimes build up within a galaxy leading to a sudden upheaval. Such a spasm would blow away much of the galaxy's substance leaving a small, dense remnant – a quasar. If this theory is correct, quasars are the galactic equivalent of neutron stars, the superdense remnants of supernova explosions. They represent, therefore, the final stage of a galaxy's life. The tremendous energy outputs of quasars suggest, however, that they are shortlived bodies which would mean that those observable from Earth became quasars quite recently. But the quasars we see are very old, in fact a few of them, at over 10 billion years, are the oldest things we can see, and what we see of them today happened long before the Sun and our solar system formed. Indeed, the quasars we observe now probably don't exist anymore.

The idea of discerning the past in quasars is the basis of the most widely held explanation of their nature: Rather than showing the end of a galaxy they illustrate the birth of one. Galaxies may have formed around black holes and the energy released from the accretion disk around the black hole would have been phenomenal, and would have varied as the hole drew in

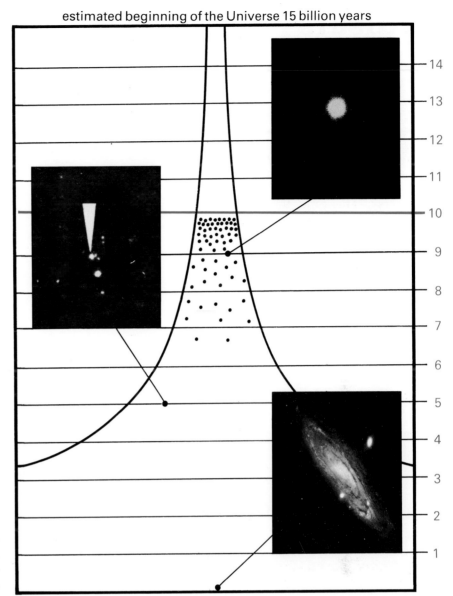

estimated beginning of the Universe 15 billion years

Above: The diagram shows how the distances of galaxies and quasars away from us places them in a time scale running back to the beginning of the universe, about 15 billion years ago. At the bottom is the Andromeda galaxy, about 1.5 million light-years away and the most distant object visible to the naked eye. In the center are the most remote galaxies so far measured, in the constellation Boötes. The brightest object in the cluster, indicated by the pointed marker in the "photo detail," is believed to be two galaxies in collision. One of the most distant objects so far detected is the quasar 3C 48 at the top of the diagram.

more, or less matter. Evidence that quasars are the cores of early galaxies was discovered recently when astronomers found halos of nebulous gas clouds surrounding quasars. It may be that quasars evolve into BL Lacertae objects, which in turn become giant elliptical galaxies. Similarly, Seyfert galaxies, also small and intensely luminous, may also have black holes at their core. Seyferts, however, are not as big as quasars and would evolve to become spiral galaxies rather than elliptical ones. Perhaps our own spiral galaxy, the Milky Way, was a Seyfert in its early days and has just calmed down as it matured.

Still, while evidence may point to quasars being the cores of early elliptical galaxies, the jury is still out, and they remain one of the great mysteries of the universe.

Pulsars

A few years after the dramatic discovery of quasars, radioastronomers were again the center of a storm of scientific controversy. This time, the fuss was about the detection of *pulsating stars*, called "pulsars" for short.

In July 1967, a small team led by British astronomer Antony Hewish were beginning research with a new radio-telescope sited just outside Cambridge, England. It consisted of an array of 2048 dipoles that covered an area of 4.5 acres. The new instrument was designed for a difficult and highly specialized task. The research team were intending to make the first detailed analysis of the scintillation of celestial radio sources, caused by the

Below: the first optical picture of a pulsar – taken at the Lick observatory in 1969. Below right shows the pulsar "on," below left the now invisible pulsar.

tenuous dust clouds in interstellar space. The phenomenon is similar to the familiar twinkling of stars viewed from Earth. This is caused by their light passing through the atmosphere. Planets do not twinkle only because their disks seen from Earth are too large. In just the same way, radioastronomers can tell that radio sources which twinkle are smaller than a particular critical size, determined by the properties of the interstellar dust clouds.

Almost as soon as they began operations the team was plagued by regular pulses of radio interference. Most radioastronomers are resigned to putting up with stray interference from local man-made sources such as radio and television stations, factory equipment, and even automobiles. But Hewish's team could not complete their observations until they had tracked down the cause of their problem and eliminated it. After systematically checking their equipment and all likely sources of interference in the vicinity, they began to suspect that their interference was a signal from space.

Examining the interference more closely, they realized they were picking up regular pulses of radio waves in bursts of about a thirtieth of a second, separated by periods of approximately one and a third seconds. Even more astonishing was the discovery that these uncannily regular pulses were coming from a fixed point in space. The signals were so regular – like the ticking of some inter-stellar clock – Hewish and his team seriously thought they may be receiving signals from another civilization circling a distant star. Soon some Cambridge astronomers had dubbed the first pulsar LGM 1, the letters standing for "Little Green Man!"

Close study of the signals, however, soon proved that they were not coming from a planet. If they were, their strength would fluctuate as the planet moved in its orbit around its parent star. The complete steadiness of the pulses suggested they were coming from some kind of stellar body, and therefore had a natural explanation.

As news of the strange discovery spread, radioastronomers round the world found it relatively easy to detect other pulsars. Forty were located in two years. Despite their success in finding pulsars, astronomers were at a loss to explain them. The early pulsars all had periods between signals of a few seconds and they were regular to an accuracy of about two millionths of a second. In astronomy a regular variation of this kind usually means that something is either pulsating or rotating. Both explanations have been put forward by those attempting to account for pulsars.

The idea of a physical pulsation in the stars seemed the most obvious explanation, but computer studies showed that none of the known types of stars could pulsate at the right frequency. Then British astronomer Thomas Gold suggested that a relatively small body rotating at great speed might create the kind of signals characteristic of pulsars. Gold showed that a neutron star – an immensely dense star made of matter collapsed into a core of solid neutrons – could produce regular bursts of radio signals. He calculated that because a neutron star is so small – perhaps no more than a few miles in diameter – it would spin very fast. As it did so it would leak

Above: part of the Cambridge Mullard Radio Astronomy Observatory's 4.5 acre radio telescope array. It consists of 2048 dipole aerials tuned to pick up radio signals at particular wavelengths. In 1967 a group of astronomers using the equipment detected mysterious regular pulses of radio energy, less than one tenth of a second long.

Below: the source of pulsar radiation is probably a rapidly rotating neutron star with a diameter of perhaps no more than a few miles. Scientists have shown that such a body would leak radiation at its poles and under certain circumstances give rise to the regular flashes of energy characterizing pulsars.

energy in the form of radio waves, emitted from the two poles of the star. Each time a pole pointed in the direction of Earth, we would observe a brief pulse of radiation.

Gold predicted that as the star lost energy it must gradually slow down. If Gold was right, therefore, the periods between pulsar signals should be gradually lengthening. Then in November 1960 American astronomers discovered a pulsar in the heart of the Crab nebula. Careful study showed that, as Gold had thought, the pulsar was slowing down. The rate of slowing was calculated as just over 36,000 millionths of a second each day.

The pulsar in the Crab nebula did more than verify Gold's theory. It became the first to be visually observed, its visible light fluctuating in exactly the same rhythm as its radio pulses.

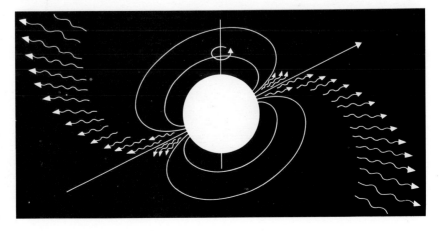

X-ray stars

The discovery of stars emitting powerful bursts of x-rays was one of the most remarkable accidents in modern astronomy. On June 18, 1962 NASA launched a rocket carrying x-ray detectors. The flight boosted the detectors above the obscuring layers of the Earth's atmosphere and pointed them at the Moon. The scientists hoped to detect x-rays produced by cosmic rays as they struck the lunar surface, and perhaps learn something from them about the structure of lunar rocks.

The experiment was a total failure. No lunar x-rays were found and none have

Below: a computer-generated correlation map, used to locate a strong x-ray source in the constellation of Sagittarius, near the center of our galaxy.

Below right: an x-ray detecting satellite in orbit. The SAS (small astronomy satellite) was launched in 1970 by an Italian space team from a mobile platform in East Africa. It carries sensitive instruments to measure x-ray sources both within and beyond our galaxy.

been detected since. But as the detectors scanned the sky around the Moon, an intense point source of x-rays was picked up in the constellation Scorpius. Now called Sco X-1, the source was soon providing astronomers with an astonishing new puzzle. No theories had ever predicted stars emitting more x-rays than our own Sun. But Sco X-1 was apparently producing millions of times as much. In fact, in terms of x-rays alone, it was emitting 100,000 times as much energy as the Sun emits across its entire spectrum of radiation.

The chance discovery of this x-ray powerhouse determined astronomers to look for more x-ray stars. And as soon as astronomers began to search for them they were detected without difficulty. Astronomers had still to explain how the radiation could be produced on such a vast scale. A major step forward came in 1966 when a joint research program involving Japanese and American astrono-

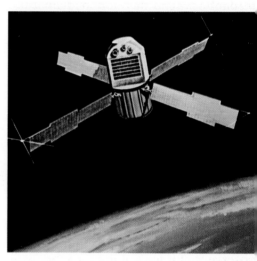

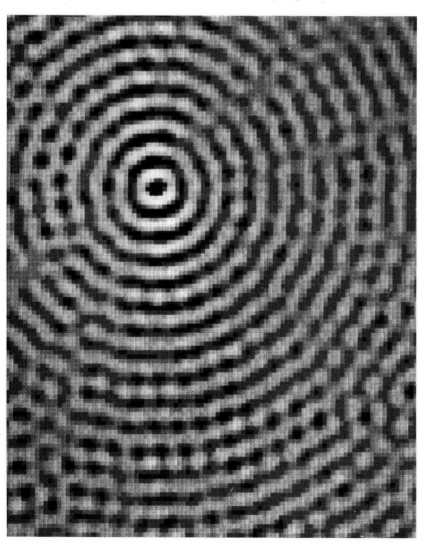

mers succeeded in identifying Sco X-1 with a visible star. This helped in two ways. First, optical astronomers could now use Earthbound telescopes to observe the star's behavior directly. Second, measurements of the spectrum of the light from the star could provide information about the processes that might be creating the powerful flux of x-rays.

Sco X-1 was discovered to be a far from normal star. It appeared to flicker rapidly and in an apparently random fashion. As a background to this rapid fluctuation, there was a longer-term flaring that caused the star to brighten dramatically, and then slowly fade to normal intensity.

What was happening to Sco X-1? Its wild variability pointed to some kind of unknown cataclysm. But what physical process could possibly produce such violent bursts of x-rays? One explanation now widely accepted suggests an awesome picture of stellar upheaval. According to the theory, x-ray sources consist of at least two stars caught in each other's gravitational fields. If one of the stars is sufficiently massive, its powerful gravitational attraction can draw gas from its companion star into its own interior. Astronomers have worked out that the captured gas would spiral downward into the star at an ever-increasing velocity. Simple physics demonstrates that an accelerating cloud of extremely hot gas will emit radiation, and calculations show that much of this would be in the form of x-rays.

Since the discovery of Sco X-1, orbital observations have made possible prolonged studies of x-ray sources throughout our galaxy. Many new x-ray stars have been located. Although their behavior still has some extremely puzzling aspects, the basic explanation of how they produce their radiation still appears to hold good. But recent evidence suggests that in some cases the massive star attracting the continuous stream of hot gas from its companion is in fact physically quite small. This means that the dominating star is probably a white dwarf or neutron star – superdense, and made of tightly compressed matter.

One system called Cyg X-1 presents an even stranger possibility. What we can see with our telescopes when we observe it is a normal star, a supergiant with a mass equivalent to about 20 of our Suns. Astronomers know the supergiant has a tiny companion. But calculations show that despite its small size, the companion is more than half as massive as the supergiant. This rules out a white dwarf or neutron star. The mysterious Cyg X-1 companion is far more dense than either. In 1973 astronomers worked out that it could be a black hole in space, the first ever detected. The idea is an awesome prospect. So dense that its gravitational field is too powerful for even light to escape it, the black hole would always remain invisible to our telescopes. But its irresistable gravitational attraction would steadily draw off

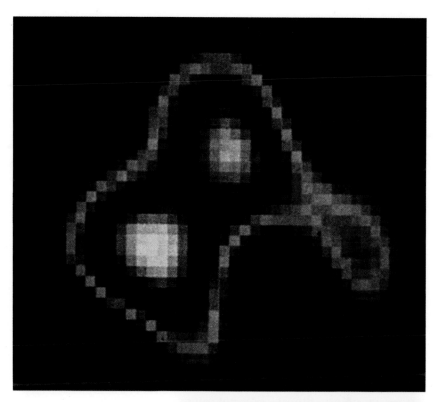

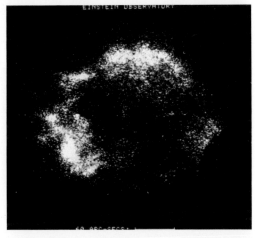

Above: an x-ray map of Cassiopeia A, the remnant of a supernova that occurred probably as recently as about 1700. The radio picture (right) shows the shell of expanding gas breaking up as it comes into contact with surrounding interstellar material.

vast quantities of hot gas from the nearby supergiant, generating powerful bursts of x-rays as it did so.

Since the idea was first proposed, the behavior of Cyg X-1 has been explained by supposing that instead of two stars at least three are involved. This would mean that in addition to the visible supergiant, there is also another normal star orbited by a neutron star.

Although the triple system could explain Cyg X-1, astronomers still do not discount the possibility of a black hole being present. The mystery will only be resolved when the technology is available to make a much more detailed analysis of Cyg X-1 and its huge outpourings of energy.

The mysteries of missing mass

In early 1985, two astronomers, Donald McCarthy Jr. and Frank J. Low announced they had detected a small, faintly luminous object near the star van Biesbrock 8 (VB8), 21 light-years from Earth in the constellation of Ophichus. The object, given the name van Biesbrock 8B, constitutes one of the most important discoveries in the history of astronomy, for it is the first time that something the size of a planet has been discovered outside our solar system.

VB8B is not a true planet. Although 10 percent smaller than Jupiter, it is 30 to 80 times as massive and is much hotter. It has been categorized as a *brown dwarf,* or failed star – an object almost, but not quite massive enough to generate the heat required for it to ignite and become a star. While providing invaluable information

Below: a photographic diagram of a neutron star binary system and the ripples in space-time it would create. Such dark, hard-to-see systems may account for some of the missing mass.

about the formation of our own solar system and the evolution of stars, VB8B may also be one piece in a vast, and mostly unfilled and unsolved jigsaw puzzle – the mystery of missing mass.

In 1933, astronomer Fritz Zwicky, while examining the Coma Berenices cluster of galaxies, noticed something strange. He had estimated a mass for the galaxies by the amount of light they emit. But then he observed the motion of the galaxies and reasoned that if they only had the mass he could see, they would be flying apart for there wasn't enough apparent mass to create the gravity needed to hold them together. He concluded there must be a lot of "dark matter" in the cluster that he couldn't see or photograph. It soon became apparent that Coma Berenices wasn't alone, and that missing mass was a universal phenomenon. In fact, it has been calculated that we can observe only 10 percent of the mass that is out there.

In truth, its not the mass that's missing, just the light from it – the mass is there, we just can't see it. What can it be? There are three main possibilities. It could be composed of conventionally dark mate-

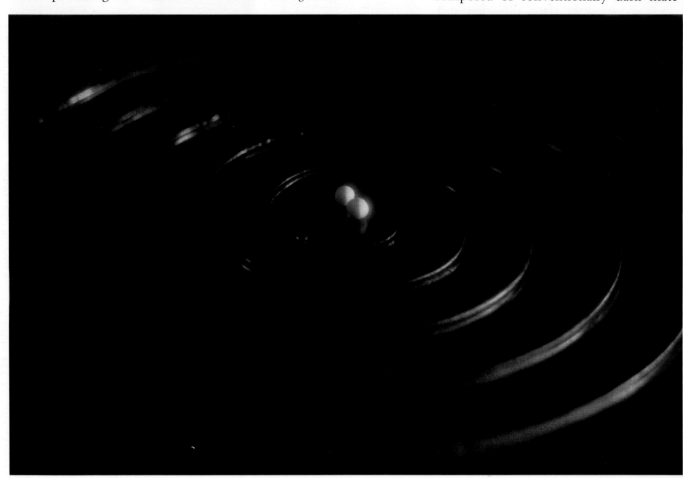

Left: an artist's concept of the Milky Way galaxy surrounded by a halo of neutrinos. Recent research indicates the supposedly mass-less neutrinos may have mass after all, which would account, to a large extent, for the missing mass.

ial, such as planets, clouds, debris and dark nebulae. A second possibility is that the missing mass is composed of black holes. Or, it could be composed of exotic cold, dark, subatomic particles.

Although the discovery of VB8B indicates that planetary formation may be quite common, it is highly unlikely that planets, along with debris, clouds and dark nebulae could account for 10 times the observable mass of the universe. Remember that the total mass of the planets in our solar system is miniscule compared to the mass of the Sun. As for black holes, that too seems an unlikely explanation of missing mass. Unless there are areas where black holes have gathered in anomalously high densities, which is highly improbable and would be easily detected, regular black holes are not good candidates for the missing mass.

But black holes are not all regular. There is the possibility of black holes that are neither mini, stellar, massive nor supermassive, but are somewhere in between. Such planetary mass black holes – or PMBHs as they are known – would be as small as an inch across. They too would be remnants of the Big Bang, and may have been the seeds around which galaxies formed. They might still be there, forming dark halos around the galaxies.

Most attention recently has been focused on the hypothesized exotic cold, dark subatomic particles. The first such particle to be considered is the neutrino, a tiny particle which does exist. The neutrino is infinitesimally small, flitting about at the speed of light and, it has been assumed, with no mass. But some suggest that the neutrino may have a tiny bit of mass after all. Furthermore, as there are 10 billion neutrinos for every atom in the universe, a neutrino would only have to have one-one-hundred-millionth the mass of a typical atom for the total mass of all the neutrinos combined to outweigh the entire universe.

Another exotic creature would be the magnetic monopole, a particle which defies common sense. When you cut a magnet in half you don't get two separate poles, but two new smaller magnets, both dipolar. However, monopoles may have been created by the Big Bang. In fact, they may have already been detected and their existence may be proven soon. Some theorize that monopoles would have mass. By one count, however, if they did have mass and are as prevalent as predicted, then there would be *too much* mass – indeed, the universe would have collapsed back in on itself after only 10,000 years.

Other fascinating particles are those postulated by the theory of supergravity which suggests that every massless particle has a superparticle that contains mass. The electron has the selectron; the photon, the photino; the graviton, the gravitino, etc. The difference between a neutrino with mass and a superparticle like a gravitino is one of energy. Even a massive neutrino would be too energetic to stick around to help gas clouds collapse to form galaxies, whereas the stodgier gravitino would remain. To confuse matters more, astronomers point to dwarf galaxies and note that even a gravitino would be too energetic to abet their formation. In order to solve the problem, supergravity theorists hypothesized a still lighter, even less energetic particle, the axion.

But no one knows if these particles really do exist, and so we are still a long way from discovering the solution to the mystery of missing mass. But one thing is certain – the day the answer is found will be one of the most momentous in the history of science. For when we discover the nature of the missing mass we will know far better how our universe began and evolved. We may also find out what its fate is, and whether it will expand forever, or whether there's enough missing mass to collapse the universe back in on itself to explode in another Big Bang and rise Phoenix-like as a new universe out of the ashes of our own.

Chapter 8
Man's Future in Space

Even before the first flights into space, people dreamed about the discoveries that were waiting for them. Today, earlier successes and rapid mastery over new technologies have encouraged a fast-growing interest in mankind's long-term future in space. The dreams are beginning to turn into reality. Already ambitious new projects are being discussed and planned – permanent space colonies in orbit around the Earth, astronomical observatories high above the atmosphere, expeditions to the planets, and even the first journeys to the stars.

Assuming society can survive social and economic upheavals on Earth, mankind's future may well extend over millions of years. It is scarcely possible to imagine what technological achievements might be commonplace in such a remote time. Supertechnologies capable of taming the energy of stars and even black holes will have evolved. Human civilization will have spread throughout the Milky Way and, with the emergence of fantastic new methods of travel, perhaps the first journeys will be made to other galaxies in the most distant regions of the universe.

Opposite: this illustration from a science fiction poster, *Disaster at Syntron*, depicts the sudden rupture of the wall of a torus-shaped space colony of the future. We are left to speculate what caused it – an incident in a galactic war, perhaps, or a meteorite impact.

The space shuttle

Shortly after dawn, on April 12, 1981, the space shuttle Columbia rose off the launching pad at the Kennedy Space Center. It was the first time in six years that Americans had gone into space. It was the twentieth anniversary, to the day, of Yuri Gagarin's historic first flight into space, and like that pioneer venture this too was a momentous occasion, for the launch of the shuttle signaled the beginning of man's future in space.

The success of the Apollo and Skylab missions was a sure demonstration that the United States had mastered the technical challenges of space flight. But the problem of justifying the huge costs of even simple missions had proved almost as difficult as developing the complex technology itself. Before more progress could be made in space some way of reducing costs was essential.

Modern space boosters are all "one-shot" vehicles. This means that once their individual rocket stages are burned out they are discarded, falling back into the ocean or breaking up high in the Earth's atmosphere. The precision-made rocket engines, their huge casings, and the complex electronic systems are lost forever. This throwaway approach to space flight is one of the main reasons why it has been so expensive up to now.

The obvious answer was to develop

Above: a 747 Shuttle Carrier Aircraft with the Orbiter 101 atop. The shuttle was developed as the first reuseable space transportation system. It has a huge central fuselage that is used as a cargo bay to take its payload into space. It is designed to be lifted into space with rocket boosters – a large external tank containing liquid oxygen and liquid hydrogen, and two smaller solid rocket boosters. The external tank and boosters are ejected before the shuttle goes into orbit. The 747 carrier is here being used during the extensive program of tests that preceded the first orbital flights.

reuseable launchers. In May 1972 the United States Congress voted funds to NASA for the development of the space shuttle, a reuseable orbital spacecraft capable of ferrying people and equipment into space cheaply and safely.

The space shuttle consists of three main elements: an orbiter spacecraft, a large external propellant tank, and twin solid fuel boosters. The orbiter is a hybrid with features common to both spacecraft and conventional aircraft. It is designed to operate for both orbital flight and maneuvers inside the Earth's atmosphere. It is therefore the first of NASA's space vehicles to have a distinctly aerodynamic shape like many orthodox high-performance aircraft. Inside the orbiter is a flight deck where the mission commander and pilot operate the craft; below the deck are the living quarters and room for passengers and the storage of equipment. Behind this on the orbiter is the large (15 x 60 feet) cargo bay which occupies much of the fuselage and is used to store satellites or other packages to be delivered to or retrieved from space. The orbiter can carry up to 65,000 pounds of payload in its cargo bay.

The orbiter has been developed by the same organization responsible for the Apollo mooncraft. The orbiter has an overall length of about 120 feet and a wingspan of 77 feet. It relies on three important propulsion systems. The most basic of these is called the space shuttle main engine or SSME for short. It consists of three liquid fuel rockets each providing over 370,000 pounds of thrust at lift off. The SSME burns for only the first eight minutes of flight and then is shut down for the rest of the mission. In many ways a conventional high thrust rocket engine, the SSME is remarkable because it can be used for 25 to 30 missions before needing to be replaced. The orbiters themselves are built to operate up to 100 missions before they are no longer considered space worthy.

In order to put the orbiter into its correct orbit and to make major course corrections including the vital maneuvers before reentry, the spacecraft also has a small engine system housed in its tail. Called the orbital maneuvering system, it consists of twin 600-pound-thrust liquid fuel engines. Minor course corrections, attitude control, and precision maneuvering is handled by the reaction control system – a bank of over 40 small thrusters mounted in the nose and tail.

The external tank of the shuttle is a large hydrogen-oxygen fuel tank which feeds the SSME during the first minutes of each launch. Over 150 feet long and 26 feet in diameter, it carries nearly 700 tons of fuel. In addition to supplying the SSME, the tank acts as a structural member linking the orbiter and its two solid fuel boosters. These twin boosters, attached to the sides of the external tank, burn in parallel with the SSME. Almost 50 feet in length, the twin boosters

separate from the external tank as soon as they burn out, usually at an altitude of about 25 miles. They are then parachuted back to a splashdown near waiting recovery vessels that tow them back to shore. Once rebuilt and checked, the boosters are ready for another launch.

The orbital flight sequence for the shuttle begins when its main engine and twin boosters burst into life on the launch pad. The vehicle lifts vertically off the pad and begins accelerating toward orbital velocity. The actual insertion into orbit is accomplished by the orbital

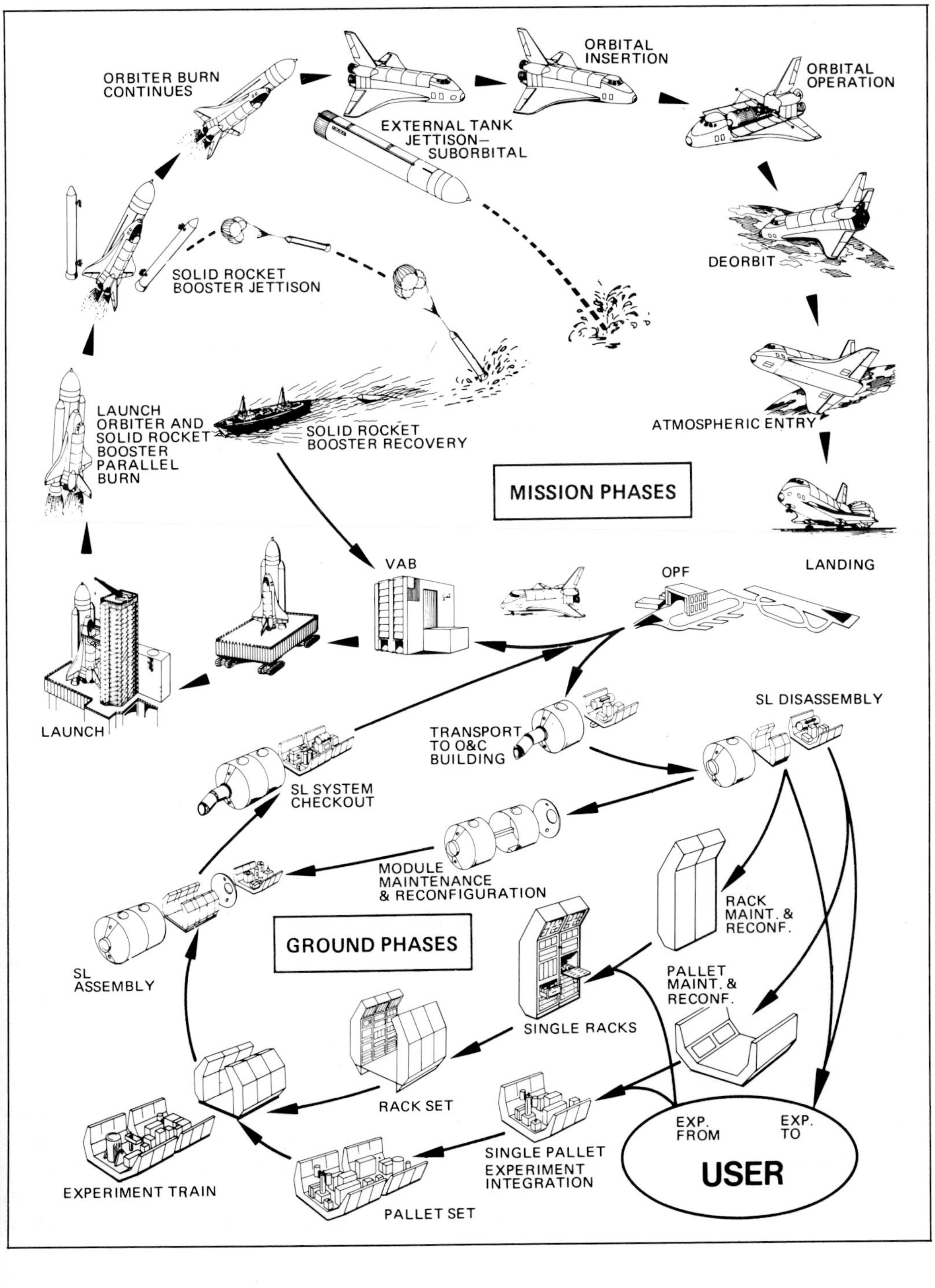

ORBITER BURN CONTINUES

ORBITAL INSERTION

EXTERNAL TANK JETTISON— SUBORBITAL

ORBITAL OPERATION

SOLID ROCKET BOOSTER JETTISON

DEORBIT

LAUNCH ORBITER AND SOLID ROCKET BOOSTER PARALLEL BURN

SOLID ROCKET BOOSTER RECOVERY

ATMOSPHERIC ENTRY

MISSION PHASES

VAB

OPF

LANDING

LAUNCH

TRANSPORT TO O&C BUILDING

SL DISASSEMBLY

SL SYSTEM CHECKOUT

MODULE MAINTENANCE & RECONFIGURATION

GROUND PHASES

RACK MAINT. & RECONF.

PALLET MAINT. & RECONF.

SL ASSEMBLY

SINGLE RACKS

RACK SET

SINGLE PALLET EXPERIMENT INTEGRATION

EXP. FROM

EXP. TO

EXPERIMENT TRAIN

PALLET SET

USER

Left: diagram showing mission and ground phases of the space shuttle transportation system. It shows the space shuttle orbiter being launched as a rocket, orbiting as a spacecraft and returning as a powerless aircraft to a runway landing. Because the cargo bay is so large, the shuttle can accommodate a wide variety of payloads such as Earth resource satellites, communication and weather satellites, and other "Earth-oriented" activities. The projected European spacelab is one such activity. The spacelab is developed on a modular basis, with its two principal components being a pressurized module that provides a laboratory with a normal Earth environment for working conditions, and an open pallet that exposes materials and equipment directly to space. The ground phases in the diagram show the assembly of the pallet segments through spacelab (SL) assembly, checkout, and Orbiter Processing Facility (OPF) at the Kennedy Space Center. From there it enters the Vehicle Assembly Building (VAB) from which a mobile launch platform transports it to the launch pad. After liftoff and before the orbiter reaches its set orbit, the solid rocket boosters (at 31 miles) and then the external tank (at 69 miles) are jettisoned. After orbital operations, the orbiter re-enters the Earth's atmosphere and lands. It is then deserviced and the payload removed.

Left: technicians inspecting the space shuttle main engine (SSME) at Rockwell International before delivery to NASA. It is part of a three-engine cluster located in the aft section of the orbiter's fuselage and provides the thrust in conjunction with the two solid rocket boosters.

Below: the Space Shuttle Transportation System shows (left to right) the space shuttle consisting of orbiter spacecraft with two solid rocket boosters for the launch, and an external tank housing the liquid fuel for the three 470,000-pound thrust main engines. The next illustration shows the solid rocket boosters at the moment of separation. They descend to a predetermined altitude where the nose fairing is jettisoned and the tanks are parachuted to Earth. They are recovered and reused. The next stage of the flight is the separation of the external tank from the orbiter. It splash-lands in the ocean and is not reused. When orbital operations are complete (fourth picture from left), orbiter fires its retro-rockets and reenters Earth's atmosphere. Special reusable surface insulation materials protect the external skins of the craft from atmospheric friction, which creates temperatures of up to 2300°F. The last illustration shows the orbiter descending toward touchdown much like a conventional aircraft.

maneuvering system which takes over from the SSME. Immediately after this vital step, the external tank is jettisoned, plunging to destruction in the Earth's atmosphere.

Once in orbit and on station, the commander is responsible for precise maneuvering while the pilot or mission specialist operates the cargo bay doors and payload handling systems. The chief handling system is a remote-controlled robot arm made in Canada and called the Canadarm. TV monitors in the cargo bay and on the robot arm itself allow the operator to see exactly what he or she is doing as the equipment the orbiter has carried into orbit is deployed.

When the mission is completed, the orbital maneuvering system slows the orbiter and starts its downward plunge into the atmosphere. As reentry begins the crew keep the vehicle in a nose-up attitude so that the full force of atmospheric friction is taken on the heat resistant (up to 2,300°F) tiles on the underside of the fuselage. Above, at an altitude of 75 miles, the orbiter handles like a spacecraft. But as it drops into the denser air below this, the orbiter's aerodynamic surfaces create enough lift for it to be flown almost like

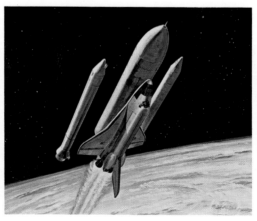

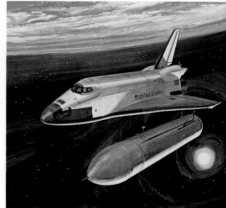

a conventional aircraft – although more like a glider than an airplane, for the orbiter descends without power.

Between 70,000 and 10,000 feet, the orbiter's speed is gradually reduced to 350 miles an hour. Once the commander and pilot start the final approach there is no turning back; without power the orbiter has only one shot at landing. As the landing sequence begins, the crew lowers the undercarriage and brings the shuttle down like an ordinary aircraft, although at the higher speed of about 208 miles an hour.

Since the first shuttle flight in 1981, there have been numerous missions. The goal of the shuttle is to be a cost effective, efficient, eventually even profitable method of launching objects into space. And so, another milestone was reached on November 11, 1982, when, with the fifth flight of Columbia, the space shuttle became operational, delivering two communications satellites (the Canadian Anik C-3 and the Satellite Business System SBS-C) into orbit. Other shuttle milestones include the maiden flights of the two other orbiters, Challenger and Discovery. The maiden Challenger flight, on June 18, 1983, was the occasion of another milestone – Dr. Sally K. Ride became the first American woman to go into space. Fittingly, aboard Discovery's initial flight, August 30, 1984, was Steven Hawley, Sally Ride's husband.

But all has not been easy for the shuttle program. In February 1984, two communications satellites launched from orbiter Challenger went awry and tumbled into useless orbits. Investors were nervous, and the insurers who had underwritten the satellites said they might no longer insure. Hence in November 1984, Discovery went on a salvage mission. Astronauts plucked the two satellites from

Above: giant test stands at Rockwell International's Santa Susana Field Laboratory. They are used for testing the space shuttle's main engines. Unlike previous rocket engines the SSME's will be reusable for up to 55 flights.

space and returned them to Earth to be used again. This use of the shuttle for service and retrieval had been demonstrated earlier in 1984 when astronauts aboard Challenger pulled in a failing NASA solar research satellite, fixed it in the cargo bay, and sent it back on its mission. No longer will malfunctioning satellites be lost in space.

But the shuttle is not used solely for commercial ventures. One of its primary duties is to carry scientific payloads, such as the joint NASA-ESA Spacelab and the Large Space Telescope into space. It has already been used as a platform to con-

duct a variety of experiments (some of them suggested by American high school students) on everything from how silkworms spin webs in zero gravity to the growth of crystals in space.

The shuttle program is also a military venture. The program was widely supported by the United States Department of Defense, in return for which the Department is guaranteed a certain number of flights each year. These missions, launched from both Kennedy Space Center and their own launch facilities at Vandenberg Air Force Base in California will carry defense communications satellites, early warning satellites and spy satellites into orbit. The first military shuttle mission was launched in January 1985, deliv-

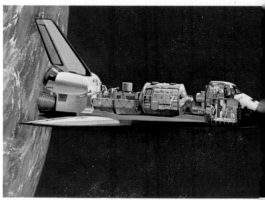

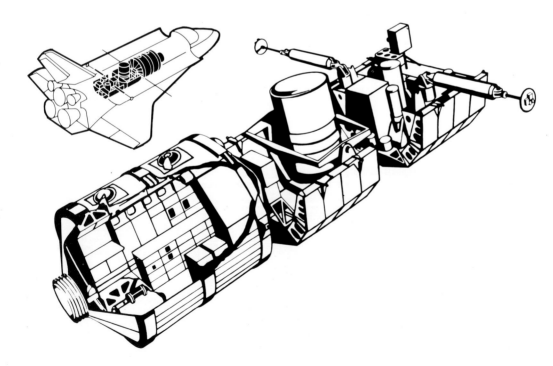

Above left: building the European-designed orbiting laboratory, spacelab. Sometime in the 1980s this laboratory module and instrument platform will be ready for operation.

Above: the orbiter shuttle with its cargo bays open exposing the spacelab.

Left: cutaway pictures of pallet segments of spacelab. On the right are the platforms for mounted experiments and equipment requiring direct access to space or uninterrupted or broad fields of view. On the left is the pressurized laboratory with instrumentation panels, a tunnel (extreme left) that gives access to the crew cabin, an optical window (top left of the module) and an airlock (top right of the module). The whole spacelab is integrated into the transporter, top left in the diagram.

ering the top secret Aquacade electronic spy satellite into space.

There have been many criticisms leveled at the space shuttle. It is behind schedule and over its $5.5 billion budget. Furthermore, as a means of getting things into space cheaply, it falls short of such expendable boosters as the ESA's Ariane. It has also been said that the cost of the program takes money from other space efforts, such as follow-ups to the Viking and Voyager missions. The sternest critics have been the Soviets, who claim the shuttle is designed purely for military purposes, and its commercial and scientific exploits are only a facade. Of course the Soviets are hard at work on their own shuttle, the Kosmolyot, a large delta-wing vehicle scheduled for service in the late 1980s.

The space shuttle truly marks the beginning of man's future in space, for it offers a simple and relatively inexpensive way to get to and from orbit, and, in terms of space ventures, that is half the battle. Once in the weightless environment, space stations, like the one NASA plans to establish in the 1990s, can be built with ease. Inside the stations, zero-gravity conditions can be exploited to create ultra-pure medicines, smaller and faster computer

chips and even perfectly round ball bearings. These space stations will also be used as home base for inter-orbital space tugs which will service satellites in the higher orbits. Most exciting is that space stations will be the launching pads for the manned missions to other planets planned for the next century.

Right: the space shuttle releasing its payload – in this case, an Earth-orbiting satellite.

Below: because of the zero gravity of orbital flight and the hard vacuum of space, the shuttle will make possible a whole range of new technological processes. Routine access to space also means that the shuttle can permit the construction of large and complex structures in space (as in the illustration here) that will be able to turn out products that cannot be produced on Earth.

Astronomy in space

In 1608 the Dutchman Hans Lippershey revolutionized astronomy by inventing the telescope. For the first time people could examine the heavens in detail. Important discoveries were quick to follow. In 1610 Galileo discovered mountains on the Moon and 50 years later Christian Huygen observed the rings of Saturn. In 1781 William Herschel for the first time spotted Uranus, one of the outermost planets of the solar system. Then in the late 1800s Giovanni Schiaparelli thought he saw a regular system of markings criss-crossing the surface of Mars. It looked as though he had discovered evidence of an alien civilization. It has become clear, however, that he had been deceived by a freak atmospheric effect.

The experience of Schiaparelli is a clear example of the dangers of relying on observations made from the surface of the Earth. The problem that misled Schiaparelli has faced astronomers for centuries. No matter how good their instruments, every observation they undertake is made through an envelope of

Below: Copernicus, the NASA orbiting astronomical observatory launched in August 1972. It is designed to conduct detailed studies of the composition of interstellar space.

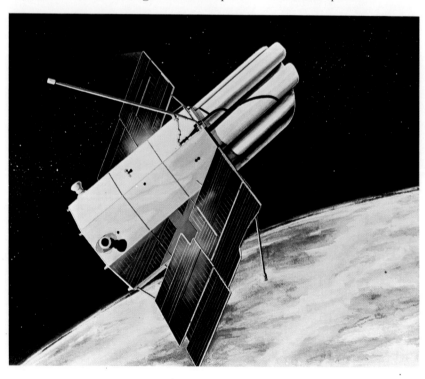

gas and dust about 200 miles thick. The Earth's atmosphere is an ever-present barrier to perfect observations, bending and distorting the light from the planets and stars. The familiar twinkling of the stars is an everyday example of how the movements of dust in the atmosphere breaks up the otherwise steady light that would reach Earth.

The age of space flight has at last provided an opportunity denied to past generations of observers. For the first time astronomical instruments – and soon even astronomers themselves – can be carried high above the atmosphere. Already the first results of space-born astronomy have led to dramatic advances in knowledge. As if a veil had been removed from astronomers' eyes, the universe is being revealed in hitherto unimagined detail. Astronomy in space represents a scientific breakthrough as significant as the invention of the telescope itself.

The first major program of astronomical observations from orbit was initiated by NASA in the United States in 1959. Work was started on a series of four ambitious orbiting astronomical observatories, called OAOs for short. Equipped with 11 optical telescopes and two scanning spectrometers to study ultraviolet radiation – a type of electromagnetic radiation normally blocked by the atmosphere – each OAO was able to send back in a matter of days more data than earthbound astronomers could accumulate in years. Only two out of the four OAOs were successful. OAO-2, orbited in 1968, studied nearly 2000 different celestial objects while making a total of 22,560 separate observations. Among its most important discoveries were mysterious ultraviolet emissions from Jupiter and the unexpected fact that many of the younger stars in the Milky Way are evidently far hotter than had ever been suspected. OAO-3 made some equally fundamental discoveries, revealing for the first time powerful sources of X-rays and ultraviolet radiation at vast distances from the Earth.

Since the pioneering days of the OAOs, more exciting observations have been made from orbit by sophisticated telescopes on board space stations such as Skylab and Salyut. These are merely the forerunners of a new generation of space-born instruments. Scheduled for launch aboard the space shuttle on August 13, 1986, is NASA's Large Space Telescope. At the heart of the LST will be its enormous 8-foot diameter astronomical mirror, which, to prevent any distortions or cracks when it was being made, was cooled slowly over a period of three years. This telescope will be so precise that were it set on the top of the World Trade Center in New York City, it could fix on a dime on top of the Washington Monument in Washington, D.C., The orbiting of this telescope will mark another revolution in the history of astronomy for it will be able to resolve planetary systems around nearby stars, and will allow astronomers to see deeper into space – and farther back into time – than ever before.

The most spectacular breakthroughs will come when the first permanently staffed observatories are in orbit above the Earth or even based on the surface of the Moon. With powerful optical and radio telescopes, the most remote regions will be studied in unprecedented detail. Nearer home, the planets of the solar system will be closely studied without the trouble of sending probes or spacecraft with crews. It will be a matter of routine to spot objects on the surface of Mars and it may even be possible to map the outer planets – Uranus, Neptune, and Pluto. Perhaps the most exciting

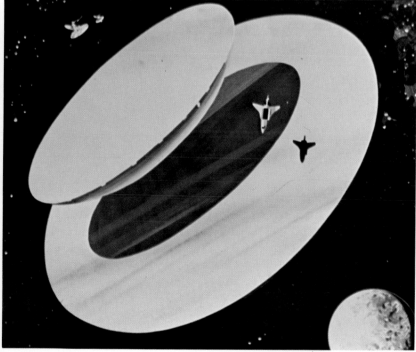

prospect of all is the real possibility space astronomy offers of picking up any detectable signals from another civilization, perhaps thousands of light-years away.

Above: an artist's impression of the huge space telescope being designed and built for NASA. It will be 43 feet long and 14 feet in diameter and will be put into orbit by the space shuttle in late 1983. The telescope is part of a Support Systems Module, which will include other optical and scientific instruments. The telescope-module will orbit some 300 miles above the Earth. It will be able to see distant objects with extraordinary clarity and might even be able to detect the planets of other solar systems.

Left: an impression of a "cup-and-saucer" space telescope of the future. The "cup" is the radio reflector, the "saucer" is a shield against radio interference from Earth. The objects at the top left are radio feeds, and a space shuttle has been included to indicate scale. Free from the obscuring effect of the Earth's atmosphere, such giant space telescopes will be able to resolve even the faintest and most remote sources of light and radiation.

Energy from space

One of the most vital resources waiting in space for mankind is raw energy. Far from being empty, the vacuum of space is constantly bathed in a powerful flux of electromagnetic waves, mainly ultraviolet and infrared radiation and visible light. In the vicinity of the Earth the source of most of this radiation is the Sun, and the idea of trying to convert it into useful energy is already being put into practice. The first prototype of solar-powered homes have been built using the energy of the Sun to heat and light them. Solar cells have been designed to power pocket calculators and wristwatches. In space, panels of them are being used in spacecraft and satellites.

It is in space that the real future of solar power lies. High above the Earth there is no day or night to limit the hours during which power can be generated. Neither is there any atmosphere to absorb and filter the radiation. With the full intensity of the Sun's rays perpetually available, space is the perfect solar energy factory.

Space engineers are already familiar with the basic technology of converting sunlight into electricity. But even space laboratories such as the United States' Skylab and the Soviet Union's Salyut cannot rely exclusively on panels of solar cells, deployed like wings, to generate enough electricity to run all their systems. Until now solar energy has been utilized on a very small scale, and the dramatic possibilities for the future depend not on a fundamental new technology with which to generate power but on a breakthrough in the scale on which solar generators are used.

The large-scale generation of solar energy in space will create a revolution in power supply – and it will probably come at a time when it is desperately needed. Already the resources of the Earth are dwindling with fossil fuels such as oil, coal, and gas running out. If the present rate of technological development is to continue, it is essential to discover new sources of energy. Atomic power plants are a part of the answer for the future, but as long as they rely on nuclear fission reactions they will always remain dangerous sources of radioactive pollution. The prospect of a cheap, safe, and plentiful supply of energy from space is of overwhelming importance for the future of mankind.

Even if space is an ideal place to generate power, how can that power be brought to Earth and distributed to the

Below right: a solar-thermal test facility at Sandia, New Mexico.

Below: Skylab; the windmill-like panels of solar cells collect the Sun's energy to provide on-board electricity.

homes and factories where it is needed? The answer is not only simple but, with recent technological advances, it is also already practical. The link between a power station in space and the Earth will be a beam of radiation known as microwaves. With a wavelength intermediate between infrared and radio waves, microwaves are a kind of heat radiation that

can be transmitted as a tightly focused beam over great distances. Recent breakthroughs have made it fairly easy to convert ordinary electrical energy into microwaves and then back into electricity. This means that electricity produced in orbit could be changed into microwaves and beamed to a ground station on Earth. The beam would be kept at a fairly low strength to make it harmless to any living creatures encountering it. Birds, for example, might experience a slight sensation of warmth but would suffer no ill effects. The ground station would turn the microwaves back into electricity and use conventional techniques to distribute the power to major centers of population.

In 1968 the American engineer Peter Glaser produced detailed plans for the world's first power generation satellite, or powersat. Like all other space vehicles, Glaser's satellite was designed with wings made up of panels of solar cells. The major difference is size. Glaser's plan, which has not yet been put into effect,

Above: a United States Army solar furnace consisting mainly of a heliostat (right) and a concentrator (left). The heliostat is a 40 foot by 36 foot framework of 355 mirrors that can be angled to any position to catch the Sun's rays and reflect the solar radiation onto the concentrator. The concentrator's 180 concave mirrors all focus this light to a single point within the test chamber, where the extremely high temperatures that can be reached are used to perform heat tests on materials under study.

is to build a pair of wings each about three miles across. The entire powersat with its giant panels of cells would be about seven miles from wing tip to wing tip and would weigh over 11,000 tons.

The wings themselves would not be completely covered by solar cells. Instead, mirrors built into them would focus the sunlight onto the cells and greatly increase the amount of electricity generated. Between the two wings, a 3000-foot diameter transmitting antenna would be deployed. Converters in the antenna would turn the solar electricity into microwaves which could then be beamed back to Earth.

Glaser knew that if his proposal were to be practical, an extremely accurate means of targeting the microwaves would have to be used. He suggested that a small part of the microwave energy be reflected from the ground station back to the powersat. This could then be used as a reference beam. Linked to attitude control systems in the satellite, the reference

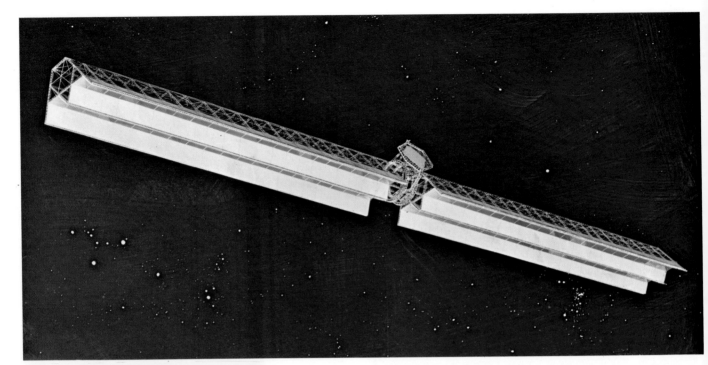

beam would make it possible to lock the microwaves onto the ground station. By placing the powersat into a geosynchronous orbit, engineers could insure that the vehicle was kept permanently over head so that the rotation of the Earth would not sweep the receivers out of line with the beam.

The main drawback of Glaser's scheme is that solar cells are not yet cheap and efficient enough to make such a powersat commercially viable. If built today, the electricity it produced would cost hundreds of times more than conventionally generated power. However, the search for low-cost cells is already underway and some engineers believe that a major breakthrough may be no more than a few years away.

But power from space does not have to rely on solar cells. Another American engineer, Gordon R. Woodcock, has shown how the Sun's energy can be used to generate electricity by familiar technology such as turbines and dynamos. Woodcock proposes to build a huge solar furnace in space. The Sun's radiation, focused by aluminum-coated films of plastic floating in space, would be used to heat a reservoir of helium gas. As the gas expanded it would turn an ordinary turbine, generating electricity in much the same way as a power station on the ground. The electricity would then be transmitted back to Earth using a system

Above: an artist's concept of a solar power satellite (powerstat) system. The satellite is equipped with large solar cell arrays that collect light energy from the Sun and convert it to electricity. This is then converted to microwaves and beamed through a transmitting antenna on the satellite to a receiving antenna on the ground.

Left: a design for a solar power satellite conceived by American engineer Peter Glaser. The huge device has vast wing-like panels of solar cells. The panels are also covered partly with mirrors, which for added efficiency concentrate the sunlight onto the electricity producing cells.

Opposite top: a powersat design by the American engineer Gordon R Woodcock. His method uses arrays of thin plastic film coated with aluminum which reflects sunlight onto a spot that becomes heated to several thousand degrees. Helium gas flows through the "hot spot" area and carries the heat energy to drive turbines that are hooked to generators.

Right: an artist's impression of a solar power satellite system being constructed in space some 36,000 miles above Earth.

of microwave converters.

In either of these two approaches, the technical problems of building powersats are enormous. Fabricating them on Earth and launching them ready-made into orbit one by one will be a prohibitively expensive process, beyond the capabilities of even the most powerful of present-day boosters. The only viable answer is to build them in orbit, and then the first step will be to construct assembly facilities in space – a major undertaking in itself. The most economic solution may well be to establish permanent colonies of engineers and construction workers in orbit. A large work force on a massive space-born construction shack could dedicate itself to a lengthy program of powersat development. If such a mammoth project were started today, a cheap and limitless supply of energy might be in use before the end of this century.

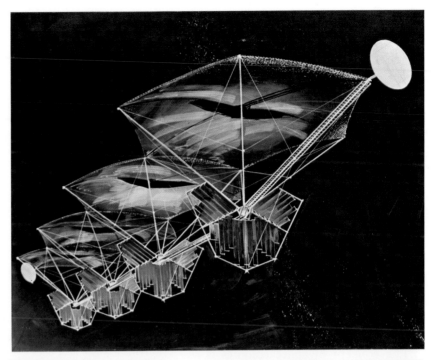

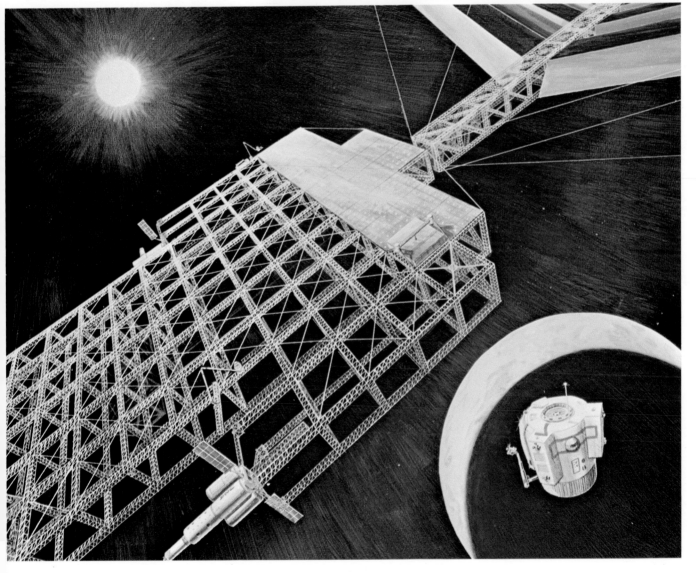

Colonies in space

By the middle of the 1990s the first permanent space stations may well be in orbit above the Earth. Serviced regularly by space shuttles and remote-control supply tankers, these large vehicles will be used as complex orbiting laboratories on which scientists can make a unique range of astronomical observations and conduct experiments under weightless conditions. Such stations will be a starting point for a bold future development that will create for people a permanent home in space.

The concept of space colonies is the brainchild of the American physicist Gerard K. O'Neill. Working at Princeton University, O'Neill led a number of study projects aimed at findings ways of solving world problems through technology. One

Right: the interior of a "Bernal sphere" space colony. It is named for the American writer J. D. Bernal, who in 1929 described a cosmic-ark method of interstellar flight. Generation after generation of people would "pilot" the ark until their remote descendants reached their destination star. In this cutaway view of such a sphere, as many as 10,000 people could live and work in the central region and there is a separate area exposed to the intense sunlight where crops would be grown.

Opposite below right: the exterior of a Bernal sphere type colony. The colony lives in the central sphere. The coiled structure is where crops are grown and the disks at either end are to radiate waste heat back into space.

Below: a space colony shaped like a cylinder. Such a colony would be a vast structure, housing hundreds of square miles of land.

of the basic problems he tried to answer was whether the surface of a planet was really the best place for an expanding technological civilization. Because people previously had not even imagined there could be an alternative, the question had never been asked seriously. Even when a few visionaries had speculated how mankind might expand its civilization beyond the Earth, their ideas invariably involved the setting up of outposts on other worlds.

O'Neill realized the drawbacks of trying to colonize alien planets. In this solar system, few of the planets could be considered likely for colonies. Those that do hold some promise – the Moon, Mars, and some of the satellites of Jupiter and Saturn – offer relatively little room for expansion, perhaps only a few times the surface area of the Earth.

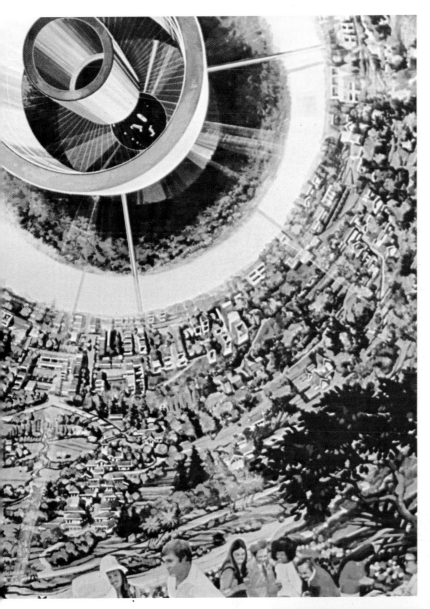

easily supply the power needs of a colony consisting of up to 10,000 people.

To build such a massive structure at a bearable cost, the starting point would be modified space stations known as construction shacks. The American engineer Gerald W. Driggers of the Southern Research Institute in Alabama has already made concrete proposals for these. The construction shack he envisages would accommodate about 2000 workers in cylindrical living modules. The shack would consist of a central sphere about 300 feet across in which most of the basic construction work took place. Below the sphere and mounted on spokelike arms are the clusters of living modules. The entire structure would rotate slowly to create an artificial gravity for the workers.

To build a space colony, at least two construction shacks will be needed – one to extract metals from their ores and the other to fabricate the parts from which the colony will be built. The ore used will be mined and sent from the lunar surface.

The problems of extracting the required metals from lunar ores will be formidable, but this will be more than offset by the overwhelming advantage of working in orbit where solar energy is cheap and plentiful. In fact, it is the energy supply alone that makes the operation viable. A secondary but important advantage is that moon rocks are known to contain a lot of oxygen that can

O'Neill decided on an entirely different approach to finding room for an expanding civilization. Instead of planning colonies on other planets, he examined the feasibility of establishing them in space itself. He soon realized that the idea was not only practical but also held many positive benefits.

By the early 1970s the first design concepts emerged. O'Neill favored a large cylindrical structure one mile long and 600 feet in diameter to house the space colony. One of the most exciting features of the proposal was O'Neill's idea for a free and inexhaustible supply of energy. He envisaged huge externally mounted mirrors oriented so that they focused sunlight onto banks of solar cells. With no day and night to restrict the generation of electricity, a system of this kind could

be released during the process of extracting metals. Enough will be produced to supply the needs of the entire 2000-strong work force.

Once the problems of fabricating the building materials are overcome, the task of constructing the colony will seem simple. In the weightlessness of orbit even large structural girders can be handled easily by small robot machines or individual workers. Most of the work will probably be done by remotely controlled units linked to a central bank of computers. Guided by a complex program of computer commands, the robots will maneuver structural members into position and weld them into place.

The construction shacks will not be pleasant places in which to live. Quarters will be cramped, and despite leisure and recreational facilities, long periods on board will place workers under considerable psychological strain. There will be little sense of romance or adventure. The crews will face long shifts of grueling repetitive work. Those who will think of the great shacks orbiting high above the Earth as little more than factories where a good wage can be earned will undoubtedly return to Earth. But a few will certainly commit their lives to living in space and will be among the first people to inherit the colony they helped to build.

What will life in the colony be like? Most experts believe that it is important to make the environment as comfortable and as much like Earth as possible. The inhabitants will not merely be living on board for a few months or even years. Most will spend their lives in the colony. One basic requirement is to create some kind of artificial gravity. The experience gained in the early years of space flight shows that during extended periods of weightlessness, people lose calcium from their bones. This process can seriously weaken the bone structure. All that is needed is to set the colony rotating around a central axis. O'Neill worked out that to create the impression of normal Earth gravity, the cylindrical colony would need to spin at about two rotations a minute.

Artificial gravity produced in this way has one serious drawback. A side effect of the spinning motion is a phenomenon scientists call the Coriolis effect, a tendency for an object to change speed as it goes inward or outward from the axis

Below: a large-scale facility – possibly a future space colony – being constructed in space. As much of the work involves routine, repetitive operations, special automated machines will be used like this assembler, which is linked to a small computer programmed to guide the assembler through repetitive operations. Human workers will occupy the roles of supervisors. Huge structural sections that on Earth would be extremely heavy and involve careful lifting, balancing, and swinging into place, can be lifted and maneuvered by quite small machines or even by individual workers. Steeplejacks will not have to cross high girders with care, merely launch themselves off in the direction they want to go. The workers themselves will live in construction shacks in space – which, while not as spacious or luxurious as the colonies they will be helping to build, will have opportunities for relaxation such as microfiche libraries, videotape centers, and television from Earth.

around which a body is rotating. As far as the colonists are concerned, too much Coriolis effect could make life very unpleasant. Besides distorting the trajectories of any objects thrown into the air inside the colony, the Coriolis effect is known to cause dizziness and motion sickness.

A less harmful characteristic of this kind of artificial gravity is that its strength would vary. Approaching the axis around which the colony was spinning, gravity would fall toward zero. Moving away, it would rise to normal Earth level.

The need to keep the spin rate as low as possible so that Coriolis effects are minimized has led to an even better basic design than O'Neill's cylinder. A team in California worked out a plan for a colony which could enjoy Earthlike gravity at a rate of only one revolution a minute. They rejected a cylindrical shape in favor of a torus – a shape rather like the inner tube of a car tire. The torus they proposed would be made of a body 400 feet wide bent into a circle four miles in diameter.

At the center of the torus and connected to it by spokes they envisaged a 400-foot diameter hub to house docking facilities for spacecraft. Inside the hub the gravity will be about one sixteenth of its value on Earth. This will make the hub an ideal place for swimming pools, gymnasiums, and any recreational facilities that benefit from a low gravity field. The hub will also become the crossroads of the colony. Instead of traveling around the torus itself, people will be able to cut

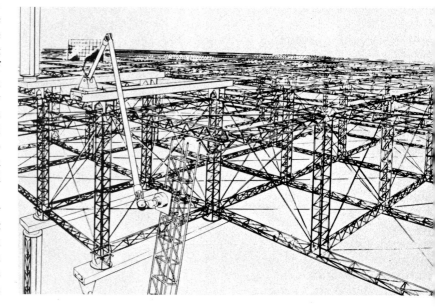

across its diameter by taking trips through the connecting spokes to reach their various destinations.

To supply electrical power, a huge mirror could be floated above the torus. Small thrusters attached to it would automatically adjust the mirror's attitude to keep it focusing the Sun's rays directly onto the colony's solar cell banks. Some of the sunshine might also be reflected by a secondary mirror system into the torus itself to make the environment more cheerful.

Would the Earth people living in space colonies eventually become Space people? That seems inevitable if the chain of events progresses as many envision it. For, as the years passed, a generation of space colonists would grow up knowing no other environment. The industry which will become the colonists' most profitable resource is solar power generation. The limitless energy of solar radiation will not only serve the colony's needs but will provide huge surpluses to the energy-hungry nations on Earth. As power demands grow, the power stations of space could take over from Earth's dwindling supplies of fossil fuel. Once this happens, the colonists will be able to wield immense political power. Initially the nations that built the colonies would receive all the energy they needed, but at some stage the colonists might declare their independence of Earth and create the first space nation. Eventually they might be looked upon in

Above: a segment of a torus-shaped space colony in the final stages of construction. The 150-yard internal diameter colony is visible through the 100-foot long strip windows encircling the wheel-like structure. The view here is of an agricultural area, and takes in a lake and a river. The louvers, shown being installed, would absorb cosmic radiation while allowing sunlight to be reflected inside. Artificial gravity would be provided to its 10,000 inhabitants by a 1 rpm rotation of the torus.

Right: the interior of a future torus-shaped space colony. At the top of the cylinder are chevron shields – mirrors that reflect the Sun's light into the colony through a series of baffles that screen out harmful cosmic rays. A typical housing complex can be seen in the top deck. There will also be some spaces for parks. The track that is visible is for part of a transit system. The second deck houses a service area.

the same way as the oil-rich nations of the world today are regarded by countries battling against a desperate energy crisis.

As increasing numbers of space colonies emerge, their populations will grow. O'Neill has already worked out plans for giant colonies housing millions of people. Eventually as congestion of space lanes became a real problem, the first colonies might be placed in orbit around the Moon or perhaps even the more distant planets. But an even more remarkable possibility exists in the far distant future. It is just possible that when the solar system is crowded with colonies, a decision may be made to fit some of them with massive engines so that they can leave their orbits and travel freely in space. Perhaps the boldest colonists would leave the solar

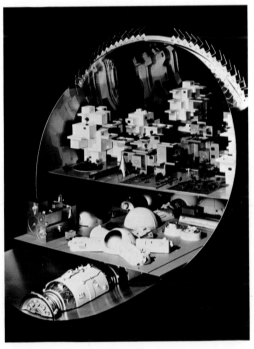

system and set a course for the nearest stars. They may even become completely itinerant, drifting forever through the interstellar gulfs without ever needing or desiring a home star.

Ultimately this strange artificial world may become a commonplace feature of our part of the galaxy. And while planet-tied communities may continue to emerge as people expand their empire from star to star, the free flying space colonies may come to be regarded as cosmic gypsies – forever on the move and acknowledging the authority of no world other than their own.

Space war

On February 28, 1959, the American military mission in space began. On that day a Discoverer spacecraft was launched into orbit from Vandenberg Air Force Base in California. What few people knew at the time was that the innocuous looking little Discoverer capsule was designed to carry cameras that could take high-resolution photographs of the Soviet Union from space.

Spy satellites have come a long way since those early days. Now both the United States and the Soviet Union, and even China, have cameras in orbit. The premier American spy satellite is the CIA-built KH-II, a phenomenal craft, equipped with infrared sensors which can see at night, an imaging radar system which can take pictures through clouds, and regular optical telescopes which can pick out something the size of a hand grenade from 100 miles

Below: *Battle Stations*, by British artist David Hardy depicts a conflict between interstellar space colonies. A giant spherical space colony has arrived in the vicinity of a planet. The inhabitants of the planets are hostile – they have possibly interpreted the very arrival of the colony as a menacing act – and have sent out space battleships to destroy it. The colonists are not unprepared, however, and have already destroyed one of the rockets with laser cannon.

in space. And now beginning testing is the ultimate spy satellite – HALO (High Altitude Large Optics), a mammouth space platform which will orbit at the geosynchronous altitude of 22,300 miles. But the military use of space is not just limited to the dozens of spy satellites now in use. There are high-speed communications satellites such as the American MIL-STAR, and military navigational systems such as the American Global Positioning System, which will allow an airplane, ship or even missile to determine its own position to within fifty feet. But the most important military objects in space are the early warning satellites, like the American Defense Support System satellites which use highly sensitive infrared detectors to register the launch of enemy missiles by the intense heat given off by their rocket plumes as they are fired.

Such watchdog satellites would be prime targets in a time of war, for if you could destroy your opponent's advance warning system first, you might be able to sneak in a devastating first strike. In 1968 the Soviet Union began testing a "hunter-killer" satellite. Called the Fractional Orbital Bombardment System (FOBS), the technique involves launching a hunter-killer

satellite into the same orbit as the target satellite. The hunter-killer maneuvers to within fifty feet of the target and is then detonated from the ground, obliterating itself as well as the target. While the Soviets continue FOBS research, the Americans are pursuing a different kind of ASAT, known as PMALS – Prototype Miniature Air Launch System. With PMALS, an F-15 fighter plane takes off and flies up to 50,000 feet where it launches a rocket carried under its wing, firing a basketball-sized payload into space. This little payload is covered with dozens of detectors that home in on the target, and jets that guide its course. When it's locked in on the target satellite it just rams into it. It doesn't need to explode, it just has to hit the target; something the size of a basketball and traveling 20 times faster than a bullet will do sufficient damage.

Both FOBS and PMALS have been tested successfully. But they mark only the beginning of plans for war in space. In March, 1983, United States President Ronald Reagan made the now famous "Stars Wars" speech, in which he proposed that America embark on an ambitious program to establish a protective shield in space against incoming missiles. Although there has been intense debate over "Star Wars," as of this writing it appears that the United States is proceeding with research into a three-tiered Strategic Defense Initiative. The first two tiers are the "Star Wars," for they involve destroying missiles as they emerge from the atmosphere and travel through space. The third tier would use conventional anti-ballistic missile defenses to knock out the missiles as they fell to their targets.

There are two main components to the "Star Wars" phases. First is the method for detecting, tracking and locking onto the missiles. The technology already exists for tracking a missile while it is in its burn phase – the period where the rockets are firing. But after the engines cut out and the missile is just arcing through its trajectory like a bullet, it is relatively cool and hard to track with infrared detectors. This is where the technology being developed by the American Defense Advanced Research Projects Agency comes into play. The project is known as Talon Gold, and when operational it will be able to track the still warm missile against the cool background

Right: a high-powered ruby laser pierces a hole in a sheet of steel.

Below: an experiment at the United States Army research and engineering establishment at Natick, Massachusets shows how concentrated rays of solar energy can be made to generate temperatures of nearly 5000°F. *A* shows the solid steel beam before exposure to the concentrated ray of solar radiation. *B* shows the four-inch ray – concentrated from 180 mirrors – flaming its way through the tough metal. *C* shows the elliptical hole that has been burned in the steel beam.

of space. Once the missile has been targeted it can be destroyed; the methods of destruction being discussed are reminiscent of the "death rays" of science fiction of years past.

The idea is to use some kind of high-energy beam that can burn up a missile within a fraction of a second. There are currently three candidates for such a beam. First is the common chemical laser, a beam of light propagated in a gas (such as helium or argon), which is so concentrated that it can burn a hole through a brick in seconds. But the time it takes for such a beam to destroy missiles may be too long. The second, and most likely candidate, is the x-ray laser. An x-ray laser satellite

Left: the electron accelerator (atom smasher) at Stanford, California. The accelerator is used for research on elementary-particle physics, but physicists have known for decades that the intense beam of charged or neutral particles such machines can produce are potentially extremely destructive. Research is now going on both in the USA and, it is believed, in the USSR to produce a new class of weapons based on accelerators.

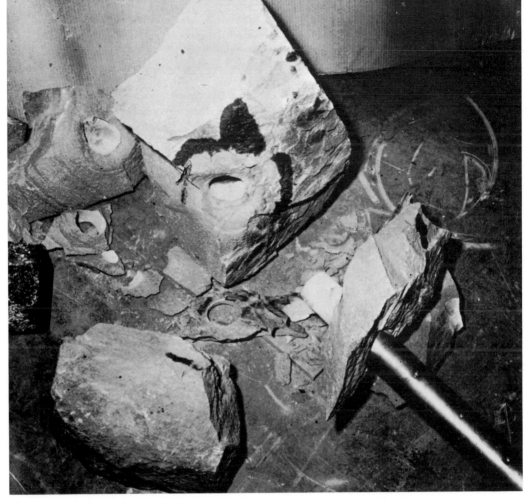

Right: a two-foot cube of rock weighing half a ton split by the action of a gun that fires microwave energy. The rays of microwave energy split the target by causing the moisture content however small, to boil and expand violently. Already scientists are looking into possible military applications

Below: an Astron magnetic fusion experiment. The device is designed for research into harnessing the energy of nuclear fusion. Its military applications are similar to those of Stanford's linear accelerator.

would have x-ray guidance rods that would track the missiles and lock on to them. When fixed, a small nuclear device at the heart of the satellite would be detonated. In the split second before the blast devastated the satellite itself, the x-ray rods would channel the high-frequency and high-intensity x-rays emitted by the blast and would focus them on the targeted missiles, destroying them in a flash.

The third "death ray" technology is the most exotic, and yet it is in use already in several places around the world. A charged-particle beam weapon would be a direct offshoot of the particle accelerators that would have been used by scientists since the 1930s to smash atoms to find out what they are made of. Essentially, in particle accelerators, particles such as protons and electrons are sped up to near the speed of light, and when they collide, the heat generated can reach millions of degrees. In these machines this intense heat is kept in place by strong magnetic fields. But what if the accelerator were open-

ended and the particles were to smash into a missile? The missile would, for all intents and purposes, disappear. Both the Americans and the Soviets are researching such beam weapons, but for the time being it seems their immense size would prohibit their use in space.

There have been many arguments against the use of a "Star Wars" defense system. There is the matter of feasibility and cost-effectiveness, for even if the technology necessary is ever realized (and some doubt it will be), will it be so expensive – up to a trillion dollars, or more than ten times the cost of the Apollo and Space Shuttle programs combined – that it will be easier for the opposition to simply produce additional fast and less targetable offensive weapons such as cruise missiles? Nevertheless, to many, the high promise of such technology – eliminating the threat of nuclear weapons – remains enticing. Whatever the final outcome, the time when space was a place of peace has passed.

Moonbase

Above: a close-up view of the soil found by the Apollo 17 crew. Analysis of such samples confirms that worthwhile mining operations could be established on the Moon.

The first Apollo astronauts to visit the Moon went as explorers and scientists in cramped spacecraft. They stayed for only a few hours – just long enough to collect soil samples, deploy scientific equipment, and raise an American flag especially designed to remain unfurled even in the vacuum of space. The Apollo adventure was a triumph of human courage and technological prowess. But despite its historic significance, it represented only a fleeting visit to another world rather than a true beginning to lunar colonization.

The next people from Earth to visit the Moon are likely to remain on the surface for months at a time. Although they will have to be trained astronauts, their main skills will be in engineering and mining. In contrast to the excitement and glamour of Project Apollo, the next generation of lunar missions will consist of a tough schedule of work aimed at establishing Earth's first permanent base on another world.

Moonbase is not planned to be an expensive luxury appreciated only by astronomers and scientists. It will have to justify its existence in economic returns to be shared by the nations of the world. The key activity will be mining the lunar rocks for valuable minerals and shipping them back to Earth.

Although lunar rocks are not rich in ores, they contain easily available stores of such useful metals as titanium, aluminum, and iron. Once mined the moon rocks can be used as the raw materials from which solar power generating satellites can be built in orbit. Orbiting construction facilities with accommodation for thousands of workers will collect the ores and extract from them the metals needed for their space-born building program. Eventually these orbiting construction sites may evolve into the first space colonies housing permanent communities in an Earthlike environment. Expected to be massive in size, they will be complete with artificial gravity and all the vital elements of a self-sufficient society. They will have factories, farms, offices, schools, houses, parks, and gardens. Moonbase will therefore provide the materials which will make it possible for people to leave their home planet and take the first permanent step into space.

To gather the ores from the lunar soil on Moonbase, engineers and miners there will use especially adapted power

Above: a lunar processor for turning ilmenite – rich in titanium and iron – into water. A container of ilmenite is heated to between 2000° and 3000°F by means of a solar reflecting mirror. Hydrogen gas is then introduced to the container, which combines with the oxygen atoms in the hot ilmenite and produces steam. The steam is cooled to water.

Left: what a future lunar mining base might look like. At the left is a "mass launcher," a sort of electromagnetic catapult that launches lunar material beyond the Moon's gravitation field to a catchment area in space. From there the ores could be ferried to an orbital position above the Earth and be used as raw materials for a space colony. The living area for the lunar base is at the extreme right. In the foreground is a ferry tug that operates between Earth and the Moon.

shovels, bulldozers, and tractors. Soil blowers will clear unwanted lunar dust by means of fierce blasts of compressed air. Once the ore is collected, the task of launching it into space will be simple and cheap. Because the Moon is much less massive than the Earth, its escape velocity is lower – a mere 1.5 miles a second compared with about 7 miles a second on Earth. It is therefore possible to launch payloads from the Moon without the powerful rocket boosters needed to escape Earth's gravitational pull. Scientists have already sketched out the practical details of a kind of electrically powered catapult known as a mass driver, which could accelerate loads of ore to escape velocity and hurl them into space.

Establishing Moonbase will not be easy. The first flights will have to carry essential survival equipment for the pioneer workers who will set up initial camps and equipment depots. Later landings will bring the mining equipment and materials with which to build the mass driver. Later still, the first relatively comfortable living quarters for the miners will be erected. But for some years they will probably have to live in modified storage containers used to carry the early loads of equipment to the Moon. Covered in lunar soil to protect them from the extreme heat of the Sun, these pressurized huts will be only rudimentary homes, lacking in basic comfort but nonetheless safe and functional.

Eventually the base will develop into an interlinked complex of huts and other quarters with nearby landing and launch facilities for supply ships. But the most striking feature of Moonbase will almost certainly be the huge mass driver. Looking something like a railroad track several miles long, its aluminum rail sections will guide bucketlike containers of ore along a carefully computed launching path. Each bucket will hover just above the rails, held in place by a magnetic force produced by powerful electromagnets. Initially the buckets will be accelerated fiercely for about three seconds. In lunar gravity and with no air resistance to slow them down, the buckets soon approach lunar escape velocity. At the exact moment the critical speed is reached, the buckets will be suddenly decelerated, making them eject their contents in a precalculated trajectory sweeping over the lunar horizon and into space. The ore will continue its headlong flight for about two days. Then after 40,000 miles of travel, it will reach a strategically positioned vehicle permanently stationed above the Moon to catch the payload. The catcher will be a simple device, most likely a large cone of some strong fibrous material fitted with small rocket engines remotely controlled from either the Moon or the Earth. The cone will be kept spinning so that payloads once caught will be kept in place by centrifugal force. The open throat of the catcher will be covered with a fine wire mesh. As the ore strikes the mesh it will be shattered into a fine mixture of small rocks and grit. When the catcher is full, ferries will carry the ore back to Earth.

Lunar mining operations could well be under way sometime in the next century because most of the technology required is already available. Though the work of the pioneer teams can only be seen as a first step toward the large-scale colonization of the Moon, a large mining encampment could well provide the impetus for a more wide-ranging lunar base. For example, the Moon is an ideal place for scientific research. Astronomy in particular would, flourish with optical telescopes unaffected by a distorting atmosphere. Power satellites of the kind that will soon orbit the Earth may one day be placed in orbit around the Moon to supply the expanding colony with its energy needs so that in the remote future the lunar plains may be crowded with complex interlinked colonies. For millions of years a dead and barren wilderness, the Moon may someday be alive with human activity.

Exploring the planets

Well before the end of this century, unmanned probes will have visited every planet in the solar system and may have even begun mapping the great asteroid belt between Mars and Jupiter. Sophisticated sensor equipment and color cameras will radio back to Earth a wealth of new detail that will keep planetary scientists busy for years. But the most exciting step in exploring the solar sys-tem will be the first manned expeditions to the planets and the difficult and hazardous task of setting up human outposts on new and distant worlds.

The first attempt at interplanetary travel will almost certainly be a flight to Mars. Blueprints already exist for an initial mission. The plan is to send a combined crew of 12 aboard two space-craft. Using a trajectory designed to burn a minimum of fuel, the shortest journey time to Mars will be 270 days. On arrival, both ships will enter parking orbits around the planet, remaining there for about 80 days. During this time small landing craft will ferry crew members between the mother ships and the surface.

Left: an Apollo 17 crewman drives an electric-powered Moon-buggy, or lunar roving vehicle. The longest distance traveled by the vehicle was made during the Apollo 17's mission – a total of 56 miles on three excursions.

Below: a Martian sunset over Chryse Planitia. The computer-enhanced photograph was begun some four minutes after the Sun had dipped below the Martian horizon. The blue to red color variation is caused by the scattering and absorption of sunlight by atmospheric particles. In the foreground (far right) is a silhouette of one of Viking's power system covers.

Despite the fact that so far the United States Congress has made no funds available for the mission, NASA officials are hopeful that the first landings on Mars will be made early in the next century. There is some speculation that the Soviets are planning a manned Mars mission to co-incide with the 100th anniversary of the October Revolution.

What kind of world will people find when they reach Mars? American and Soviet probes that have already orbited and landed on the planet tell the story. Their cameras reveal a bleak and hostile landscape pitted with craters and covered with dune fields, dusty plains, and boulders. The atmosphere is much too thin to sustain human life and the first explorers will have to wear fully pressurized suits. Radio will be needed for communication because the rarefied air barely carries sound waves. Even the loudest shout would be inaudible more than a few yards away.

Although there should be no difficulty in keeping in touch with the Earth, widely separated bases on the planet's surface could face real problems. The difficulty arises because the Martian atmosphere does not contain the same kind of reflective layer – called the ionosphere – which enables people on Earth to bounce radio signals right around the planet if necessary. One possible solution could be to use the twin moons of Mars, Phobos and Deimos, as natural relay stations; but neither rises high enough in the sky to be really effective and the ultimate answer will probably have to be to place a number of artificial satellites in orbit around the planet.

Getting around on Mars will present no major problems. Although the terrain is rough, vehicle designers have already learned how to cope with boulder fields and dust bowls on the Moon, where their self-propelled machines were used with great success during Project Apollo. For long journeys, however, far more complex vehicles that are pressurized and air conditioned will be needed.

If the first scientific bases on Mars are to evolve into large-scale colonies, they will have to become as self-sufficient as possible. Every natural resource on the planet will have to be fully exploited. It is unlikely that plants will ever grow in the open, so agriculture will have to take

Left: a hydroponic greenhouse flourishes in an Israeli desert area. High-intensity agriculture, including multiple cropping, interplanting, and the breeding of high-yield varieties of cereals, fruits and vegetables will play a vital part in any space colony of the future.

place under cover. Instead of using soil as a growing medium for crops, special nutrient solutions will be used in the form of cultivation usually called hydroponics. One long-term problem colonists will have to accept is the effect of low gravity on physique. Gravity on Mars is only about one third of the gravity on Earth. Evidence gathered during long space missions suggests that low gravities can eventually weaken bones and muscle. Children born in the colony will probably adapt well but may never be able to face the strains of high gravity environments. Without special training, a visit to Earth would be dangerous. Later generations may become so well adapted to Martian gravity that no amount of training will enable them to visit their home planet. In a very real sense, these men and women will have become a new Martian race.

Perhaps the most fundamental question the first settlers will answer is whether there has ever been life on Mars. Imagine the excitement of discovering traces of fossils or even a few simple organisms still alive. The first contact with traces of extraterrestrial life – even if it is at a very primitive level – may well revolutionize human understanding of biological science. However, such major breakthroughs need not be confined to Mars. Even in more remote and alien environments – the sulfurous atmosphere of Venus or the frozen and turbulent

clouds of Jupiter and Saturn – especially adapted life forms may well have evolved.

Mercury and Venus are least attractive for interplanetary exploration in the fairly near future. Their environments are so utterly hostile that landings will be almost out of the question.

Surprisingly, the most likely possibility for human flights after trips to Mars is in the farther reaches of the solar system. Planets such as Jupiter and Saturn are no less inhospitable than Mercury and Venus but they have the advantage of ready-made orbiting observatories high above their atmospheres. The first missions to Jupiter and Saturn will not attempt landings on the planets themselves but on the larger of their moons.

Perhaps the most exciting possibility of all is a mission to Titan, largest of Saturn's moons. Titan is thought to be unique in the solar system. Orbiting over 700,000 miles above Saturn, Titan is actually a larger body than the planet Mercury. What makes it particularly remarkable is that astronomers believe it has a fairly thick atmosphere with pressure at the surface being perhaps 10 times as great as on Mars. Titan is nonetheless a far from inviting place. Temperatures on its surface seldom rise

Below: one of the crewmen of Skylab 3 demonstrating the effects of weightlessness. The Skylab 3 team spent a record 84 days in space and demonstrated that humans could survive long exposure to weightless conditions and still readapt easily to normal gravity.

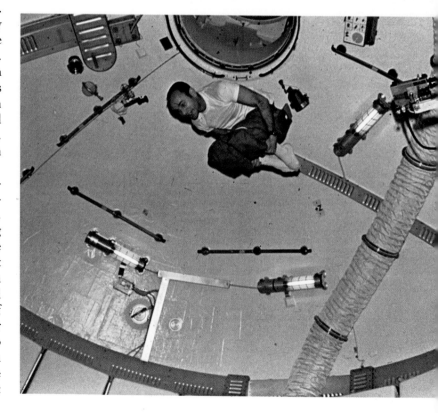

above −290°F, and although the atmosphere is thick, it is predominantly nitrogen with an unpleasant mix of ethane, acetylene, and hydrogen cyanide.

Titan's surface is probably a spectacular place. Methane would rain from the sky, forming rivers of liquid natural gas, flowing into methane lakes and oceans. Most awesome of all must be the sight of the planet Saturn dominating the sky.

A base on Titan would be a lonely and remote place, a single voyage to or from Earth taking several years. Because Saturn is nearly 800 million miles from Earth, the team on the Titan base will have to commit themselves to a long stay with no hope of rescue if an emergency occurs. Eventually, however, Titan could become an important stepping stone for the first flights to the outermost limits of the solar system. Titan's remoteness may also be reduced by breakthroughs in propulsion technology, perhaps bringing it within a year's flight of Earth.

Jupiter, largest of the planets, has 13 known moons. Bases on the two largest of Ganymede and Callisto will present similar technological problems to that of Titan, but the scientific rewards could be even greater. The fearful cauldron of gases that is Jupiter's atmosphere has for long mystified astronomers. Strange and violent meteorological effects seem to be at work whipping the Jovian gas clouds into a dazzling display of multicolored loops and swirls. Best known is the Great Red Spot, a feature now thought to be a gigantic storm which has already lasted at least 300 years. Some scientists have suggested that despite its turbulent nature, the Jovian atmosphere could nonetheless support especially adapted life forms, possibly ones that drift like balloons among the swirling clouds.

Perhaps the most logical step in interplanetary exploration will be the first flights into the asteroid belt, a region between Mars and Jupiter. In it thousands of minor planets, ranging in size from large boulders to miniature worlds up to 650 miles across, move in regular orbits around the Sun. Although permanent bases on any of them is a possibility only for the remote future if at all, building beacons on them to radio data back to Earth is an exciting possibility, achievable even with present technology.

Above: a germanium selenium crystal being "grown" in the weightless conditions aboard Skylab. It was found that crystals could be grown bigger (this one is about one inch long, compared to a typical one tenth of an inch on Earth) and more pure than on Earth. Scientists believe that other industrial processes could be conducted with far greater efficiency in the absence of gravity.

The really attractive feature of certain asteroids is the irregularity of their orbits. Although most are confined to the main swarm, some wander beyond Jupiter or close to the Sun. Icarus, for example, passes well inside the orbit of Mercury, almost brushing the Sun at a distance of 17 million miles. A transmitting station on Icarus could give much valuable information as well as conveying a fantastic close-up glimpse of the solar atmosphere without the immense problems of trying to build a space probe that could survive the searing heat.

Asteroids may also be rich in metals and minerals. If they are, it may only be a matter of time before the first prospectors are sent to explore them. Eventually fleets of specially equipped mining ships may range through the asteroid belt looking for sources of the materials people need most for their civilization.

An even more remarkable prospect is the possibility of using whole asteroids as building material. It may become feasible to capture suitable asteroids and, using space tugs, tow them out of their normal orbits. They could be taken back to orbits around Earth, the Moon, or Mars and there used as a handy source of metal ore or as a starting point for a space station. Hollowed out and equipped with life-support systems and appropriate scientific equipment, asteroids could provide stations that would need little external maintenance and would last almost indefinitely.

Star voyages

By everyday standards, many of the planets of the solar system are unimaginably remote. Pluto, for example, orbits on average more than 4000 million miles from the Sun. But eventually it will be possible to reach all of them using more or less conventional space vehicles. Already the first probes are moving among the outermost planets and, although their journeys take many years, it requires only patience before mission controllers on Earth receive the data for which they have been waiting.

The difficulties of flying to the stars are immeasurably greater. The basic problem is distance. It is almost impossible to grasp the vastness of the gulfs separating the Earth from even the closest star. Try to imagine the solar system shrunk so that the 93 million miles between the Sun and the Earth becomes a mere six feet. The outermost planet, Pluto, would be orbiting on average 240 feet from the Sun; but even on this vastly reduced scale the nearest star is more than 3000 miles away. When we use the gigantic measuring rod of light-years, the closest star – Alpha Proxima in the Centauri system – is 4.3 light-years away.

How can people hope to span these huge distances? Clearly, the technology being used to explore the solar system is hopelessly inadequate. The Pioneer 10 spacecraft that visited Jupiter recently is now headed out of the solar system at a cruising speed of around 25,000 miles an hour. Even if it were aimed toward Alpha Proxima, it would take over 100,000 years to reach its objective.

Current space technology is only the starting point for the design of the first practical starships. But despite the awesome problems, some scientists and engineers have already made serious studies of interstellar flight and have suggested a range of design possibilities. Although the means of building the first starships may not be within grasp until the middle of the next century, scientists

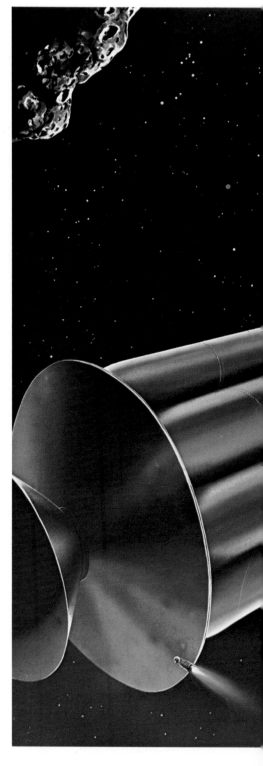

Right: a space ark has arrived at the planet of another star. The cylinder is about 30 miles long by about 6 miles in diameter. It has a mobile precurser shield to protect it from erosion during its long journey through interstellar space. This is now being maneuvered away. A small space tug can be seen inspecting the drive unit. There are three reflectors used to reflect light into the ark's long windows when it is in the vicinity of a star. In interstellar space the reflectors are closed up and the ark's thermonuclear reactors supply all its energy requirements. The ark is now in orbit thousands of miles above its destination. In the foreground is a beacon satellite, perhaps placed there by the crew of one of a later generation of faster starships that were developed hundreds of years after the ark departed and yet arrived before them to begin the task of terraforming the planet.

already have a good idea of the kind of propulsion system they will use and the kinds of missions they will undertake.

Starship designers are faced with two basic approaches to interstellar flight – fast or slow. A fast trip might take only a few years but would need highly sophisticated drives capable of accelerating the starship to velocities approaching the speed of light. The more leisurely ap-

proach to starflight has been dubbed the Space Ark concept. Although traveling at much higher speeds than our current space vehicles, the Space Ark would still take hundreds or perhaps thousands of years to reach its objective.

The attraction of the Space Ark concept is that it requires no fundamental breakthrough in propulsion systems. Although the huge vehicle would have to

be accelerated by powerful engines up to a practical velocity, the problems are not as complex as those involved in achieving near light speeds. But the Ark presents some unique difficulties that may never be satisfactorily resolved. The concept requires an entire colony of human beings to commit themselves to a journey that will take generations to complete. The men and women leaving Earth would have to resign themselves to death in space far from their homes and knowing that perhaps only their remote descendants would ever see an end to the interminable voyage. Generation after generation would be born, live, work, and die in a strange artificial environment on the way to the final goal.

The Space Ark is as much a sociological challenge as a technological one. The size and composition of the Ark's population would have to be carefully planned. The people chosen would need one or more of a range of vital skills. There would have to be engineers, scientists, doctors, and teachers to pass on knowledge from generation to generation and social psychologists trained to ease some of the complex tensions likely to build up in the colony during the flight.

Below: a starship of the future. If high-speed starships become a possibility in the future Einstein's Special Theory of Relativity – especially that part of it that states that a moving object ages more slowly in relation to a stationary one – will have important practical effects. On very fast journeys, those that rely on velocities fractionally short of the speed of light, a journey to a star might take a matter of months as far as the astronauts are concerned. But back on Earth years or even centuries may have passed.

Most modern estimates suggest that about 10,000 people would be needed initially to create a stable, self-sustaining community. This means that the Ark itself would have to be a huge vehicle weighing perhaps 4 million tons or more. Complex on-board systems would be needed to keep the passengers alive and well throughout the voyage, and formidable primary engines would have to be relied upon to accelerate the Ark to its cruising speed.

The unanswered question about the Space Ark concept is whether human beings can remain sufficiently motivated to insure the success of a mission lasting centuries. If a generation loses interest,

no one can predict the social disaster that might follow. Perhaps succeeding generations would forget Earth and mistake the Ark itself for the only real world. New religions and mythologies might arise. In the end, the people arriving at the ultimate destination could easily prefer the environment of their Ark to a landfall.

One interesting possibility of slow voyages to the stars is that colonists might be overtaken in flight by fast starships developed after the Ark's departure from Earth. This could mean that when they eventually arrive at their destination, Earth people might already be waiting to receive them.

Although Space Arks are a real possibility for the future, the main hope of interstellar exploration today lies in high speed starships. Speeds of perhaps a fifth of the speed of light make the traveling time to many of the nearer stars only a few years long. But at such high speeds it is necessary to define exactly what is meant by traveling time. According to Albert Einstein's theory of relativity, the measurement of time depends on how the person making the measurement is moving. Time as far as the crew of a fast starship are concerned would appear to pass more quickly than on Earth. This means that a mission during which the crew ages by three years may take over 10 years for flight controllers on Earth. The faster the starship travels, the more pronounced the effect becomes. Close to the speed of light, ship-time is moving at a crawl compared to Earth-time. After just a few years of flight at this kind of speed, returning crews might find that thousands of years had passed on Earth.

Can we ever hope to reach such fantastic speeds? The secret lies in keeping up a small acceleration for long periods of time. At present rocket engines can only do the opposite – produce high accelerations for fairly short periods. If a system could keep accelerating a spacecraft steadily and gently for several years or more, we would be able to explore the entire Milky Way in a matter of years. For example, imagine an acceleration equivalent to the pull of the Earth's gravity – about 32 feet per second per second. If a starship traveled under this acceleration for a year it would reach a velocity of nearly 60 million miles an hour. Just 12 years of ship-time would be needed to cross the galaxy.

Space engineers have already proposed a number of possible propulsion systems for the starships of the future. One of the most remarkable is known as the laser galleon, a concept first suggested by the Canadian scientist Philip Norem. His idea avoids the problem of installing complex drive systems aboard the starship, proposing instead that the means of propulsion be left on Earth. Norem's plan is to build a powerful bank of lasers either on Earth or in orbit around it. The 1000-ton galleon would have a huge parachute-shaped sail made of aluminized plastic about one micron thick attached to the spacecraft by a cable up to 20 miles long. When the

Below: an interstellar ramjet, viewed from a small shuttle craft traveling at the same speed. The large structure in front of the craft generates a magnetic scoop which sucks up ionized hydrogen gas to be burned as fuel in a fusion rocket. The ramjet is accelerating and has reached about two thirds the velocity of light. At such speeds the light from the stars has been shifted, so that they appear redder than they really are. Already the cooler stars have disappeared because their light has been shifted into the infrared part of the spectrum. Some of the very hot blue stars can be seen because their emissions in the ultraviolet now appear as visible light.

intense beam of light from the lasers is directed on the spacecraft, its energy exerts a gentle but steady pressure on the sail to pull the spacecraft behind it. Eventually speeds of over 60,000 miles an hour would be reached.

Norem has actually worked out a flight plan for an imaginary voyage to Alpha Proxima. The first phase of the journey would be an eight-and-a-half-year cruise directly toward the planned destination. The ship would then veer off-course, making a leisurely 180-degree turn to approach the target star from the far side. The turn could take anything from 10 to 30 years, but once complete the ship will be moving into the laser wind and will lose speed coasting toward its target. Once the crew members landed and established a base on a suit-

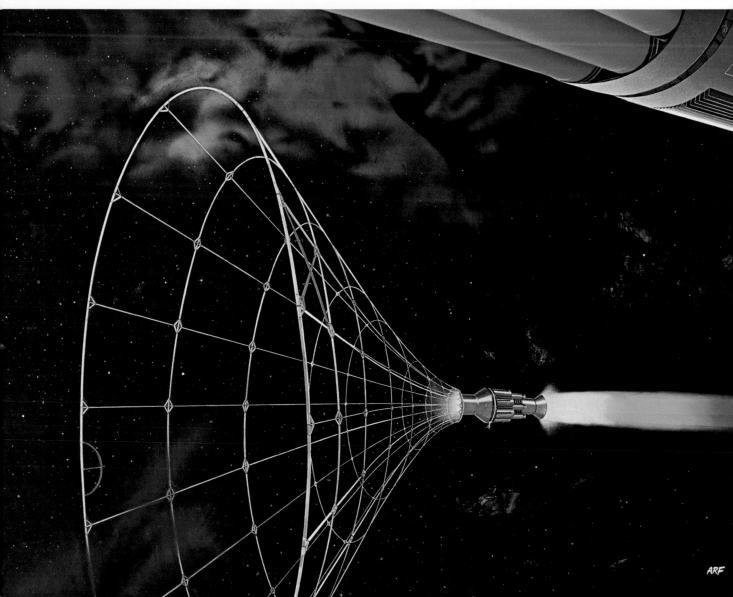

ARF

The laser galleon

Right: Canadian scientist Philip Norem's "space galleon" was designed to use the propulsive force of an array of laser beams. The array, set up on Earth, would drive the galleon at a speed of 62,500 miles a second toward a planet of the nearest star Alpha Proxima, 4.3 light-years away. The huge parachute-shaped sail that can be seen in the illustration is made of aluminized plastic and is only one micron thick. It is attached to the galleon by a 20-mile long cable. The laser beam "wind" would drive the galleon along its course for 8.5 years after which it would make a wide 180 degree turn. This maneuver, shown in the diagram below, would take between 10 and 30 years to execute, by which time the galleon would be approaching Alpha Proxima from the opposite side to the Earth and it could use the laser wind as a brake. The craft would then be able to coast down to its target planet, taking another 8.5 years. The astronauts, by then much older, would set up a laser array on the planet to act as a brake for future craft. Subsequent journeys would take 10 years instead of the 60 or so taken by the pioneers.

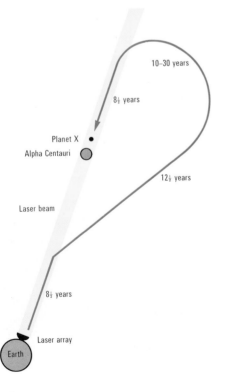

able planet, they could build another laser bank. This would subsequently be used by future galleons as a space brake, eliminating the need for the lengthy 180-degree turn. This could cut flight times by half.

The simplicity and potential of the laser galleon has great appeal, but it does not offer a very flexible kind of starflight. The ship is entirely dependent on the laser beam from Earth and could not easily make changes of course during the voyage. Perhaps the most disturbing feature of the laser galleon is that if anything goes wrong on board, the crew would be unable to call off the mission and return to Earth.

A far more versatile system sharing some of the advantages of the galleon is the interstellar ramjet suggested by the American engineer R. W. Bussard. Like the galleon, the ramjet is basically simple. Its most remarkable feature is that it carries no fuel of its own. It literally picks up what it needs as it goes along.

The ramjet relies on the fact that the space between stars is far from empty. It is filled with a tenuous mixture of gases, mainly hydrogen and helium. Although densities probably average less than two atoms in each cubic inch of space, there may be regions in which the local density is much higher. Bussard worked out a way of using the hydrogen in the interstellar gas as fuel for a specially designed nuclear ramjet. The principle is very simple. As the ramjet moves through space, a huge scoop deployed in front of the vehicle gathers hydrogen, directing it into a small nuclear fusion engine. Inside the engine, the nuclei of hydrogen atoms are fused together into heavier nuclei of helium. As this process takes place, vast amounts of energy are liberated and the ramjet is driven forward by the hot exhaust of waste helium produced by the fusion engine.

Because the ramscoop would need a diameter of about 1800 miles to collect enough fuel, building one is clearly out of the question. Instead the scoop would consist of a projector capable of throwing out a huge scoop-shaped magnetic field. The atoms of hydrogen in space being largely in the form of electrically charged ions, this magnetic field would attract them and draw them toward the mouth of the ramjet engine.

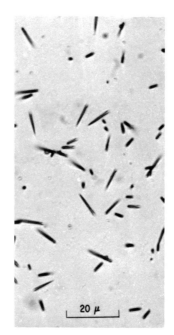

Above: a photograph made by Professor P B Price of California University showing radiation damage in the form of fission fragment tracks in a crystal from an ancient meteorite. It is from evidence such as this that scientists build up a picture of the locally violent environment that existed in our galaxy immediately before the birth of our slar system.

The ramjet would need to be traveling at high speed before it could scoop up enough fuel for its engine. A secondary engine will therefore be needed in the first phase of flight. But once the ramjet starts working, the starship would quickly accelerate to extremely high velocities.

As the starship approached the speed of light, effects of relativity would begin to work in favor of the ramjet system. Einstein's theory says that at high velocities distances appear to shrink. This means that as far as the ramjet was concerned, the distance between atoms of hydrogen in space would drop. The faster it traveled, the denser the source of its fuel would become and the more efficiently its fusion engine would work.

Once the ramjet reached velocities close to the speed of light, how would it reduce speed when the crew nears its destination? Bussard has pointed out that the ramscoop could be turned into an ingenious space brake. With the fusion engine switched off, the magnetic field could be reversed. Instead of scooping up interstellar gas, it would reflect it, turning the ramscoop into a kind of invisible parachute. Final braking maneuvers would be handled by the secondary rocket engines.

Formidable technical problems have to be overcome before Bussard's interstellar ramjet could become a reality. Some experts believe it will never be practical. Even if a small fusion engine could be built, they argue, the scoop could never collect enough gas to keep it going. But despite the unresolved questions surrounding the ramjet concept, the exciting potential of the idea is enough to sustain the interest of many starship designers. With a fleet of Bussard ramjets, people could go anywhere they wished in the galaxy and, if enough fuel were available, they might press on into intergalactic space itself.

Bussard envisaged a very sophisticated fusion engine. Other designers have been investigating a much more crude way of using the energies of nuclear fusion for starflight. As long ago as 1891 the German inventor Hermann Ganswindt suggested that a space vehicle could be propelled by detonating a series of bombs under it. The impulse from each blast would, he thought, hurl the spaceship forward. In the 1950s an American team of research engineers led by Dr Stanislaw Ulam updated the concept by suggesting the same principle be applied using small fission bombs. Ulam's team proposed to explode a large number of atom bombs in a continuous sequence behind a large "pusher" plate fitted to the base of a spaceship. The plate would transmit the force of the detonations as a smooth and steady acceleration. But Dr Freeman J. Dyson at the Institute for Advanced Study in Princeton University rejected the idea, pointing out that atom bombs are much too inefficient for a system of the kind proposed. Because they release a good deal of their energy as deadly radiation, the spaceship would have to be fitted with massive shields to keep the crew alive.

Dyson argued that it was much more practical to apply the same principle but use hydrogen bombs instead. His suggestion is now known as the fusion pulse rocket and he has worked out some detailed plans for a test vehicle. He envisaged a spaceship with an initial mass of about 400,000 tons, two thirds of it in the form of 300,000 individual hydrogen bombs. Each of these would yield an explosive force equal to about one million tons of TNT. Only about 45,000 tons of the initial mass would be available for payload but this would probably be enough for a reasonable size space colony. Dyson calculates that to reach a cruising speed of about 5000 miles a second, one bomb would have to be detonated every three seconds for the first 10 days of flight.

The fusion pulse rocket would take about 300 years to reach targets 10 light-years away. However, there is another complication. If half the bombs were kept for use in decelerating the spaceship when it reached its destination, the flight time might be more than doubled.

Below: the table shows the effect of time dilation on the crew of a starship under a steady 1g acceleration. As far as the crew are concerned, after 0.9 years of flight they will have traveled 0.4 light-years. But on Earth a year will have passed. After an apparent 9.7 years aboard the ship the Earth will have aged a staggering 10,000 years. By this time, the crew will be about 10,000 light-years away, deep in interstellar space.

Dyson's proposals, therefore, can only be seen as propulsion for a slow Space Ark journey.

The idea of using hydrogen bombs to reach the stars sounds extremely far-fetched, but recent studies suggest that the nuclear pulse rocket in a highly refined form may be the best hope of reaching the stars within the next century. This is certainly the conclusion of the first full-scale starship study carried out from 1973-1978 by the British Interplanetary Society. The result of their work is the first practical proposal for a mission to the stars, complete with full details of the spacecraft and its flight plans.

The British scheme is called Project Daedalus. The key to the mission is a highly efficient nuclear pulse engine. The principle of the engine is to induce fusion reactions in small pellets of fuel by blasting them with laser beams. Experiments in the United States have already shown that fusion can be started in this way as long as the laser pulse is sufficiently powerful. To keep ignition temperatures as low as possible, the Daedalus team recommended using as fuel a rare lightweight isotope of helium known as helium-3. They worked out that about 30,000 tons of it would be needed for the mission.

On Earth helium-3 does not exist in huge quantities. The only way of gathering the amount needed would be by mining it in orbit from the helium-rich atmosphere of Jupiter. The Daedalus spaceship itself would also be built in orbit around Jupiter. The team envisages a large two-stage crewless vehicle which would carry a number of small probes to be dispersed in the vicinity of the target star.

For such a project, it would be essential to guarantee the maximum scientific

EARTH TIME IN YEARS	SHIP TIME IN YEARS	DISTANCE COVERED IN LIGHT-YEARS
1	0.9	0.4
10	3.0	9
100	5.2	99
1000	7.5	999
10,000	9.7	9999

returns after the immense efforts needed to make the mission possible in the first place. The group therefore settled on Barnard's Star after a great deal of thought, largely because there is strong evidence suggesting a planetary system around it.

The six light-years to this star would be covered in about 50 years. For the first four, the spaceship would be gradually accelerating toward its cruising velocity of about 23,000 miles a second. The mission would take the form of an undecelerated fly-through with the small probes being sent into orbit around the star to carry out an investigation of its planetary system.

Extremely accurate navigational systems will be vital if Daedalus is ever to come close to its target. The long interstellar trajectory will be constantly checked by on-board computers which will judge the spaceship's position in space by observing certain pre-chosen stars and using them as navigational beacons. Delicate sensors will have to be able to spot planets around the target star while still several years from the closest approach. The computers will need all of this time to work on the complex trajectories needed to make certain that the various probes are launched at the right moment as Daedalus hurtles past its target.

If the nations of the world decided to go ahead with Project Daedalus now, it would take about 20 years to design and build the vehicle. Allowing for 50 years in transit and at least six years for the transmission of data back to Earth, the Project would take nearly 80 years overall. Unfortunately, present knowledge of fusion technology is not yet advanced enough to even contemplate Daedalus, and would probably not be ready much before the end of this century. This still means that people could reach the stars before the year 2100.

Project Daedalus is the first real blueprint for interstellar flight. It demonstrates conclusively that by building on existing technology and known scientific principles, missions to the stars are practical. But the remote future may well hold possibilities that at present we can hardly imagine.

One remarkable propulsion system has been popularized by a television

series called *Star Trek*. It works by mixing matter with antimatter. When the two meet they instantly annihilate one another in a blaze of energy. If the violence of the reaction could be controlled, the matter-antimatter annihilation would be the most efficient energy source imaginable. But despite the fact that the process represents 100 per cent conversion of matter into energy, only a fraction of the energy appears in a usable form. Much of it is generated as gamma rays and tiny particles called neutrinos, neither of

which can be deflected into a concentrated exhaust beam. But if only a fraction of all the energy produced could be turned into propulsive power, the engine would be enormously powerful.

A major drawback to the system is the difficulty of handling and storing antimatter. At present no one has any idea how antimatter can be prevented from interacting with the ordinary matter around it. But the problem may be solved within a few centuries.

The future will no doubt hold many

Above: the Daedalus spacecraft blasts off from the vicinity of Jupiter on its 50-year journey to Barnard's Star, about 6 light-years away.

other remarkable new breakthroughs and as long as mankind survives social and economic crises on Earth, voyages to the stars will probably one day be a routine part of human activities. People can only speculate how the exploration of the galaxy will affect civilization. Much will depend on what is found when the stars are reached. But given long enough, mankind's empire could extend throughout the Milky Way – and the Earth may become merely one of a multitude of inhabited planets.

Supertechnology

It is much easier to predict the fate of stars than to describe how civilization may develop in the centuries to come. The natural world obeys more or less fixed scientific laws, and once they are understood, scientists can make confident statements about how nature will behave at any given time in the future. But intelligent life is a more unpredictable element in the universe, and when people try to imagine the ultimate future of mankind, they can only speculate, building on what is known of present-day trends and capabilities.

Technology is the tool people use both to change the world around them and to utilize natural forces to make their life more comfortable. Any kind of technology needs fuel. In most basic terms, technology consumes energy and the more technologically developed a civilization becomes, the more energy it uses. The last few thousand years of technological progress is only a flicker of time on a cosmic scale and so far people have been able to make do with the resources of this planet. In recent years, however, with ever-growing demands for energy these resources have begun to dwindle.

The time must come when the riches of a single planet are not enough, and if the pace of technological progress is to continue then mankind must seek new fuels and new sources of energy. Scientists studying technological development have mapped out an evolutionary future linked to mastery of energy resources. In the present state of relying on a single planet, society can be described as a Phase One civilization. Phase Two begins with the exploitation of the resources of the solar system. If the second phase is ever successfully completed, a third phase remains – the exploitation of the entire galaxy. A Phase Three civilization has probably never yet evolved in the history of the universe. If it had done so, there ought to be visible evidence of it. But Phase Two is an essential development if technological growth is to continue.

How can people ever hope to utilize the entire resources of the solar system? With present-day technology the challenges seem too formidable to contemplate. But given the continued survival of mankind, there are many millions of years in which to advance skills and understanding of scientific principles. Even the most far-fetched ideas being considered today may be commonplace realities in the distant future.

Society is already taking the first steps toward the second phase of development. Plans to utilize solar energy are underway and, although current proposals are only the most minor beginnings of future grand-scale exploitation of the Sun, it is a vital start toward avoiding a devastating energy famine before the end of this century.

There are already many ideas about how to mine the wealth of other planets, but they involve working in artificial environments, which is very restricting. An exciting prospect for the future is to go several steps farther and change the total climatic conditions of another

Right: a modern technological triumph. This map of cloud-shrouded Venus was produced by radar astronomers at NASA's Jet Propulsion Laboratory. The area indicated is a 910-mile wide zone in the equatorial region that reveals huge shallow craters up to 100 miles across. The black band through the center of the picture depicts an area that cannot be accurately scanned from the radar tracking station in the Mojave Desert. The map at the bottom gives the position of the heavily cratered area (circled) on the planet's face. Other bright areas on the face, marked by letters, were mapped in previous radar scans of Venus.

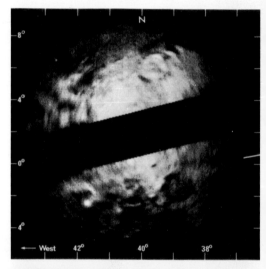

Far right: an ultraviolet picture of Venus, showing (in the crescent shape) atomic hydrogen both in the upper atmosphere, and (in the vertical bars) reaching far above the atmosphere. Combined with oxygen, this atmospheric hydrogen could become a source of water to irrigate the planet.

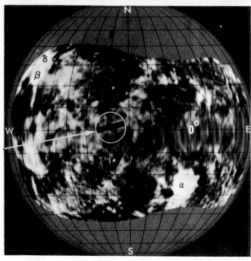

world, making it almost as suitable for human beings as the Earth itself. This kind of advanced technology is called *terraforming* and there are two candidates in the solar system on which the process could be started within the next 100 years.

One of the proposed planets for transformation is Venus, and a plan has already been worked out to convert its hellish conditions into a livable environment. To do this it will be necessary to remove the high concentrations of carbon dioxide in Venus' atmosphere so that the temperature and atmospheric pressure of the planet will drop.

The American astronomer Carl Sagan has suggested how this can be done. In 1961 he described how the clouds of Venus could be seeded with a tough variety of algae, a genus called *Nostocacae*. Once established in the upper atmosphere, the algae would begin photosynthesis – the basic life process of all

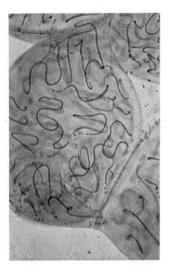

Above: a highly magnified (×40) photograph of a *Nostoc*, a genus of blue-green algae which, in the presence of light, synthesizes organic matter from carbon dioxide and water.

green plants. The important feature of photosynthesis is that it extracts carbon dioxide from the atmosphere, releasing greater volumes of oxygen.

As this happened, temperatures and pressures near the planet's surface would drop. But more important, water vapor in the Venusian atmosphere would begin to fall as rain. The initial cloudburst would never hit the ground, but instead would turn to steam while still high in the atmosphere. After repeated attempts, though, the downpour would eventually reach the surface. Within a few years the impenetrable clouds of Venus will part, leaving an atmosphere rich in oxygen with surface temperatures of about 70°F. As the sky clears and the Sun can at last be seen, the most hellish planet in the solar system will have been changed into a temperate world in which hardy plants and animals from Earth could flourish. Human beings will be able to settle a new world without any protective equipment.

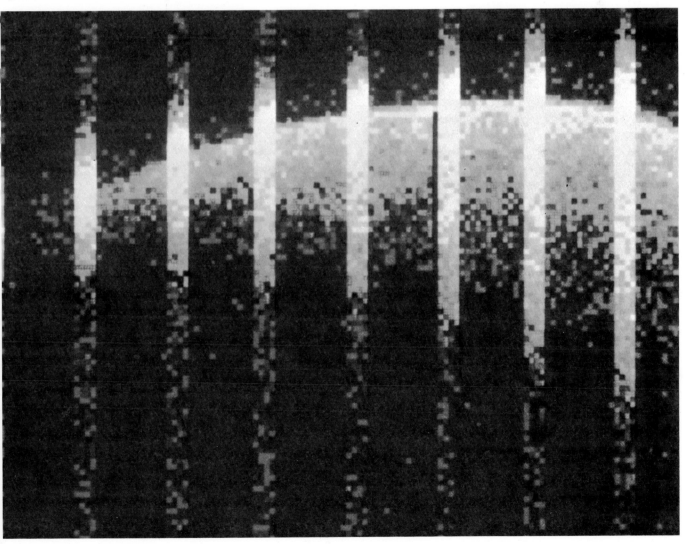

Left: the descent to the surface of Mars of a future spacecraft. With the prospect of a lunar base being established in the 1990s, manned expeditions to other nearby planets are a strong possibility for the end of the century.

The other candidate for terraforming is Mars. There is strong evidence suggesting that in the remote past, conditions on Mars were much more like Earth than they are today. It is known that locked in the Martian ice caps and in the permafrost covering much of its polar regions are vast quantities of carbon dioxide and water. To release these life-giving substances the planet has only to be warmed up. The Earthlike conditions of the past may well have been caused by the 50,000 year cycle during which Mars tilts toward the Sun in the course of its orbit. Some early terraforming proposals have suggested that the tilt could be induced artificially by altering the positions of the twin moons of Mars or by towing a large asteroid into the vicinity of the planet. But juggling with gravitational fields is a complex and unreliable process. Much simpler ideas include stationing a huge reflector above the ice caps to focus the sunlight on them to gradually melt them. An even easier technique might be to sprinkle large quantities of black powder over the poles. This would make them absorb solar energy more strongly and could, over a period of many years, raise polar temperatures enough to release the frozen water and carbon dioxide. The final stage of terraforming would be to use algae to photosynthesis the carbon

dioxide into oxygen and thereby turn the planet into a habitable world.

The creation of new worlds congenial to people would be a tremendous boost for mankind but even these will not be enough to go on supplying the material needs of civilization. Ultimately the eyes

Below: a circular spaceship soft-lands on a planet with a Venus-like atmosphere.

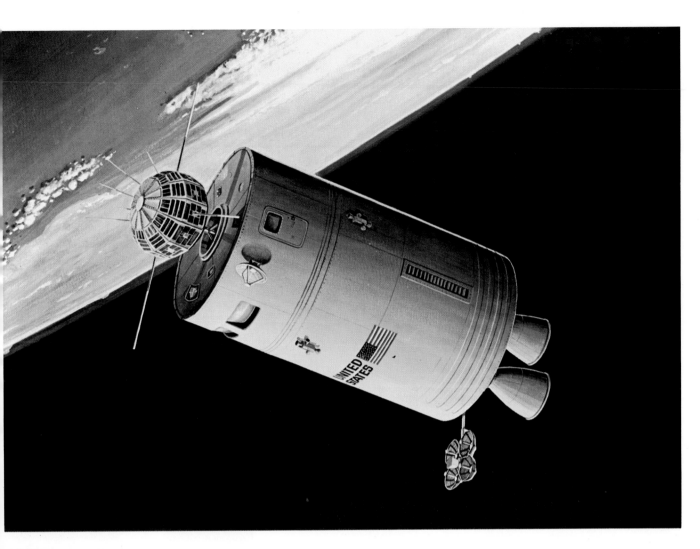

of the supertechnologists will be turned to the Sun.

The Sun is the chief source of energy in the solar system, but most of its radiation is wasted. Even systems of space-born power stations designed to convert sunlight to electricity would be harnessing only a minute fraction of the Sun's overall output. The energy-hungry civilizations of the remote future will look for ways of catching all the radiated energy from the solar system's richest resource.

In addition to being hungry for energy, the fast expanding civilization will also be in desperate need of living space in the future. One awesome technological possibility answering both problems together will be to build space colonies englobing the Sun as suggested in 1959 by the American physicist Freeman J. Dyson. There now seems no reason why it should not become a practical reality within a few hundred years.

The Dyson Sphere, as it is called, will

Above: an artist's conception of a space tug, a multi-use space vehicle of the future, designed to operate around the Earth, Moon, or in support of interplanetary missions. The tug is shown here retrieving an Earth-orbiting satellite. Its other uses might include assembly of space station modules, rescue of disabled craft, and the transport of payloads from one altitude to another.

be the supreme achievement of a Phase Two civilization. But virtually enclosing the Sun in a gigantic artificial sphere with little of its radiation escaping the solar energy converters of the colonists presents one serious problem. Where can the enormous quantities of building materials needed for its construction be found? One possible answer is to dismantle Jupiter, the largest planet in the solar system – an idea even more fantastic than the building of the Sphere itself.

Jupiter is a world with a spherical core of metallic hydrogen 75,000 miles across. Above this is about 9000 miles of semi-solid and gaseous hydrogen, methane, and ammonia. But recent analysis of the light from Jupiter suggests that although the planet is about 78 percent hydrogen, it also contains huge quantities of oxygen, nitrogen, carbon, silicon, aluminum, iron, and other heavy metals. If all, this material could be broken up into manageable fragments, there would be enough material to build 38 new worlds with the same mass of the Earth or hundreds of thousands of smaller space colonies – enough in fact to construct the Dyson Sphere.

For years scientists protested that to destroy Jupiter would be to upset an important gravitational balance, with potentially disastrous consequences for the rest of the solar system. The gravitational field of the largest planet is known to affect the orbits of the others and even slight changes in an orbit could bring catastrophe to life on Earth. But recently astronomers have made the first thorough investigation of the likely effects of Jupiter's removal and have concluded that the event would have no appreciable effect on the orbits of the rest of the planets.

How could such a feat ever be achieved? Dyson was the first to show that dismantling planets was a practical possibility. In 1966 he worked out how Jupiter could be taken apart simply by applying today's technological knowledge. At present the planet rotates roughly once every 10 hours, complicated by the fact that different cloud layers actually rotate at varying rates. He proposed encircling selected latitudes of the planet with gigantic metal grids. Electricity generated by solar energy and passed through the grids would cause Jupiter's speed of rotation to increase from once every 10 hours to about once each hour, and the centrifugal force created by the spinning motion would rip the planet apart. The drawback of Dyson's proposal is that even using the most optimistic estimates, the task of speeding the planet's rotation to the right level would take about 40,000 years.

Dyson's plan at least demonstrated that today's technology is equal to the task. The vast superiority of future technology will therefore certainly make planetary engineering of this kind achievable in a few years of concentrated effort. The British astronomer Iain Nicolson has even suggested that within a few hundred years scientists may be able to devise bombs capable of shattering Jupiter in a matter of hours. The explosive power required would be equivalent to 1000 million million million hydrogen bombs each of 40 megatons.

Whatever technique is eventually used,

there is little doubt that our highly advanced descendents will be able to use the materials making up Jupiter to construct the Dyson Sphere, and building the Sphere itself will be relatively simple. A particular solar orbit will be chosen, perhaps just beyond Mars, and suitably sized pieces of material will be towed into position. Millions of different sized worlds could then be built ranging from platforms housing a colony of a few thousand people to structures the size of the Earth and providing living space for millions. Some, such as science fiction writer Larry Niven, have suggested that as a precursor to building a sphere, ringworlds could be built, with a band of material completely encircling a star.

At present only one millionth of one percent of all the Sun's radiated energy is

Right: the Stanford Torus as conceived by the British artist David Jackson. It is depicted against a background of the planet Venus which is in an early stage of terraforming. The colony's central, non-rotating "south pole" is uppermost. The huge, half-mile-square main radiator is positioned with its surface edge-on to the Sun, radiating the waste heat of the colony into space. Below the torus a giant mirror reflects sunlight into the interior of the colony. The three spheres between the torus and Venus are man-made habitats. To the right of the torus a small rocket-powered utility craft directs an asteroid toward a new location for mineral extraction.

captured by the Earth. The Dyson Sphere by contrast should collect nearly 90 percent of the energy. Life on the surface of the Sphere would be oriented around the efficient use of solar energy. Some worlds would be specialized industrial centers, others agricultural areas, others especially set aside for leisure and recreation. Within a few centuries life on the inner planets – Mars, Earth, and Venus – will seem hopelessly outmoded. The real focus of human activity will be the ever-moving surface of the Dyson Sphere. Dyson has suggested that we may well find evidence of Phase Two civilizations through our telescopes, for a star, completely englobed by an artificial world, would give off a dim, characteristic radiation.

In the more distant future, civilization will not be satisfied with the resources of a single star. As the centuries pass steps will be taken to extend the Phase Two civilization exploiting the whole solar system, to Phase Three. Plans will be laid to utilize the resources of the Milky Way itself. Dyson has estimated that a fully developed Phase Two civilization could move from star to star in 1000-year stages. In about 10 million years – a short time by astronomical standards – mankind will have spread throughout the galaxy.

One problem facing a Phase Three civilization is the fact that the galaxy contains too many stars and not enough planets. Where will the supertechnologists find the raw materials to build their Dyson Spheres and other great structures? One solution will be to actually mine some of the older stars, extracting the heavier elements locked within their

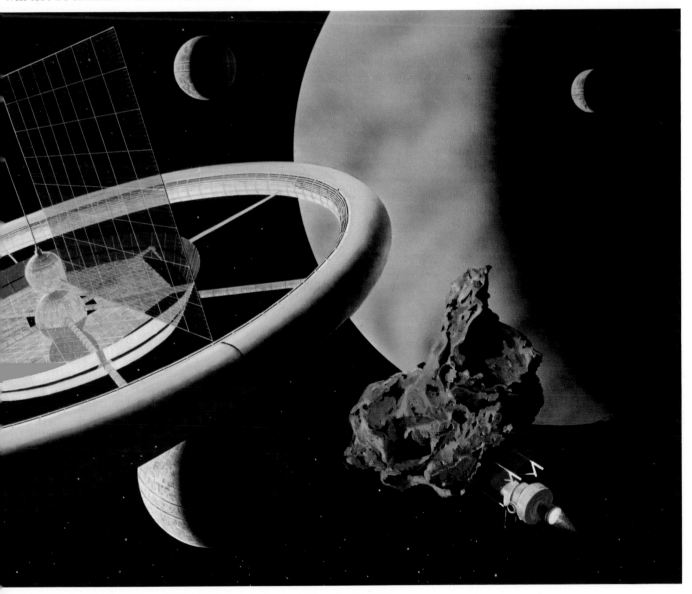

inner cores. Several thousand years from now the concept of dismantling a star will be seen as a straightforward problem in stellar engineering.

How will it be done? One mechanism already exists in Nature – the supernova. At a particular stage in a star's evolution, usually when it is hundreds of times the size of the Sun, processes within its core reach a crisis point. Either it ends up as a tiny dwarf star or a cataclysmic explosion sends fragments of its core hurtling away into space. A star undergoing a supernova can shine briefly with the brilliance of a 100 million Suns.

The problem for the technologists of a Phase Three civilization is to find a star approaching its crisis and then to tamper with its internal structure to insure a supernova. One technique might be to build a kind of gigantic laser cannon. Instead of producing a powerful beam of light, the laser would work at much higher frequency and generate gamma radiation. It would have to be able to irradiate an area about six miles in diameter with a fantastically powerful blast of energy, enough to weaken the star's outer layers at the very moment that its inner stresses and pressures reached a maximum. The sudden laser attack would unleash the explosive forces within the star's core and trigger a supernova.

A few years later, when the lethal radiations accompanying the upheaval had died away, the Phase Three technologists could begin sweeping through the clouds of stellar debris, collecting the vast quantities of heavy elements previously hidden in the star.

This kind of stellar engineering may someday be a commonplace technique. Even more staggering feats may be undertaken involving whole clusters of stars. It may become possible to construct vast interconnected Dyson Spheres weaving an intricate network of artificial planets and space colonies around groups of hundreds, even thousands of stars.

Perhaps the most awesome possibility of all is the eventual use of black holes. A rotating black hole could contain as much energy through its spinning motion as the Sun emits over a period of 10 million years or more. But while the Sun radiates its energy into space, a black hole retains its rotational energy more or

less indefinitely. A black hole therefore could become a perfect energy store for a Phase Three civilization. The problem is to find safe and reliable means to deposit energy in the hole and retrieve it whenever necessary.

Recently the British mathematician Roger Penrose discovered that it is theoretically possible to simultaneously deposit and release energy from a black hole. The technique is quite straightforward. A lump of material is dropped into the hole, but just as it reaches the event horizon it is exploded into two fragments. One is lost into the hole. If the explosion is timed exactly right, the other piece is spun violently away into space, picking up a good deal of the hole's rotational energy. A similar approach can be used with stationary black holes except that the energy drawn off is

Right: a method of obtaining energy from a rotating black hole as proposed by the British mathematician Roger Penrose. Here, a glowing mass is torn into two parts near the event horizon of a rotating black hole in space. One part falls beyond the horizon and is lost into the hole, the other part is flung violently away with more energy than it originally contained.

gravitational rather than rotational. The process is much less efficient but since spinning black holes may be rare, the loss of efficiency could well be an unavoidable drawback.

Penrose's findings are only one possible basis for exploiting black holes. A developed Phase Three civilization may have a far better understanding of the way they work and may discover many ways of exploiting them. The technological problems involved in actually using the Penrose mechanism are unimaginably complex, far beyond even the most extravagant dreams of present-day civilization. But if scientific progress continues unchecked for the next few million years, black hole technology may become a routine engineering skill.

It is even possible that a Phase Three civilization will tame black holes to such an extent that people will be able to use them as a means of instant travel around the galaxy. Current knowledge of rotating black holes suggests that if travelers could survive the journey into one, they might be able to escape again. But on leaving the black hole they would find themselves in a totally different part of the universe – or, as some scientists postulate, perhaps even a different universe altogether.

At present there is only the vaguest idea of how this process might operate and no one can even be sure that the black hole theory will actually be borne out in reality. To our descendants millions of years from now, however, black holes may be both rich sources of energy and convenient gateways making possible instantaneous travel to the most remote regions of the galaxy.

Chapter 9
Life in the Universe

Most scientists are convinced that life must have evolved on planets other than the Earth. In the immensity of the universe it seems certain that the chemistry that created life here must have been repeated, perhaps thousands of times over. But what do we mean by life and what remarkable chemical process converted nonliving substances into the first organisms?

Perhaps the most challenging problem is both to unravel the origins of life and to find out how intelligence has developed. In the search for a complete understanding of the universe, scientists are faced with not only the likelihood of one day finding alien life forms, but also with the far more awesome possibility of contacting extraterrestrial intelligence. If contact is made, will scientists know an alien intelligence when they meet it? Can humans really hope to communicate with beings from worlds light-years away from Earth?

Opposite: conceived like a monstrous ironclad Gulliver towing Lilliputian ships, this robot creature of some distant planet illustrates typical human fantasies about life on other planets – mainly that it will be (as the ancients once imagined their gods) "like ourselves writ large." Most writers depict other life forms as more advanced, more awesome, and invariably hostile. But need they be so?

Life and intelligence

could not hope to recognize them as living simply because the creatures would lie outside the human definition of the term. Recognizable life, therefore, can only be "life as we know it."

There are six basic requirements for something to be recognized as living. First, it must be able to take in food from outside itself, a process called nutrition. Secondly, it must be able to convert this into more of itself. In other words, it has to be able to use food for growth. Thirdly, it must break down part of the food to produce energy. This is called respiration. Fourthly, it has to eliminate waste materials by excretion. Fifthly, it must be able to reproduce likenesses of itself. Sixthly, it must react to changes around it such as fluctuations in temperature, light, and violent stimuli such as physical attack. The most common reaction is movement.

If something exhibits these six features it is certain to be alive within the human definition of the term. It does not matter whether the creature is a microscopically small, single-celled protozoa or a large, multimillion-celled animal such as man himself. But such a simple checklist is unlikely to reveal anything about a creature's intelligence. While human beings and the protozoa are both alive and share a common biological heritage, they represent vastly different evolutionary stages as far as intelligence is concerned.

There is no complete answer to the question "what is intelligence?" Biologists agree that the physical origin of intelligence is the nervous sytem – a complex network of electrical linkages that usually connect a control center or brain with the cells that make up the living creature. At its simplest level, nervous systems control reflex responses

As exploration of the solar system gets underway and scientists look farther afield at the stars and galaxies, one of the most exciting possibilities to be faced is the chance of encountering extraterrestrial life and intelligence. But will the new forms of life and intelligence be recognized even if they are met? The answer depends on how well humans themselves understand what life and intelligence are.

The nature of life is a mystery that science may never fully unravel. But modern biology has made remarkable advances in identifying the basic properties that seem to distinguish the living from the nonliving. Life – as far as the scientist is concerned – can be defined in terms of the essential things plants and animals have to do in order to live. But this kind of definition is necessarily limited to this particular corner of the universe. After all, life can only be observed and analyzed as it exists on Earth. The fact remains that if scientists encountered living creatures that conformed to quite different criteria, they

Above: the American biologist James Watson (right) and the British biophysicist Francis Crick with their model of the DNA molecule. The discovery of the structure of the DNA molecule in 1953 was one of the most important biological discoveries of the 20th century.

Right: a seven-day chick embryo floating in amniotic fluid. The embryos of most of the higher animals and man develop from an egg that grows and divides into thousands of millions of cells before eventually becoming a fully formed young animal or baby.

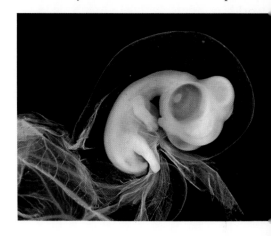

Left: a spider uses a web to catch its prey. Each filament of the web is covered with a sticky substance so that any flying insect that lands on the web becomes entangled and immobilized. The spider bites and paralyzes the insect – a housefly, for instance – before devouring it. Its behavior is not learned, the insects and lower animals are more akin to computers, with every aspect of their behavior "programmed" into their genes.

Below: Frank Whittle (right), the British test pilot and aeronautics engineer whose jet-engine design revolutionized every aspect of the aeronautics industry from military aircraft and guided missiles to passenger aircraft and helicopters. Such creative intelligence as shown by inventors such as Whittle is one of the main distinctions that marks man out from the animals.

such as those in plants and lower forms of animals. In more complex organisms such as the apes, dolphins, and humans, nervous responses are so advanced and complex that such creatures are said to think.

In itself there is little to commend such a high degree of intelligence. Insects, for example, are a very successful class of animals and have flourished for millions of years without the intelligence needed to build railways, automobiles, or even spacecraft. Indeed, a visitor from another world, watching the activities of ants or bees might easily mistake them for highly intelligent creatures. Conversely, the same alien could well draw the opposite conclusion about mankind by watching some human activities.

To avoid making this kind of mistake, there must be some clear definition of intelligence. Like the alien visitor, conclusions must be drawn through observing behavior. But the whole process of observation becomes very complicated when dealing with a creature never previously encountered, and therefore

just the kind of creature for which it would be necessary to make an assessment of intelligence. For example, how would a human react when faced with an organism physically quite different. In such a case, completely inaccurate assumptions might be made about what its behavior implies. The upturned corners of human lips and the offering of an outstretched hand is a gesture of friendship and welcome. But what might it mean to a creature without lips or hands? As far as we are concerned, it would be easy to mistake a raised tentacle as a movement of blind aggression rather than a sensitive appeal for understanding between two alien but intelligent species.

Perhaps the best way of measuring intelligence is by judging how well creatures communicate. A Canadian anthropologist, R. B. Lee, believes that human intelligence reduced to its essentials is the same thing as human language. Using this idea in a wider context, intelligence can be regarded as the ability to transmit complex and, in particular, abstract information from one individual to another. In practical terms, if a creature can communicate its ideas to humans, it must have a fairly well-developed intelligence. If those ideas are themselves fairly complex and abstract, then it can be assumed that it is a highly intelligent being – whether it comes from some region on the Earth or the farthest reaches of interstellar space.

The chances of extraterrestrial civilizations

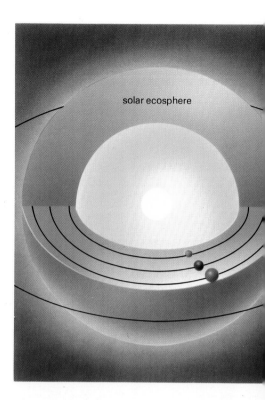

Few people today still believe that life is unique to Earth. Among the countless millions of stars in the galaxy to which the solar system belongs, there must be many with planets capable of supporting life. Although life is a rare and remarkable biochemical accident, living organisms must surely have evolved on some of the vast numbers of possible host worlds.

But the mere fact of life is not necessarily a way of measuring the likelihood of a civilization emerging. Civilizations require a high degree of intelligence and social organization. If life is a com-

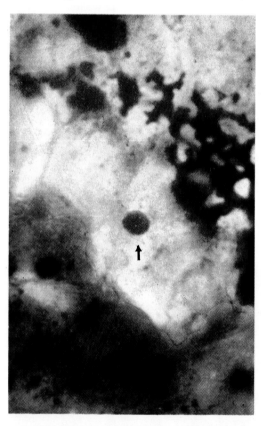

paratively uncommon phenomenon in the universe, extraterrestrial civilizations must be very rare indeed. But just how rare are they? And most important of all, how many of them in the Milky Way have developed the technical ability to travel in space and communicate over great distances? These, after all, are the

Below left: this greatly magnified meteorite fragment has a patterned area of rock (arrowed) like that produced by certain algae found on Earth. There has been much speculation that meteorites may play an important part in distributing life-forming substances throughout planetary systems. Confirmation of the theory must await the recovery of meteors from space before they have entered the Earth's atmosphere and become contaminated.

Above left: the solar ecosphere showing the four planetary bodies that probably evolved from the same cloud of gas and dust. Of the four only two could possibly have supported life – Earth and Venus, both of which have similar masses. If Venus had an orbit some 15 million miles further from the Sun, its atmosphere might have been almost identical to the Earth's. The masses of both Mars and Mercury are too small and their gravitational fields too weak to have retained atmospheres that would support advanced forms of life.

Above right: a cross section of a region of space in a sphere around the Sun with a diameter of 33.4 light-years. Even in this relatively small region there are at least two other stars which may have planets capable of supporting life.

Right: the atomic bomb. It seems that the more technologically advanced a civilization becomes, the greater the possibility that it will develop the means to destroy itself.

civilizations with whom mankind may one day make contact.

Scientists have already made estimates of their number. The calculation is a multiplication of a number of different quantities. The total number depends first on the average rate at which stars are formed in the galaxy. Astronomers know this number fairly accurately. It depends next on the number of stars that have planets around them. This is less well known but a reasonable estimate can be made. It depends on the proportion of the planets that can support life and on the number of those on which life actually evolves. Next it depends on the further fraction of planets on which an intelligent species appears among the emerging life forms. It then further depends on the fraction of these intelligent species that create a technically advanced civilization. Finally – since we are only interested in civilizations that are in existence at the same time as that on Earth all the previous numbers have to be multiplied by a further factor that takes account of the average lifetime of a technically advanced civilization.

Obviously, most of the various factors can only be estimated very roughly. A good deal of intelligent guesswork has to be involved. But the best modern estimate of the number of advanced civilizations in the Milky Way turns out to be about

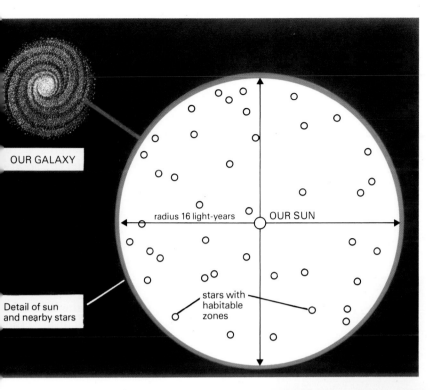

OUR GALAXY

radius 16 light-years — OUR SUN

Detail of sun and nearby stars

stars with habitable zones

why have scientists seen no signs of them? Why has Earth not been colonized by aliens?

The chance of advanced civilizations existing in the Milky Way is one tenth of a number that can only be guessed at. If it is very small and technical civilizations have only a brief lifetime, the total number will be even smaller. But if an optimistic estimate of millions of years can be accepted, the number must be in the order of hundreds of thousands.

Perhaps the most interesting and important fact that emerges from the calculation is that however large or small the answer may turn out to be, it cannot be zero. In other words, if our basic assumptions are correct, there must be some advanced technical civilizations in the galaxy. But it must be admitted that the answer to our impossibly difficult sum may be that there is just one such civilization, our own.

one tenth of the average lifetime in years of such a civilization.

It is impossible to be sure of what the average lifetime might be. We know of only one example of an advanced civilization that has definitely evolved – the one here on Earth. Mankind is at such an early stage in its development it is difficult to make any firm prognosis about future prospects. Some pessimistic estimates do not rate mankind's chances very highly and there is a school of thought that argues that any technically advanced civilization carries within it the seeds of its own destruction. Certainly the dangers that threaten civilization can already be sensed – nuclear war, social disintegration, energy shortages – but it does seem possible to contain all these and deal with them successfully.

If technical civilizations can survive for very long periods, however, they will presumably become increasingly advanced and sophisticated. Human civilization is only about one hundred thousandth of the age of the galaxy and it is reasonable to expect civilizations in other star systems to have reached the stage of building interstellar empires by now. Enrico Fermi, the Italian physicist who helped to build the first nuclear reactors, summed up the dilemma in the simple question: "Where are they?" If civilizations that are millions of years old exist,

Origins of life

For centuries people believed that life on Earth developed spontaneously. The 6th-century BC Greek philosopher Anaximander of Miletus is said to have taught that living creatures first emerged from some moist substance as it was dried out by the Sun. Aristotle believed that animals and plants sprang from dirt, slime, and decaying matter.

The concept of life as some kind of instantaneous event was popular until the 17th century. Only then did the first scientific investigations begin to reveal the fallacy of the idea. In 1668, Italian scientist Francesco Redi performed an experiment showing that maggots were not, as then commonly believed, spontaneously created in rotting meat. He placed one dead fish in an open jar and another in a sealed jar that kept out the flies. Only the specimen on which the flies crawled and deposited their eggs became infested with maggots. In this way Redi showed that the maggots were not spontaneously formed in the fish but were merely the progeny of the flies able to lay their eggs in the rotting flesh of the uncovered specimen.

By the 1890s scientists were beginning to speculate about life having chemical origins. A British physicist, John Tyndall, pointed out that the human body, remarkable though it was, consisted of the same elements that make up rocks, soil, and other inorganic material. Tyndall concluded that life was just a special arrangement of the common natural elements that are found everywhere.

It was a bold idea and theologians attacked it bitterly. It seemed to them an unpardonable heresy to reduce the origin of life to an arrangement of chemicals. It was also a direct challenge to the Scriptures, which held that a divine hand had created life. At the time it was too difficult to look beyond Tyndall's idea and see perhaps a divine hand manipulating the chemistry by which life was formed.

But the real breakthrough was made by the English naturalist Charles Darwin. Taking Tyndall's ideas an important step further, he wrote: "But if we could conceive in some warm little pond, with all sorts of ammoniac and phosphoric salts, light, heat, electricity, etc., present, that a protein compound was chemically

Left: lightning flashes in the atmosphere when the Earth was still in a primitive stage of evolution could have supplied energy to form the amino acids and nucleotides from which living organisms are made.

formed ready to undergo still more complex changes, at the present day such matter would be instantly devoured or absorbed, which would not have been the case before living creatures were formed.''

Darwin's idea is the key to all modern ideas of the origin of life. The conditions on Earth at around the time life first formed must have been similar in some respects to Darwin's "warm little pond." In fact the reality was much more dramatic. The ancient seas that must once have covered much of the Earth were not merely salt water. They were a complex watery soup of chemicals. Nei-

ther was Earth's atmosphere the breathable blanket of gases it is today. Instead it was a turbulent mixture of carbon dioxide, nitrogen, hydrogen, ammonia, and water vapor. Earth in this primordial era was a terrible and spectacular cauldron of lightning storms and volcanic eruptions.

In 1953 two American chemists, Stanley Miller and Harold Urey, tried to simulate some of the features of this ancient environment. For a week they passed electrical discharges – representing lightning bolts – through a mixture of ammonia, methane, hydrogen, and water. By the end of the experiment they had a deep-red solution, rich in large organic molecules, including some amino acids, the basic building blocks of life. In 1965, Cyril Ponnamperuma, a biologist working for the American space agency NASA, using techniques similar to those pioneered by Miller and Urey, succeeded in synthesizing the basic ingredients of the DNA molecule, a fundamental part of cellular material, and essential to life.

Today most scientists agree that these experiments show that life must have evolved in the ancient seas. Under the stimulus of lightning, heat from volcanic activity, and ultraviolet radiation from the Sun, chemicals combined and recombined to form increasingly complicated molecules. After millions of years, a particular configuration of molecules emerged that was able to replicate itself. Following this first step, still more com-

Below left: a 15th-century engraving of the barnacle tree. Many people believed that barnacles grew on trees and that as they fell, geese hatched out. Others held that living things could arise from nonliving matter – a process known as spontaneous generation.

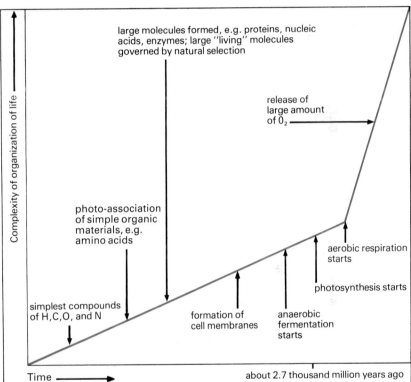

Above: a graph showing the complexity of living things increasing with time. The timescale is only very approximate. The curve rises sharply with the appearance of photosynthetic plants. The large amounts of oxygen they produced made the atmosphere suitable for the evolution of animal life.

plex molecules appeared with growing numbers of the attributes we now associate with living matter. Ultimately the first tiny living thing formed out of countless successive syntheses of other molecules. It must have reproduced quickly, devouring the less advanced ingredients of the primordial soup covering the Earth. Its progeny will have included slight mutations, different types of living organism which, if successful in the battle for survival, spawned others of its own kind. The 4000-million-year story of evolution, which culminated in the emergence of mankind, had finally begun.

Evolution of Life

The long road from the first organic molecules to human beings has taken around 4000 million years. For most of that period, life on Earth was restricted to simple algae and marine organisms. It is only in the last 500 million years that animals have moved over the surface of the world and the first land plants have grown. For most of this time, reptiles dominated the Earth. The first mammals did not appear until only 150 million years ago and man himself did not emerge until about 5 million years ago. Human civilization is a mere few thousand years old.

The awesome timespan of life on Earth is the very reason why there is such a fantastic diversity of living things. The keynote of the last 4000 million years has been change. Infinitesimally slow but remorseless, the slight changes from generation to generation among plants and animals have brought about the rich variety of organisms that flourish today.

The now accepted ideas of how life has evolved are comparatively recent. Only 100 years ago most scientists believed that the bewildering variety of plants and animals showed the enormous versatility of the divine Creator. To suggest otherwise was regarded as a challenge to basic religious doctrines. To biologists, zoologists, and botanists of the mid-1800s, the past history of life was simple because it was static. They believed that all living things had been created exactly as they were in the world of the 19th century. Although this concept seems implausible today, at the time the Earth was believed to have existed for only a few thousand years. The true immensity of past geological ages was not realized and it is this very idea of long periods of evolutionary time that makes our modern concepts reasonable in the first place.

The origin of the modern theory of evolution lies in the remarkable pioneering work of the Swedish scholar Carolus Linnaeus. In the mid-1700s, Linnaeus

Below: God creates Adam, the first man. The account of the creation of the first man from the clay of the Earth – and from him the first woman – is found at the beginning of the Book of Genesis in the Bible. Until modern times, Jewish and Christian believers accepted this account as the historical origin of human life.

worked out a means of naming and classifying organisms according to the physical relationships between them. By considering the degrees of likeness between varieties of plants and animals he was able to distinguish distinct plant and animal types. Linnaeus' work provided biological science with a fundamental tool. For the first time a systematic framework could be devised within which plants and animals could be seen as members of distinct families or species.

Once scientists began to see the living world in this ordered and logical way, a few began to speculate more deeply about the diversity of species. In the late 1700s French biologist Jean Baptiste Lamarck made probably the earliest suggestion that species could from time to time alter and so produce new species. He had no idea by what mechanism such change could take place and it was not until the remarkable work of the British naturalist Charles Darwin that Lamarck's concept of evolutionary change was given a firm rational basis.

During the 1830s the British Navy despatched HMS *Beagle* on a voyage of exploration. Young Charles Darwin sailed as ship's naturalist, entrusted with

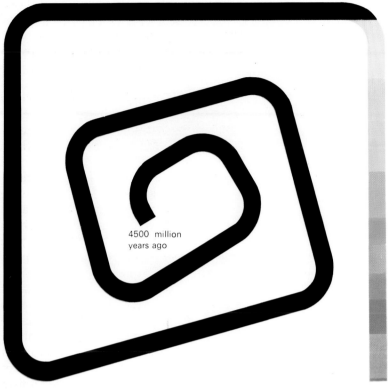

Millions of years ago	
600	Cambrian
500	Ordovician
440	Silurian
400	Devonian
350	Carboniferous
270	Permian
225	Triassic
180	Jurassic
135	Cretaceous
70	Tertiary
1	Quaternary

4500 million years ago

Left: Earth's 4500 million-year history, represented here by the snaky line, with the column of geological eras in color, spanning less than one seventh of the total length.

Below: two of the 14 different species of finch that evolved on the Galapagos Islands off South America. These volcanic islands have never been connected to the mainland so that the living things on them must have flown or floated across the 600 miles of sea between them and the nearest land. They then evolved into new species isolated from all mainland influences.

the job of collecting specimens of flora and fauna from all over the world. During the five years of the *Beagle*'s voyage, Darwin became convinced that the great diversity of life was not a sign of a single act of inventiveness on the part of a divine Creator but that species were constantly, if slowly, changing and giving rise to new species.

Darwin found the most impressive evidence for evolution on the Galápagos Islands off the Pacific coast of South America. There he found that creatures such as giant tortoises and ground finches had clearly evolved on the Islands. The finches – today called Darwin's finches – were particularly important to Darwin's ideas. He assumed that the Islands had originally been colonized from the South American mainland by just a single species of finch. Yet on the various Islands, Darwin observed more than 12 different kinds of finch. Individual species had evidently evolved from the original colonizing species, developing distinct physical and behavioral characteristics to suit them to their environment.

But although Darwin was convinced that evolutionary change was slowly altering species, he could not work out

by what natural mechanism it operated. Why did the change take place? Wrestling with the problem after his return to England, in 1837 Darwin read a new work by the celebrated sociologist Thomas Malthus on survival factors in human populations. It was the key for which Darwin had been searching. The driving force of evolution became clear. It was the need for living organisms to survive, prosper, and thereby reproduce.

Darwin knew that when a member of a single species reproduces, the offspring sometimes has physical characteristics slightly different from other members of its species. An offspring of this kind is a mutation. In some cases, the mutation – however minor – might enable the new plant or animal to prosper in its environment more successfully than other unmutated members of the same species. The more successful an organism is, the more likely it is to reproduce in large numbers. This means that over a long period of time, mutations within a species that make the mutated individuals more successful in surviving, will tend to occur with increasing frequency until the mutant strain within the species is sufficiently large to form an independant species or subspecies of its own.

This process – known as natural selection – is the cornerstone of the modern theory of evolution. It is a process in which all living things are constantly adapting to their environment or finding new ecological niches within which to flourish. The single common battle that links them all is the struggle to survive. Living things are constantly competing with each other. In forests, trees com-

Above: an ice fish, also called a "bloodless' fish because, unlike all other vertebrates, it has no oxygen-absorbing red bloodcells. Ice fish appear to take up all the oxygen they need through their gills and possibly directly through their skin. Most of the 17 species of this fish live in the Antarctic, feeding on crustaceans.

Right top: the rock ptarmigan and, right, the willow ptarmigan are two species of game bird that occur in the Arctic. They are largely brown or gray in the summer, but both species are mainly white in winter. Their change of color gives them some protection against birds of prey and predatory animals. Their extremely dense plumage also protects them against the cold.

pete for light and sustenance from the soil. In the animal world, creatures fight for food and territory within which they can safely rear young. But nature is harsh and unyielding. Most animals do not survive long enough to breed. Only the fittest can stay alive to pass on their own inherited features and so perpetuate the physical and behavioral characteristics that make them successful.

A spectacular example of natural selection comes from the 20th century. The growth of industry meant that many trees became blackened by the soot and smoke from factory chimneys. This change in the environment favored the previously very rare dark mutation of the peppered moth, *Biston betularia*. The once-common light form of the moth was preyed upon by birds so extensively that it became a comparatively rare sight. The black moth – perfectly camouflaged against the soot-blackened trees – became much more successful in surviving and breeding. After the widespread clean-air legislation in many European countries since the 1950s, the lighter colored moth has staged a comeback. It is now the dark peppered moth that is conspicuous against the cleaner trees.

Once Darwin had grasped the basic idea, he set to work to justify it on the evidence of his numerous observations, recorded during the epic voyage aboard the *Beagle*. Despite the encouragement of his friends, Darwin delayed publishing his theory, insisting it was still not fully documented. Then in 1855 he received a sudden shock. He was sent a paper written by a traveler and naturalist called Alfred Russel Wallace. Writing to Darwin from the Malay Archipelago, Wallace told him about a theory he had recently devised and asked for Darwin's views about it. Darwin was astounded to read in Wallace's paper the very ideas about natural selection that Darwin himself had recently worked out.

The following year at a meeting of the Linnaean Society in London, Darwin presented his own and Wallace's views. In 1859 Darwin published the fully documented account of his theory of evolution. Called *On the Origin of Species*, the book is today regarded as one of the most important textbooks ever written. It caused a sensation at the time and was branded as blasphemous.

The controversy was inflamed still further by Darwin's suggestion that mankind itself was the result of a long chain of adaptations by which humans had evolved from the ape. The wife of one eminent clergyman is reported to have protested at the idea saying ". . . let us hope that it is not so. But if it is, let us hope it does not become widely known!"

Today Darwin's basic ideas of evolutionary changes are universally accepted. Present studies of fossil records and appreciation of the true scale of geological time has made it possible for scientists to verify how life has changed over the millennia. They can look back over thousands of millions of years and trace the emergence from the oceans of the first marine species to adapt to life on the land. From these early animals and plants the immense range of different species gradually developed and adapted to form the types of organisms seen today.

But even now evolution continues. The story of life's development can only end when life itself ends.

Below: drawings of two fish by the British naturalist Alfred Wallace, made in 1848 while on a voyage up the Negro river in South America. Wallace is noted for having worked out the theory of the origin of species through natural selection independently of Charles Darwin.

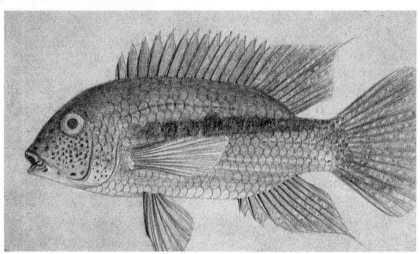

Life in the solar system

Until quite recently many people seriously believed that advanced life forms might well have developed on Earth's nearest neighbors, Mars and Venus. In the popular imagination, Mars was often regarded as harboring a technologically superior and possibly malevolent civilization that had ambitions to take over the Earth. In fact, in 1938 mass hysteria gripped the United States when a radio dramatization of British author H. G. Wells' classic story *War of the Worlds* started with a fictional news bulletin announcing that Martian invaders had landed. The American public needed no convincing. Whole populations reported seeing the invaders and panic swept the country.

By contrast Venus was often regarded as a young world, much as Earth must have been millions of years ago. Amidst steamy jungles, people imagined dinosaurs to be living – monstrous but essentially primitive life forms that could threaten only those intrepid enough to visit the planet.

But what is the reality behind all the non-scientific speculations of the past? Today we know vastly more about planetary conditions elsewhere in the solar system. And almost everything we have learned has suggested that only on our own world are conditions right for life. The first manned expeditions to the Moon, remote-controlled landings by probes on Mars and Venus and close fly-bys of Mercury, Jupiter, and Saturn, have all failed to reveal any real evidence for extraterrestrial life, even at the simplest level of bacteria or algae.

Venus – once thought to be covered by lush jungles and oceans – is now revealed as a scorched wilderness of dry rocks and dust. The atmosphere is a choking mixture of sulfur and carbon dioxide. The only rain that ever falls on this inhospitable world is dilute sulfuric acid. No one seriously expects life to have evolved under these extreme conditions.

Below: Earth invaded by Martians. The red planet Mars, named for the god of war, has invariably been featured by science fiction writers as harboring a hostile civilization. This picture illustrates the British author H G Wells's story *The War of the Worlds*, in which the Martian invaders are eventually quelled by Earth's bacteria, against which they have no immunity.

Mars still remains an enigma. On July 20 and September 3, 1976 two remote-controlled laboratories called Viking 1 and 2 touched down on the Martian surface. The probes were amongst America's most ambitious space projects and they carried equipment designed to conduct sophisticated experiments, including three designed to detect the presence of life. Of the three, the experiment known to space scientists as the labeled release test, revealed the first promising result. The idea of the test was to try to spot evidence of metabolic activity in soil samples gathered from the Martian surface. A collector arm extending from the Viking probe picked up some soil and delivered it to the experiment container inside the spacecraft. The soil was then bathed in a nutrient solution so that any bacterial life would be encouraged to increase its metabolic activity. An easily identifiable by-product of such activity is carbon dioxide. But to make identification even easier, the nutrient sap fed to the soil sample was "labeled" with radioactive carbon 14. Four experiment cycles were

run at one of the Viking landing sites and five at the other. Radioactive carbon dioxide was detected on every occasion that the soil was not first heated to a sterilization temperature. The gas failed to appear only when the sample was first heated.

The results were greeted with considerable excitement. But when the other life-seeking experiments aboard Viking consistently failed to detect any supporting evidence, doubts began to grow. Today most scientists believe that Viking has not shown conclusively that life exists on Mars. The results of the labeled release experiment have been explained convincingly by supposing a purely inorganic chemical reaction in the Martian soil.

But although Viking has provided no firm evidence it has certainly not ruled out the possibility of life on Mars. After all, only a few ounces of soil have been examined taken from two specific places on the face of the planet. The search for Martian life will certainly continue and may yet be successful. The real test will

Above: Jupiter, as seen from one of its satellites, according to a 19th-century artist. The vegetation is clearly based on that found on Earth since little was then known of other celestial bodies.

come when the first mobile laboratories are landed on the planet.

One of the most startling discoveries of strange life forms in the solar system occured quite recently – on Earth. At the very bottom of the ocean, where no sunlight penetrates, scientists have discovered bizarre plants and fish that thrive around vents in the deep-sea floor that allow heat to bubble through from below the Earth's crust. These are the first life forms discovered that do not, in some way, depend on the Sun for survival.

Elsewhere in the solar system, the chances of finding life seem remote. However, there is evidence that in the turbulent atmosphere of Jupiter there are some fairly temperate regions. Although the atmosphere is a poisonous mixture of gases to humans, its constituents – methane, ammonia, hydrogen, water vapor – are actually the basic ingredients from which life on Earth formed millions of years ago. Fantastic though it seems, scientists do not rule out the existence of some primitive life forms evolving high above the surface of the giant planet.

Communication with extraterrestrials

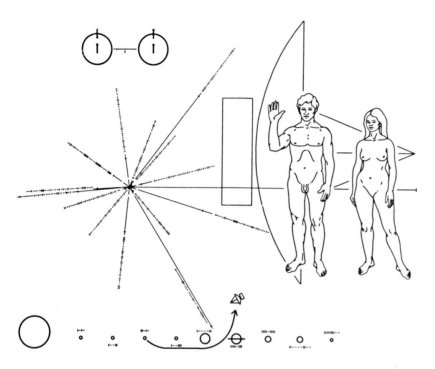

Mankind's first attempt to communicate with extraterrestrial civilizations began on March 3, 1972. On that day American space agency NASA launched Pioneer 10, a probe designed to pass close to Jupiter. But the project was unique not only because of its planetary objective but because its trajectory would take it far beyond Jupiter, eventually carrying it outside the solar system and into interstellar space itself.

Fixed to the body of the spacecraft is a six-inch by nine-inch gold-anodized plaque that carries a message that should remain intact for hundreds of millions of years. It is written in the only language we believe we might share with other civilizations – the language of science. Together with graphics depicting the human form, there is scientific data related to the atomic structure of hydrogen. In an effort to give an unambiguous location for the Earth, its position is indicated by a radial burst of lines which should tell an advanced civilization where the Earth is in relation to 14 pulsars located in the Milky Way.

The plaque aboard Pioneer 10 is more of a gesture than a serious effort to con-

Above: a salute from Earth. This drawing was etched on an anodized gold plaque aboard the Pioneer 11 space probe. From the top left are the symbols of a hydrogen atom, the position of certain pulsars seen from Earth, a woman and a man in front of Pioneer, and our solar system (bottom) indicating the planned trajectory of the craft into interstellar space.

Right: Dr Frank Drake with other senior staff members at the Arecibo Observatory, Puerto Rico. It was Dr Drake who, in 1960, set up Project Ozma, the first attempt to listen to broadcasts from extraterrestrial sources. Not surprisingly, as it is estimated that only one in a million stars might be the source of purposely directed signals, this pioneer effort was unsuccessful.

A much more ambitious project was outlined in 1971 at a conference organized at NASA's Ames Research Center. Astronomers and physicists discussed the possibility of a massive follow-up to Ozma called *Project Cyclops*. The idea is to deploy a vast array of about 1000 individual radiotelescopes extending over a huge area of land. The sole objective of this complex would be to mount a systematic and long-term search for signals from other civilizations in the galaxy.

But the problems of undertaking such an elaborate search are formidable. The galaxy is vast and even if millions of civilizations are beaming signals into space, there is still only a very small chance of Earthbound scientists choosing the right moment to listen, the right place to beam the telescopes and the right wavelength to tune into.

The problem is deciding the best wavelength for the search is especially daunting. There are an enormous number of choices. It is a real possibility that Cyclops could choose the right star to investigate but miss the message beamed from it by being tuned to the wrong radio frequency. What frequency does one choose? Some scientists have made the simple suggestion that if another civilization wanted to attract attention it would make detection of its signal as easy as possible. Faced with the choice of wavelengths and frequencies, the alien civilization might choose one of universal significance. This is why many SETI ventures focus on what's known as the "water hole" of fre-

tact other civilizations. Even if the spacecraft were aimed at the nearest star to Earth, it would not arrive there for at least another 80,000 years. But Pioneer 10 is in fact bound for a region of space where there are no nearby stars. Scientists estimate that it may pass close to the star Aldebaran in about 2 million years time and probably will not actually enter another star system for at least 10,000 million years. By that time mankind may no longer be living in the same solar system and could conceivably have already made contact with other nearby civilizations by manned interstellar expeditions.

Long before Pioneer 10 was launched, scientists were already making efforts to pick up messages sent to Earth from distant civilizations. The emergence of radioastronomy in the 1940s at last provided astronomers with the tools to receive such messages. In 1960 the first systematic SETI – Search for Extraterrestrial Intelligence – experiment was attempted. Called *Project Ozma,* the experiment was master-minded by an American astronomer, Dr Frank Drake, using an 85-foot radiotelescope at Greenbank in West Virginia. He spent a total of one week listening for signals from two of our nearest neighboring stars, Tau Ceti and Epsilon Eridani. Current estimates suggest that only about one in a million stars might be the source of artificial signals, and it is not surprising therefore that Drake's efforts were unsuccessful.

Above: the proposed moon-based Cyclops interstellar search system, from a high altitude aerial view. In this artist's concept the central control and processing building can be seen together with an array of 150 100-meter antennae.

Below: the highly sensitive radio telescope being erected in Kharkov Oblast in the USSR. It is designed to detect cosmic signals from 10,000 million light-years away.

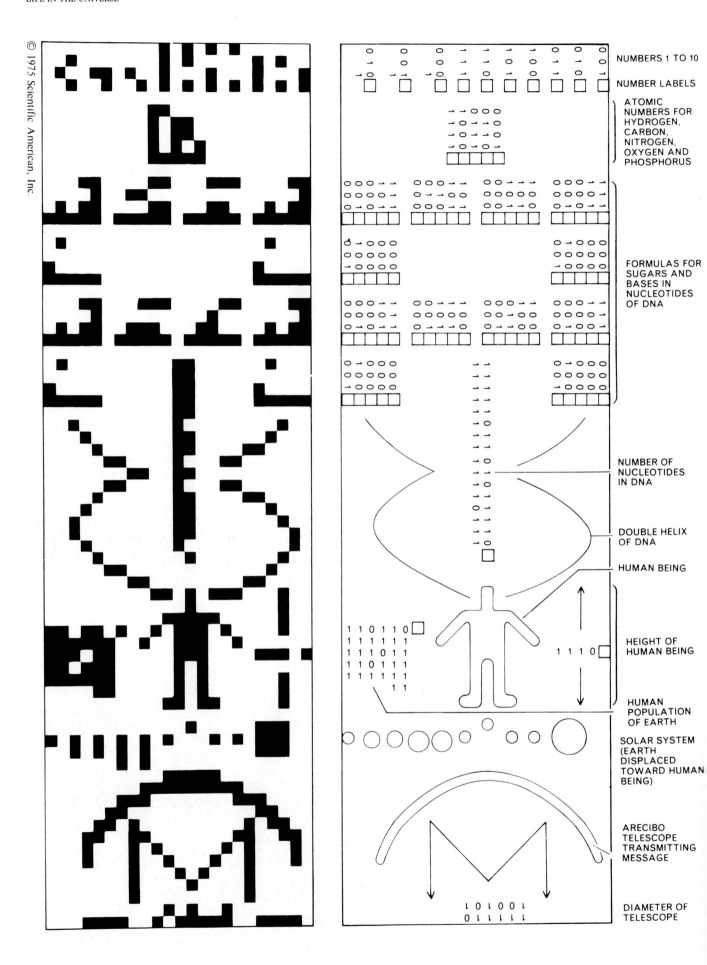

NUMBERS 1 TO 10

NUMBER LABELS

ATOMIC NUMBERS FOR HYDROGEN, CARBON, NITROGEN, OXYGEN AND PHOSPHORUS

FORMULAS FOR SUGARS AND BASES IN NUCLEOTIDES OF DNA

NUMBER OF NUCLEOTIDES IN DNA

DOUBLE HELIX OF DNA

HUMAN BEING

HEIGHT OF HUMAN BEING

HUMAN POPULATION OF EARTH

SOLAR SYSTEM (EARTH DISPLACED TOWARD HUMAN BEING)

ARECIBO TELESCOPE TRANSMITTING MESSAGE

DIAMETER OF TELESCOPE

quencies between 1.4 gHz (the natural radio emission frequency of hydrogen) and 1.7 gHz (the frequency of radical hydroxyl (OH)), which is related to water, the source of life. This water hole also happens to be radio quiet, that is, used little on Earth so that it is uncluttered and easy to monitor. Other scientists question concentrating on this range however, noting that an alien civilization might not want its communications detected by emerging civilizations and might use much higher frequencies.

So far there have been over 35 SETI investigations around the world since Project Ozma. They have been launched in the U.S.A., the U.S.S.R., Canada, France, Germany, the Netherlands and Australia. They have examined specific stars near and far, and stars at random (all told, several thousand), over a variety of wavelengths and frequencies. And still not a peep. But, there are millions of stars to listen to (even after one has disallowed stars either too big, too old or too volatile to allow life on their planets) and a vast spectrum of frequencies to be monitored. In effect, the search has just barely begun.

One of America's most enthusiastic proponents of the search for extraterrestrial intelligence is Carl Sagan, Professor of Astronomy at Cornell University. In 1973 he wrote "It is an astonishing fact that the great 1000-foot-diameter radio-telescope of the National Astronomy and Ionosphere Center, run by Cornell University in Arecibo, Puerto Rico, would be able to communicate with an identical copy of itself anywhere in the Milky Way." A year later Sagan's enthusiasm was translated into action and the giant Arecibo telescope began beaming a carefully worked out message into space.

Behind the experiment lay the realization that even if there were millions of civilizations capable of communicating, no one would be aware of the others unless someone decided to stop listening and start signalling. Choosing a cluster of stars in the Milky Way as a target, a coded message was sent from Arecibo describing the basis of life on Earth and giving simple details about the solar system and the telescope transmitting the signal.

Even if such a message is picked up by another civilization, it may be centuries

Opposite left: the message beamed by the radio telescope at the National Astronomy and Ionosphere Center, Arecibo. The message is in binary code, which is decoded by breaking up the characters into 73 consecutive groups of 23 characters, and arranging the groups one under another. This results in this visual message, in which each 0 of the binary code represents a white square, and each 1 a black square. The translation on the right shows the message decoded.

Below: a "Sun-pumped" laser – so called because it is powered by the Sun's rays collected in the parabolic mirror.

before we can hope for a reply. Interstellar radio communication is bound to be a slow process with conversations taking thousands of years. But it is just possible that at some time in the future, we may discover a much quicker way of sending signals. At present this idea seems to contradict one of the basic laws of nature that states that information cannot be transmitted from one place to another at a velocity greater than the speed of light. But modern astronomy is already revealing mysteries that challenge many of our basic concepts about the universe and we cannot discount the possibility of future breakthroughs that might make practical two-way conversations spanning the galaxy in minutes rather than centuries.

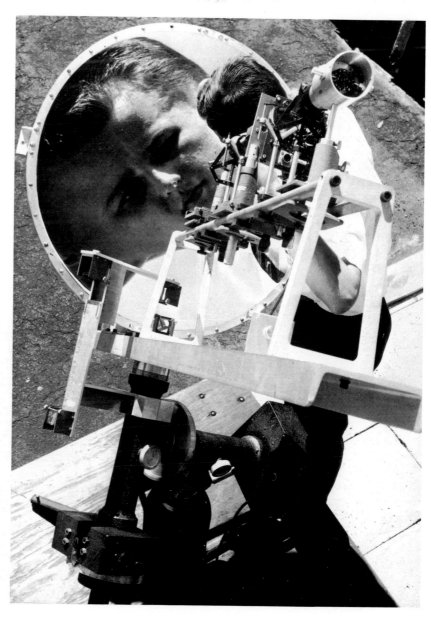

Chapter 10

The UFO Enigma

No one would deny that there are such things as unidentified flying objects. Disagreements begin when people start trying to explain them. Most are later identified as aircraft, weather balloons, even planets seen under unusual conditions by untrained observers. But a few sightings cannot be dismissed so simply. Among all the incidents, ranging from genuine misidentifications of natural phenomena to fabrications by cranks and publicity seekers, there remains a residue of inexplicable sightings.

Despite the real mystery that UFOs represent, little has been done to seriously investigate them. Only recently have scientists gone on the offensive. Instead of bemoaning the lack of reliable data, groups dedicated to actively seeking out UFO phenomena are being established, using the latest technology to record what they find. These latest efforts could put an end to all the speculation and stories surrounding UFOs. They may even lead to the greatest breakthrough in human history – mankind's first contact with extraterrestrial intelligence.

Opposite: two British UFO enthusiasts enlist modern technology to help them crack the mystery of unidentified flying objects.

Ancient myths – has earth been visited?

One of the most startling theories concerning extraterrestrial life to emerge in recent times suggests that Earth has been visited by aliens on a number of occasions in the past. Research has uncovered a wide range of ancient records and artifacts that some people interpret as evidence for encounters between humans

and extraterrestrial beings.

One of the best-known discoveries was made in the Peruvian Andes in the 1930s. On an arid 37-mile-long plain in the Nazca region of southwestern Peru, archaeologists discovered a system of strange geometrical patterns formed by deep furrows in the ground exposing the

Below: the strange markings on the plain near Nazca, Peru, seen from the air. Because the full design of these ancient lines are only clear when viewed from high above, it has given rise to a theory that the plain was in fact an airfield built by space visitors to Earth.

yellowish subsoil. Some of the lines run parallel for miles. Others cross or link together to make distinct geometrical shapes. In some places the figures of animals are also outlined in the soil. One of the most usual explanations of the Nazca mystery is that the markings are the remains of road systems built centuries ago by the Incas. Another suggestion is that the patterns have an astronomical and religious significance.

But the really perplexing aspect of the furrowing is that the animal outlines and the complex interrelationship of lines become fully apparent only when viewed from the air. There are not even any nearby hills from which the supposed Inca builders could admire their handiwork.

Why should a technologically primitive people such as the Incas go to such trouble in tracing out patterns they could never themselves see? Surely not in order to build a system of roads leading nowhere. And if there was an astronomical significance, no modern astronomers can even guess what it might have been.

The boldest explanation has been proposed by the Swiss author Erich von Däniken. He believes that the markings were designed to be seen from the air by people the Incas regarded as gods. His breath-taking hypothesis is that the plain of Nazca is covered with the remains of an ancient airstrip, a landing field for alien spacecraft. Von Däniken even speculates that the airstrip might have been built by the aliens themselves, the animal figures being later elaborations added by awed and respectful local tribes as a kind of appeasement for the gods.

It certainly seems reasonable to assume that visiting space people would be regarded as gods by primitive tribes. But von Däniken takes his hypothesis one step further. Perhaps his most sensational claim is that the ancient space visitors are actually our forebears. He suggests that *homo sapiens* may well be the result of an artificially induced mutation, a kind of cross-breeding between the extraterrestrial beings and the apelike creatures they found inhabiting the Earth. This idea supposes that the landings at Nazca were not the first and that the extraterrestrial beings had come to Earth perhaps millions of years earlier.

To justify his ideas, von Däniken uses

a wealth of archaeological evidence. Almost every ancient figure depicted with strange headgear is said to be a spaceman. Artifacts with the slightest suggestion of a winglike shape are regarded as representations of spacecraft. To support his case, von Däniken cites rock drawings from all over the world. In the Tassili mountains in the Algerian Sahara, for example, there are sketches that von Däniken is convinced show astronauts wearing round helmets floating in the weightlessness of space.

If Earth has ever been the temporary home of alien visitors with great technological powers, we would expect to find much more unambiguous proof than stylized cave drawings and curiously shaped sculptures. Nazca is offered as one important item of evidence as is the Chilean plateau of El Enladrillado where 233 huge, ancient stone blocks have been

Left: a rock painting in the Tassili mountains of the Algerian Sahara, in which the main figure seems to be floating in space.

Right: a group of the monumental stone heads that stand on Easter Island in the Pacific.

Opposite below: another view of the Nazca plain markings.

Above: a figure that appears on rock paintings in Australia and is known as Wondjina by the Aborigines. Believers in the theory that the Earth was visited long ago by extraterrestrial beings say that the head ornament is a space helmet of some kind.

arranged in a primitive amphitheater. The plateau is extremely difficult to reach and the task of transporting and erecting the blocks must have been a desperate labor for primitive tribesmen. How were the stones manhandled into position and what was their purpose? Von Däniken believes that extraterrestrial beings helped the tribes to build what was to become a visual beacon to help pilots land their spacecraft on the plateau.

Von Däniken also suggests that the famous statues of Easter Island, over 2000 miles off the coast of Chile, are further signs of alien activities. On the island are hundreds of strange statues, some between 33 and 66 feet high. No one is certain about their origins or purpose. Von Däniken claims that the ancient islanders could not possibly have carved and erected so many great

monoliths and argues that alien visitors must have been responsible for them.

Although their mystery cannot be denied, the general appearance of the statues is typical of religious carvings found on many of the Polynesian Islands. Anthropologists such as the famous explorer Thor Heyerdahl have also reported that present-day islanders still possess records explaining how the giant figures were first put into position.

But although some of von Däniken's evidence can be quickly dismissed, he is always ready with numerous other mysteries to support his theories. One of von Däniken's most extraordinary tales is an account of a visit he made to Equador to meet the discoverer of a huge complex of underground passages and halls. Argentinian Juan Moricz claims to be the legal owner of this strange underground world. Acting as von Däniken's guide,

Moricz showed the writer some of the remarkable treasures he had found.

According to von Däniken's description of the place, the passages are not in a roughly hewn state. The walls and ceilings are smooth and form perfect right angles with each other. Moricz led von Däniken into a vast hall "as big as the hanger of a jumbo jet" with galleries branching off it. They continued through a passage into another large and perfectly proportioned hall. Von Däniken says that in the center were a table and some chairs apparently made of plastic yet as heavy as steel. All around the furniture were figures of animals cast in solid gold — elephants, crocodiles, lions, bears, wolves, even crabs and snails, like the inhabitants of some fantastic zoo.

The most astonishing artifacts, however, were housed nearby in the same hall. Moricz showed von Däniken a collection of metal plaques each only millimeters thick but nonetheless extremely strong and unbending. Von Däniken estimated that they were 3 feet 2 inches by 1 foot 7 inches in size. Each was covered in a strange writing that appeared to have been stamped onto the metal by some kind of machine. Von Däniken agreed with Moricz that the plaques were the pages of a library left by extraterrestrial visitors centuries ago.

If the underground world discovered by Moricz really does contain these remarkable artifacts, modern archeologists would be faced by a profound mystery. But so far von Däniken seems to be one of the few people other than Moricz himself who has seen the treasures at first hand. Systematic investigations of the passages and hallways may someday provide us with the first indisputable evidence about the truth of von Däniken's claims.

There is no doubt that von Däniken's theories are popular among non-scientists. His books have been world-wide best-sellers and von Däniken himself is a much-sought-after lecturer. But many people have criticized his rather reckless methods of research and the highly subjective way he has argued his case. Critics say that he has plundered almost every

Left: a dramatic glimpse of how spacecraft could have been used to erect the huge stone heads of Easter Island.

Left: a traditional representation of Ezekiel's vision of a chariot from heaven, recounted in the Old Testament. Some modern UFO enthusiasts claim that the vision was actually a sighting of a craft from outer space.

Below: an ancient Mexican stone figure in strange clothing and posture. One theory argues that the person is lying in just the way an astronaut would recline in a spacecraft and that the clothing could be interpreted as a kind of spacesuit.

his ideas, strange accounts in biblical writings had been identified apparently suggesting sightings of alien craft in the skies. One of the best known accounts appears in the Old Testament when the prophet Ezekiel tells of a vision in which he seems to witness a landing by a strange and spectacular aircraft. One translation of Ezekiel's words is as follows:

"The appearance of the wheels and their work was like the color of beryl. And the four had one likeness. And their appearance and their work were as it were a wheel in the middle of a wheel. When they went they went upon their four sides and they turned not as they went. As for their rings, they were so high they were dreadful."

culture to find mysteries he can use to fit his ideas and that he deliberately ignores any facts that tend to disprove his claims.

Scientists have repeatedly uncovered fallacies in his arguments and many of his books contain surprising errors of fact. One academic who has investigated von Däniken's work in detail is Professor Basil Hennessy of Sydney University in Australia. He describes von Däniken's work as "a fascinating collection of unsupported, undigested, disconnected and often inaccurate claims." He illustrates just how seriously von Däniken errs in his facts by referring to the great Pyramid of Cheops in Egypt. According to von Däniken, each block of stone involved in its construction weighs 12 tons. In fact, the best modern estimate is that they each weigh under 3 tons. Moreover, von Däniken claims that the Egyptian slaves and engineers had no ropes or wooden rollers to maneuver the stones during the building work. In reality, there are actual examples of ancient Egyptian rope still in existence and records clearly show that enough wood was imported by the Phaorohs to make the building of rollerways a simple matter. By just assessing easily available facts, it becomes obvious that there is no need to argue for extraterrestrial help in the building of the Cheops or any other pyramid.

Von Däniken's ideas are undeniably exciting. But too often he appears to prefer the most outlandish explanations for phenomenon rather than more prosaic natural possibilities. However, even before von Däniken had begun publishing

Right: a photograph of ball lightning taken in August 1961. Although rare, ball lightning is a well-recorded natural phenomenon. Almost certainly many supposed UFO sightings involving bright lights seen moving in the sky have been no more than a glimpse of such lightning.

Not surprisingly, von Däniken seems convinced that such a description could only be an account of an alien spacecraft. However, Donald Menzel, an American professor of astrophysics and an extremely eminent figure in scientific circles, has pointed out that Ezekiel's vision is in fact a very good description – albeit in symbolic language – of a well-known meteorological phenomenon called *parhelia*. The apparition produced looks like one or more bright circles in the sky illuminated by the sun. Sometimes the circles cross each other, creating a spectacular and awesome sight.

False interpretation of data – whether by Ezekiel thousands of years ago or by men like von Däniken today – can provide

Above: the megalithic monument of Carnac in Brittany, France, has avenues lined with stones. Like Nazca, Carnac has been explained by some as a series of landing signs for alien visitors from space.

a very distorted view of some of the genuine mysteries of the natural world. Perhaps no other field of space phenomena is more prone to this than the sightings of unidentified flying objects and the extraterrestrial hypothesis that accompanies them. Nonetheless, most scientists agree that we are far more likely to reveal a genuine mystery by investigating the enigma of UFOs than by examining von Däniken's remarkable ideas.

Flying saucers

In June 1947, an American pilot, Kenneth Arnold, was flying over the Cascade Mountains in Washington State. Arnold was searching for the wreckage of a commercial aircraft thought to have crashed in the area and was hoping to claim the $5000 reward for the discovery. Arnold never found the wreckage but a quite different kind of discovery made him a world-famous celebrity overnight.

Suddenly, from high above the Cascade peaks, Arnold sighted nine shining disks weaving about the sky in close formation. Describing what he saw to reporters, Arnold said that each of the disks moved "like a saucer skipping across water." An inventive journalist elaborated on Arnold's description and coined the phrase "flying saucers" to describe the mysterious disks. The term soon became a household word, even appearing in a number of dictionaries. Ever since, it has been used to describe a wide range of unidentified objects reportedly seen in the skies. But recently a more accurate term has taken over in the popular imagination – UFO, an acronym for the

Right: the photograph that is probably one of the most famous of all UFO pictures. It was taken in California in 1952, identified by the photographer as a Venus scout ship.

Below: Kenneth Arnold, an American businessman whose sighting of a UFO in 1947 was the first to get wide publicity. It was in writing up Arnold's story that a reporter coined the term "flying saucer" to describe UFOs.

words Unidentified Flying Object.

Kenneth Arnold was not the first person to report seeing strange flying objects. As long ago as January 1878 a Texas farmer claimed he had observed an inexplicable fast-moving object in the sky. He even described it as "looking like a large saucer." But it was Arnold's sighting that attracted a blaze of publicity. Flying saucers suddenly caught the public imagination and became an instant talking point all over the world. As interest mounted, ever-increasing numbers of reports of other sightings flooded in. Psychologists have suggested that the almost hysterical reaction to an idea that was by no means a new one was linked to human attitudes following the trauma of a devastating global war. In the closing years of World War II people of all nations had glimpsed the terrifying

Above: a group of glowing objects in the sky over Salem, Massachusetts, in a photograph taken by a Coast Guard on July 16, 1952.

Below: E J Smith, a commercial pilot, uses a plate to illustrate his sighting of five UFOs during a trip on July 4, 1947.

potential of high-technology weapons such as rocket-powered missiles and atomic bombs. Perhaps there was some kind of reassurance in the idea that there were wiser creatures in the universe, possibly keeping a benign eye on mankind's activities from the flight decks of their saucer-shaped spacecraft.

Only a matter of days after Arnold's well-publicized sightings, the first reports came in of UFOs seen by trained military observers. Four days after Arnold had spotted the strange disks, two United States Air Force pilots and two intelligence officers saw a bright light performing "impossible maneuvers" over Maxwell Air Base in Montgomery, Alabama. On the same day, another Air Force pilot reported five or six objects over Lake Meade in Nevada. A few days later the pilot and co-pilot of a commerical DC-3 watched a formation of five "saucers" for about 45 minutes. When the first group flew out of view another formation of four appeared. A few days after this incident a whole series of sightings were reported in the vicinity of a secret Air Force Test Center in the Mojave Desert of California. The first sighting was by a test pilot from Edwards Air Base in an experimental aircraft. He reported seeing a yellow spherical object moving across the sky against the prevailing wind. Several other officers at the base had seen similar UFOs 10 minutes earlier. Two hours later, technicians at nearby Rogers Dry Lake spotted a spherical object, apparently made of a metal such as aluminum. They watched for about 90 seconds before losing sight of it. Later that same day, a jet pilot flying 40 miles south of Edwards Base saw a flat reflective object high above him. He tried to investigate but could not safely climb high enough in the aircraft he was flying.

Some months later, on January 7, 1948, the UFO craze claimed its first victim, an Air Force pilot so intent on getting a close-up view of a UFO that he pushed his aircraft too far. The alert that sent Captain Thomas Mantell to his death came after a telephone call from state highway patrol officers to Godman Air Base in Kentucky. The police reported that local townspeople were claiming to have seen a strange object in the sky. Soon after the call, the UFO was seen from the Base itself. Three Mustang F-51 fighters were diverted from a training exercise to investigate. Before long Mantell, flying the leading Mustang sighted the object. Outdistancing his wingmen, Mantell radioed to the Base: "I see something above and ahead of me and I'm still climbing. . . ." When asked to describe the UFO Mantell said: "It looks metallic and is tremendous in size. . . ." Seconds later came his last transmission: "It's above me and I'm gaining on it! I'm going to 20,000 feet."

Later that same day Mantell's body was found in the wreckage of his aircraft, 90 miles from the Godman Base. The UFO was never seen again.

Despite the Air Force's efforts to describe the cause of the disaster in rational terms, journalists reported that Mantell's plane had been attacked and destroyed by the UFO. The more reasonable explanation was that Mantell had in fact seen a large Skyhook weather balloon, at that time a new kind of research device that would have been unfamiliar even to a veteran flyer such as Mantell. The direct cause of the crash was that Mantell, in his excitement, had simply flown too high. Taking an F-51 above 15,000 feet without oxygen was extremely dangerous. Pushing his aircraft beyond its normal operating ceiling and probably affected by oxygen starvation, Mantell evidently lost control of his aircraft.

But the idea that Mantell had been attacked by the UFO was a much better news story and the tragic incident soon became one of the best-known "confirmed" encounters with a flying saucer. Even before Mantell's death excitement had been caused by several well-publicized hoaxes. In July 1947, for example, some alleged fragments of a UFO were discovered on Maury Island in America. Despite the fact that the affair was quickly shown to be a fraud, many people seemed to want to believe the story genuine at almost any cost. To complicate matters, two military investigators were killed when the plane carrying them to Maury Island crashed. Rumors immediately spread that there was something sinister about the deaths. Perhaps some extraterrestrial influence had caused the disaster in order to forestall an official investigation of the UFO report.

The situation was further inflamed by

articles written by Kenneth Arnold published at around this time. During the military investigation of his own reports, Arnold became convinced that the officials who interviewed him already knew the truth about the UFOs he had seen. In his articles Arnold implied that a high-level government conspiracy was going on aimed at keeping the truth from the public.

The Air Force were certainly doing

Above: UFO and United States Air Force Mustang in combat? On January 7, 1948 during a training exercise led by Captain Tommy Mantell, three planes were asked to investigate a strange craft reported over a town in

Kentucky. Mantell left the others behind and climbed dangerously high in pursuit of the UFO – only to plunge out of control to the ground. Many people still believe that Mantell was attacked by the UFO and sent crashing to his death.

little to counter Arnold's claim. Anxious to keep UFO hysteria damped down officials may well have given the impression of some kind of conspiracy of silence. But their reasons were far from sinister. The nationwide security network responsible for keeping watch on the skies was already being clogged up with innumerable reports of UFO sightings. From the Air Force viewpoint, too much time was being wasted checking

out the claims of hysterical or publicity-seeking citizens. They wanted to be able to get on with the main job of watching the skies for enemy aircraft or missiles. The cloak of secrecy about UFOs was really a reflection of concern about national security. But the official approach was a major error of judgment. Instead of damping down UFO reports, the theory about the government being involved in a conspiracy actually heightened public interest.

But the Air Force was facing greater problems from within its own ranks. The myth of a top-level conspiracy was even accepted by a number of Air Force officers in the official UFO investigation team. This began to show itself in the results of their work. For example, in 1952 members of the team together with Navy analysts studied the first cine-film of a UFO ever taken. It was a fairly amateur sequence shot by a Navy officer in Tremonton, Utah. The investigators quickly concluded that the film showed "fast-moving, self-luminous objects." A group of scientists studying the same film showed that the images were quite consistent with nothing more mysterious than a flight of birds.

Despite the counter-productive effects, the Air Force maintained an air of secrecy and kept all its UFO investigations classified under the code name Project Blue Book. By persisting in this approach they actually added fuel to public speculations and made many otherwise balanced people receptive to even the wildest UFO stories.

Right: a United States Navy photograph of a skyhook balloon being inflated. When they were first used, few commercial or military pilots were familiar with their distinctive shape. Many early UFO reports are thought to be due to sightings of such balloons at high altitudes.

Opposite top: a photograph at first thought to be of a UFO in flight over the Mount Palomar Observatory. On later investigation, it proved to be a defect in the film.

Opposite bottom: the Lubbock Lights, named after the town over which they were sighted on 14 occasions in 1951, and photographed by a student at the technological college there. An official statement following the sightings said that the origin of the lights remained a mystery. In other words they were genuinely unidentified flying objects.

Left: Harry Barnes, chief of radar at a United States government agency, working over the radarscope that picked up the flight of some strange and unexpected objects over Washington, DC in July 1952. The sensational sighting still remains the subject of considerable controversy.

In 1951 a movie was screened called *The Day The Earth Stood Still*. It began with a flying saucer being tracked as it approached and landed in Washington DC. Just a few months later, radar operators at Washington's international airport convinced themselves that the same thing was happening. The result was the biggest UFO scare of all time.

At 10.30 pm on July 26, 1952, radar operators picked up several slow-moving targets approaching the airport. They were definitely not scheduled aircraft and calls were sent out alerting the Control Tower and nearby Andrews Air Force Base. By this time, the mysterious targets had been spotted on their radarscopes as well. Jet interceptors were requested but it was not until midnight that the aircraft arrived on the scene. Just as they arrived the radar signals

ceased. The pilots reported no unusual objects and turned back to Base. Soon afterward the radar images reappeared. This time they remained even when the interceptors returned. Visibility was very poor because of low cloud and heavy rain and the reports made by the pilots were vague. Guided toward the UFOs by ground controllers, the pilots thought they saw an occasional distant glow in the sky but despite an extensive search not a single flying saucer was seen.

Because the sightings took place over a period of several hours, many of America's top UFO experts were able to gather at the airport to watch operations. The Press ran sensational stories demanding a public enquiry into the mystery. Under fierce public pressure, the Air Force convened a press conference. It turned out to be the longest

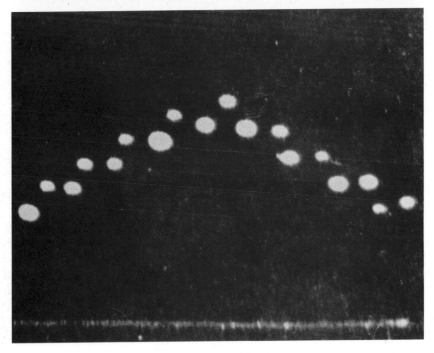

and best-attended since World War II. The official view of the incident was explained by Major-General John Samford II, Director of Intelligence at the Pentagon. Backed by expert advice from scientists on the atmospheric conditions over Washington, he said that the most logical explanation of the puzzling pheno-

mena was an effect called anomalous propagation. Radar experts had known about this for some years. Under certain kinds of air conditions dozens of images can be picked up by radar that are false images of objects often hundreds of miles away. The effect, therefore, is the radar equivalent of a mirage.

Above: one of the publicized UFO sightings of 1954 was reported by a pilot and copilot of a British passenger plane. They said that their airliner was accompanied by a large elongated object and 12

smaller ones for about 80 miles on their journey from New York to London. They thought that the small ones entered the larger one just when a scout plane sent out by the Air Force was approaching.

of trained radar operators being fooled are also on record.

However, the operators from Washington airport rejected the official conclusions. They insisted that they knew a true image from a false one. Some scientists also pointed out that conditions were by no means ideal for the mirage effect to take place. Arguments raged but to the public it looked once again as if the military establishment was trying to hush-up the truth.

What was needed was an impartial study of this and other incidents by an independant team of scientists with access to Air Force Blue Book files. UFO enthusiasts had demanded such an enquiry for years but it was not until late in 1965 that the Air Force finally agreed to the idea. A team of scientists at Colorado University under the direction of Dr Edward U. Condon was allowed access to the Blue Book files and supported by a large government grant began sifting the innumerable documented UFO reports.

At first, UFO organizations were jubilant believing that an unprejudiced view of the UFO enigma could only be in the best interests of nations throughout the world. But by the time the Condon Committee published its findings in January, 1969, most of its credibility had been destroyed. Only a few months after the Committee began work Condon himself was reported to have said in public: "It is my inclination right now to recommend that the government get out of this business. My attitude right now is that there's nothing to it . . . but I'm not supposed to reach a conclusion for another year." Condon's statement came only months before his Committee made a request for a further $250,000 in order to continue researching a mystery about which Condon himself had already made up his mind.

A further blow to the credibility of the Committee came when a private memorandum written by Committee administrator Robert Low became public. It was apparently written while the University of Colorado was still considering whether it should involve itself in the enquiry. It seemed the University authorities were anxious both to gain publicity for themselves while ensuring that no damage was done to the University's prestige among

One of the most notorious incidents involving anomalous propagation of radar waves occurred during World War II. An American battleship shelled and sank a target located by radar. It was later discovered that they had been shooting at a false image of the island of Malta. Many other less spectacular illustrations

academics. Low – then Special Assistant to the Vice-President and Dean of Faculties – wrote: "The trick would be, I think, to describe the project so that, to the public it would appear a totally objective study but to the scientific community, would present the image of a group of non-believers trying their best to be objective but having an almost zero expectation of finding a saucer."

It came as no surprise when the Condon Committee finally recommended that "further extensive study of UFOs probably cannot be justified in the expectation that science will be advanced thereby." The Committee's report is a massive 965 page book. Some scientists who have studied it in detail are extremely critical of the techniques of analysis used and have concluded that much of the Condon Committee's work is glaringly unscientific. Moreover, with over 25,000 well-documented UFO reports to study, the Committee chose to examine only about 90. It reached its conclusion that UFOs are not a fruitful area for further investigation despite conceding that more than 30 of the cases it dealt with had to be catagorized "unexplained."

But the Condon Committee's report appears at least to have satisfied the United States Air Force. In December 1969 it officially terminated its investigations into UFO reports, depositing the

Above: this photograph taken in 1950 remains one of the best authentic records of a UFO sighting.

Below: a drawing of the green fireball seen by Dr Lincoln La Paz. La Paz was convinced that the phenomenon was not caused by a meteor burning up in the atmosphere.

hitherto classified files of the Blue Book in publicly accessible archives. Subsequent close study by UFO enthusiasts now shows that many UFO reports were treated very casually by Air Force investigators. But despite the generally poor standard of analysis, there is nothing to suggest a deliberate attempt to conceal or play down important facts. The late

James McDonald, a distinguished American scientist and UFO expert, made a comprehensive examination of Blue Book case histories. He concluded: "I see a 'grand foul-up' but not a 'grand cover-up'." His work suggests that a good deal of the analysis carried out by Blue Book investigators, viewed scientifically, is almost completely meaningless.

One example of the Blue Book's lack of rigor cited by McDonald concerns a report made by two air traffic controllers at the Kirtland Air Base in New Mexico. On November 4, 1957 the two men saw an illuminated egg-shaped object descend toward the runway area, hover, and then climb swiftly away into the overcast sky. The Blue Book's conclusion was that the object was probably a light aircraft, off-course and probably confused by the runway layout at Kirtland. When McDonald himself spoke to the two men he discovered that neither had been informed about this conclusion. They found the idea of the object being a light aircraft somewhat amusing and made it quite clear that they were certain that they had not seen a light aircraft that day. Indeed it is hard to imagine any kind of light aircraft that could perform the hovering maneuvers they observed. Moreover, the two men had scrutinized

Above: a disk-shaped object seen and photographed over the south of England in 1965.

Below left: another photograph of a UFO taken in 1965, this one in Oklahoma.

Below right: British policemen warily inspect one of six bleeping disks found in fields across southern England in 1967. They were found to be a hoax – and two students later admitted to having fabricated them.

the object using 7-power binoculars and were sure that it had no tail or wings. It was about the size of an automobile and had a single white light in its base. They recalled that it hovered in full view for about 20 seconds before climbing away from the airfield at extremely high speed.

Even more disturbing than the Blue Book's casual dismissal of such a remarkable and well-authenticated sighting, is the logic behind the conclusion. The official investigator at the time was Captain Pat Shere. At the end of his report he wrote that the object was most probably an aircraft because: "(a) The observers are considered competent and reliable sources. . . . (b) The object was tracked on a radarscope by a competent operator. (c) The object does not meet identification criteria for any other phen-

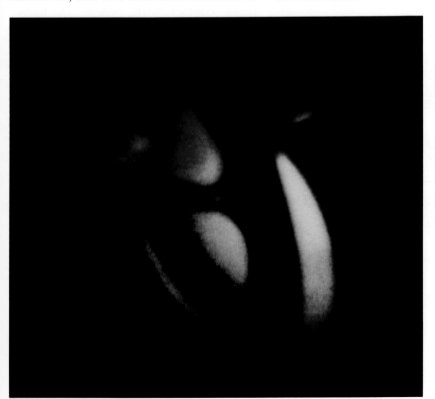

omenon."

In other words, because the witnesses were highly reliable and because what they saw could be explained in no other way, the object must have been an aircraft. Clearly if serious investigators used this kind of reasoning more widely, nothing could ever be classified as "unidentified" and the UFO enigma would be resolved for ever.

Aliens have landed!

The American astronomer Dr J. Allen Hynek is one of the world's top UFO authorities. He was formerly a scientific adviser to Project Blue Book's investigators, and is now one of the Project's most outspoken critics.

In the course of a systematic study of UFO phenomena, Hynek has shown that there are four basic ways in which UFOs seem to present themselves: as nocturnal lights, as daylight disks, as radar targets, or in detail at ranges of less than 1000 feet. Hynek argues that the greatest potential for scientific study must lie in the latter category – sightings that Hynek calls "close encounters." He subdivides them into three types. A close encounter of the first kind is a close-up sight of a UFO but one lacking in any real detail. A close encounter of the second kind is far more interesting, involving actual physical evidence of the event such as scorch marks or indentations in the ground. But most dramatic of all is the third kind of

Right: a sketch of a Venusian, made by one of the people who claimed to have seen him in the California desert.

Below: a sketch made from a description of a UFO landing by two men in Venezuela. The hairy short creature was said to have been hostile and aggressive.

close encounter. In this comparatively rare category of UFO experience, a human being actually makes contact with the UFO's crew.

Stories of human encounters with aliens are not new. Probably among the most well-known are the accounts written by a Polish-born American, George Adamski. He not only believed that UFOs were spacecraft from other worlds, he was also convinced that their pilots were humanlike creatures who wanted to make contact with people on Earth. Adamski resolved to try to obtain closeup photographs of the flying saucers and to make contact with their crews.

On November 20, 1952, Adamski and six colleagues went into the California desert looking for a region where it was said that saucers were often seen. Adamski and his team were rewarded remarkably swiftly. Adamski says he saw a flash from the sky and spotted a small, saucer-shaped craft drifting down toward him. It landed just out of sight about half a mile from where he stood. Soon afterward, Adamski reports he was approached by a young man with unfashionably long hair dressed in ski-type trousers.

In his account of the meeting Adamski writes: "Now, for the first time, I fully realized that I was in the presence of a man from space – a human being from another world. . . ." He learned that the stranger came from Venus and that other space people like himself were growing

increasingly anxious about atomic weapons being developed on Earth.

Adamski's later accounts of his contacts with aliens grew steadily more fantastic, and when Adamski died in April 1965, few people were prepared to take his claims seriously any longer. They had grown too wild for even the most ardent enthusiast.

There are many other examples of encounters with aliens that UFO enthusiasts can study. Some of the best stories involve dramatic abductions. One such case was reported by two Americans, Betty and Barney Hill. On September 19, 1961, the couple were driving home after a short vacation. During the drive they suddenly noticed a light in the sky. It grew brighter and seemed to be following their automobile. They

Above left and right: two pictures taken by Adamski of what he called a Venusian spacecraft. The first shows two scout ships leaving the cigar-shaped mother ship. The second shows the mother ship and six scout disks in a brightly glowing cluster.

Below: Barney and Betty Hill, a couple who claimed that while on a car trip between Canada and New Hampshire in 1961 they had been taken aboard a UFO for examination by space beings. They are shown in a trance, having undergone hypnosis in an attempt to find out what had happened to them while aboard the alien spaceship.

stopped and Barney Hill studied the object through binoculars. He saw some kind of vehicle with two rows of windows through which he could see figures looking out at him. The couple drove on but were soon overcome by an unaccountable drowsiness. About two hours later they roused themselves to find that they were now in a location 35 miles south of the place where Barney had stopped and scrutinized the UFO.

The full extent of their strange experience was not evident until three years later. Barney Hill was then undergoing some psychiatric therapy that involved treatment under hypnosis. During one session the missing two hours were reconstructed. When Betty Hill was hypnotized subsequently, she also described the same events her husband had revealed.

According to their accounts they had been kidnaped by aliens and taken aboard their spacecraft. They had been laid on separate operating tables and subjected to a detailed physical examination during which samples of their hair and toenails were removed. Before being released the aliens made it clear to them that they would never afterward remember their ordeal.

An even stranger abduction occurred in Brazil in October 1957. A young farmer named Antonio Villas Boas had twice seen UFOs. On the second occasion he claimed that a huge egg-shaped luminous object had touched down in a field

to have been exposed to some strong source of radioactivity. He also commented on some odd puncture marks on Boas' chin – the place from which Boas claimed blood had been taken. On close examination, Fontes found evidence of two needle marks around which the skin looked smoother and thinner than elsewhere, as though it had recently been renewed.

Perhaps the most chilling possibility suggested by some people is that alien beings are already living amongst us, posing as ordinary humans. It is obviously fairly easy to make such a claim. Producing supporting evidence presents no difficulties when you can photograph a man or woman and argue that they are in fact from another world.

One of the stories of this kind to receive most publicity was told by a New Jersey signwriter, Howard Mengel. According to Mengel, he had met space people on and off since he was a young man. Some time later after making a radio broadcast about his strange experiences, Mengel met a strikingly beautiful blonde girl called Marla who, he claimed, was from another planet. The already fantastic tale took an even more bizarre turn when Marla published a book that revealed that Mengel himself was also from another planet. It appeared that unknown to Mengel, his true origins were on far-off Saturn!

right beside the tractor he was using for plowing. Although he ran away he was pursued and caught by three humanoid creatures who lifted him off the ground and carried him into the spaceship. He was stripped naked and sponged all over. Then a blood sample was taken from Boas' chin and he was left alone in a room empty except for a couch. Soon afterward a beautiful naked woman entered. She walked over to Boas and embraced him. During the passionate sexual encounter that followed, Boas recalled that "some of the grunts I heard coming from that woman's mouth at certain moments nearly spoiled everything, giving the unpleasant impression that I was with an animal." Afterward, Boas says he was released unharmed and the spaceship took off and was soon lost from sight.

Boas' story is easy to dismiss as the sexual fantasy of an over-imaginative young man. But soon after the incident Boas was examined by Dr Olvao T. Fontes. After the examination Fontes reported that the young farmer seemed

Above: a drawing made from the Hill's description of the beings who kidnapped them.

Right: Howard Menger, the New Jersey signpainter whose stories about contacts with Venusians, Saturnians, and other creatures from outer space gained much publicity for him and his wife.

Opposite right: one of the pictures Menger claims to have snapped on a tour of the Moon arranged for him by friendly aliens. The figure is said to be that of the Venusian pilot in front of his spacecraft on the Moon.

A more sinister series of events was described by UFO investigator Alfred K. Bender, founder of the International Flying Saucer Bureau. Despite the popular success of his organization, Bender suddenly closed it down in 1953, only two years after its formation. For more than six years Bender remained silent about why he had wound up the organization. Then in 1962 he published a book called *Flying Saucers and the Three Men*. In the book Bender recounted a number of extraordinary experiences including mysterious visits by threatening aliens and telepathic messages warning him to

Above left: Albert K. Bender, founder of the International Flying Saucer Bureau.

Above center: one of the three men in black who, according to Bender, forced him to close his bureau in a cloud of secrecy.

Above right: a corpse said to be a crew member of an alien spacecraft that crashed in the 1950s.

stop "delving into the mysteries of the Universe."

Most serious investigators today dismiss Bender's book as a work of pure invention. But whether any stories of contacts with aliens are genuine is impossible to determine without scientifically assessable evidence. Despite the profusion of colorful accounts, little is available that can be objectively investigated. A major problem is that the field of UFO research attracts many people who have no scientific knowledge or training. This means that whether or not their experiences are real, imagined, or invented, they are unable to objectively assess and evaluate what they see. They are rarely even able to record their experiences with any accuracy.

However one of the most promising developments in UFO research has been efforts in the early 1970s to establish genuinely scientific approaches to investigations, using traditional scientific skills alongside up-to-date technology. This new and hopeful development could lead to the first real breakthroughs in UFO studies that may yet reveal remarkable new discoveries about the universe and the life-forms that populate it.

Science and UFOs

How can a scientist begin studying UFOs? His objective is easily defined. He must try to determine in each reported case what the UFO actually was, thereby turning it from an unidentified to an identified flying object. There is, after all, no doubt that UFOs exist – there are obviously many objects seen in the skies that observers are unable to identify. This is no more than a result of terminology. One person's UFO is someone else's weather balloon or experimental aircraft, depending on the skill and knowledge of the person concerned. The job of the scientific investigator, therefore, is to apply a wide range of knowledge to the best UFO reports and pick out the balloons, aircraft, and other everyday airborne objects. Once the investigator has eliminated these and so identified a residue of genuinely mysterious reports, a reasonable hypothesis to explain them must be worked out.

But before a scientist can reach the stage of forming a hypothesis, he or she must examine the reliability of the data. This is one of the most crucial parts of UFO investigation. The scientist has only secondhand information on which studies can be based. The investigator must therefore attempt to gauge the relative credibility of different reports and even different aspects of a single report. This places most scientists in a quite unfamiliar situation. Usually in scientific research, electronic or mechanical instrumentation registers and records data. A researcher can therefore be reasonably sure that the measurements made are objective and fairly reliable. But UFO research is necessarily based not on the data acquired by dispassionate machines but by human beings.

The fallibility of human powers of observation is notorious. Even trained observers can be completely fooled by unexpected events. For example, some years ago an American police officer – a reliable, well-balanced individual – telephoned Air Force experts with an

Above: Dr Charles Moore, a meteorologist, with a weather balloon. Once when tracking such a balloon in preparation for launching a skyhook balloon, he and four Navy enlisted men sighted a flying saucer.

Left: the V-173, an experimental aircraft without wings known as the "flying pancake." The United States Navy offered the idea that this was one of the strange disks being sighted widely throughout the United States.

Right: a huge bubble of burning gas produced by a plane to test how people would react if they saw it. The purpose was to see if people not only reported the bubble as a UFO, but whether they would elaborate the sighting probably even claiming contact with alien visitors. The test, planned by the US Air Force, was never carried out.

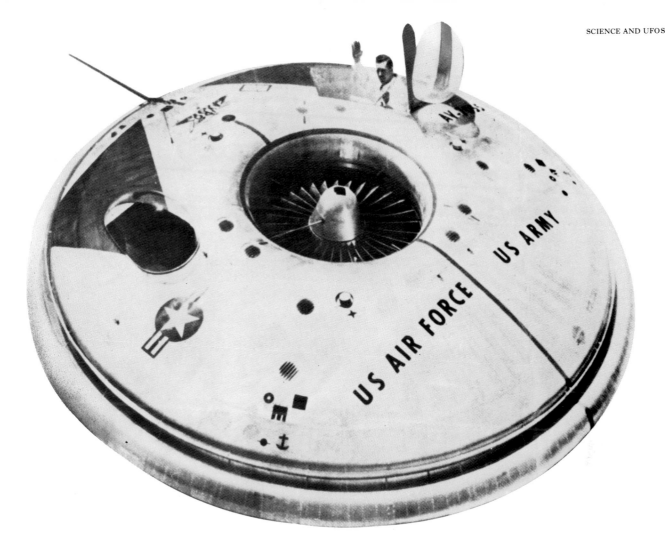

astonishing message. He said: "I have in sight a flying saucer that has just landed. It is a couple of miles off, blinking in the sun on the horizon. It's disk-shaped and made of some kind of metal." A few hours later the experts reached the patrolman who was still keeping watch on the UFO. He told them that there had been no sign of movement since the landing. The men advanced on the UFO with understandable caution. Then as

Above: the AVRO, a kind of hovercraft built by the A. V. Roe Aviation Company and being developed by the American armed forces in 1959. When a freelance photographer got a picture of the disk as he flew over the Roe plant, it was publicized as a possible secret weapon – one that might account for many of the UFO sightings.

they closed on it they realized they were actually stalking a large aluminum waste tank, dumped by the side of the road. The police officer was a reliable man and a trained observer. Everything he had reported was accurate – except for one important item. He had not seen the "UFO" land. He had driven his patrol car through a bend in a road that he knew very well and had suddenly seen in all that familiarity a new and unknown object. Unconsciously he had made the automatic assumption that the object had appeared there at that very instant. In other words, it had just "landed."

The experiences of untrained observers can be even more bizarre. In March 1968, three people, including the Mayor, all living in a small town in Tennessee, gave a convincing and detailed description of a UFO that they saw together in the early evening. They noticed a light moving in a slightly curved trajectory trailing an orange-colored glowing tail. It came straight toward them and passed right overhead. One of the observers said: "It was shaped like a fat cigar . . . it appeared to have square-shaped windows . . . I thought I caught a metallic look about the fuselage . . . two-thirds or three-quarters of the fuselage toward the front end had windows that were lit up." Other people in different parts of the United States also appeared to have seen the same UFO and the Air Force were flooded with detailed reports from reliable sources.

What had caused the apparition? Obviously something had hurtled across the sky that night. But speculations came to a slightly disappointing conclusion when the Soviet Union announced that one of its spacecraft, Zond IV, had de-

Left: Dr James Harder, an engineering professor, and Dr Allen Hynek, an astronomer and space scientist are both leading experts on UFOs.

veloped a fault and had just reentered the Earth's atmosphere. From the trajectory of the vehicle it was soon apparent that the supposed UFO had in fact been the blazing fragments of a disintegrating spacecraft, burning up in the atmosphere 75 miles above the heads of excited observers on Earth.

This unreliability of witnesses who are themselves genuinely convinced by what they have seen, is a great obstacle to the scientific investigation of UFOs.

There does appear to be one category of UFO experience that shows consistent features, however. UFO expert J. Allen Hynek calls them "close encounters with physical effects" and describes a typical case that can be taken as a prototype of many such reports. Hynek writes: ". . . a bright light suddenly appears and seems to seek out the witness's car. As it stops to hover over the car, the car lights dim or fail and the engine dies. Often the occupants of the car report feeling hot and prickly. After a few minutes the appari-

tion leaves and the car returns to normal operation."

Close encounters such as the kind described by Hynek offer the best hope for investigators looking for objective evidence with which to investigate the mystery of UFOs. Physical effects of the kind Hynek describes are at least measurable by instruments. This exciting possibility has been a factor in the recent formation of a number of groups in different parts of the world trying to apply high-technology instrumentation to UFOs and the physical effects that they leave behind them.

In the United States in 1968, during a Congressional meeting on science, one of America's top aerospace experts gave possibly the first clear and authoritative statement of the direction in which UFO investigations should be moving. Dr Garry Henderson, Senior Research Scientist in Space Science at General Dynamics, said that qualified observers with the right instruments and planning should be able to devise schemes to reveal the truth about UFOs. He insisted: "What we need to establish concerning the existence or non-existence of UFOs is not merely a review of sighting incidents, but an implemented plan to acquire hard facts."

Right: a physicist shows how an illuminated object can be formed by temperature inversion. This is one of the explanations often given by official agencies to explain UFO reports.

detection system capable of converting into sound or TV pictures any complex sequence of light signals that a UFO might produce as a response to the laser beam. Sophisticated computers will accumulate and process the data that the detectors will provide. The remarkable network of instruments will offer a degree of objectivity in the analysis of events that has never before been possible.

Project Starlight International is only a beginning. Its director, Ray Stanford, hopes that as other groups are established in other parts of the world, they will co-operate with each other to build up a common base of hard facts derived from rigorous scientific investigations.

It is over 30 years since the modern era of UFOs began. Only now are the first efforts being made to examine scientifically the mystery UFOs represent. For the first time the phenomenon is being studied by researchers who make no unwarranted assumptions about what they discover and are as ready for negative results as they are for some dramatic breakthrough. But inevitably in the mind of even the most sober researcher is the fascination of being involved in work that could culminate in the most exciting event in human history – mankind's first contact with extraterrestrial intelligence.

In the United States in particular, subsequent developments have been rapid. In 1977 a non-profit-making research corporation was established known as *Project Starlight International*. Its permanent base of operations is an isolated site about 20 miles Northwest of Austin in Texas. In two laboratory complexes elaborate and highly sophisticated instruments are used to monitor, record, and correlate a wide range of physical effects that could help in UFO investigations. Although the main base is the fixed site near Austin, research teams are on standby to quickly load essential equipment into especially designed transporters in order to visit areas where UFO activity is said to be taking place.

The kind of equipment used by Project Starlight is the very latest in up-to-date technology. A system that may be used to communicate with UFOs has been built using a helium-neon laser capable of transmitting voice, code, or video pictures. There is also a light pulse

Above: Major Hector Quintanella, one-time head of Project Blue Book, shown with some of the bogus items he received from UFO fanatics during his investigation. Among them are radio parts and pancakes.

Right: a picture of the UFO sighted on the coast of Brazil near Rio de Janeiro in May 1952 by a reporter and a photographer. Authorities proved it was a fraud by showing that the shadow on the disk could only have been produced if the sun had been shining directly up from the water instead of down from the sky.

The UFO enigma

What are the unidentified flying objects that are reported from time to time? In the majority of cases the answer is simple. Often they are nothing more mysterious than a misidentified aircraft, weather balloon, or even a planet. But a small number of incidents seems to defy explanation. Even if only one of these is both completely genuine and completely inexplicable, it presents a profound and exciting mystery.

There are many concepts that people want to believe in – such as fairies, immortality, reincarnation, or the ability to read minds. In recent years the idea of superbeings visiting Earth from distant worlds has joined the list. But precisely because the idea of alien beings landing on Earth is so appealing, it is all the more reason to view the possibility with skepticism. Whenever there is an emotional stake, however small, in an idea, people are likely to deceive themselves into believing it to be true.

Almost as soon as the first UFOs were reported they were identified in popular imagination with flying machines piloted by space people from other worlds. Because the machines could apparently perform amazing maneuvers so as to evade even the fastest of our aircraft, the space visitors were assumed to come from a technologically superior civilization. Indeed, the very idea of them visiting the Earth was obvious proof of their technical prowess.

This so-called extraterrestrial hypothesis used to account for UFOs is one of the most exciting – and one of the most abused – ideas in popular science. In true scientific terms, the concept is hardly anything as grand as a hypothesis. It is based on few real facts and is largely an emotional response helped along by sensationalized press reporting. Until recently those who put forward the hypothesis argued that the alien visitors were coming from one or more of the planets in the solar system. But today space re-

Right: Dr Carl Sagan, director of planetary studies at Cornell University, New York. He is one of the foremost proponents of exobiology – a new field of study dealing with the possibility of extraterrestial life and ways of detecting it.

Below: Dr Frank Drake, director of the National Astronomy and Ionosphere Center run by Cornell University in Arecibo, Puerto Rico. He initiated an attempt using the huge Arecibo radio telescope to communicate with intelligent beings elsewhere in the universe.

Opposite: a brightly colored and glowing UFO sighted in France on March 23, 1974. Reports about unidentified flying objects have come from every part of the world and from all kinds of people – sometimes from the skeptical and questioning as well as the credulous.

search has shown that none of them could support the kind of advanced life forms that might want to fly to Earth in their spaceships. In order to preserve the concept of extraterrestrial beings it became necessary to look beyond the solar system and speculate that the visitors came from planets revolving around some of the nearby stars. Once this possibility is admitted, however, the idea of true superbeings has to be taken into account. The technological problems of flying between star systems and entering the atmospheres of alien planets – even landing on them from time to time – are immense. By contrast, mankind is at least 100 years away from mounting even a crude fly-by mission using an unmanned probe targeted on one of the nearest stars.

Any worthwhile hypothesis must be tested before being either accepted or discarded. Einstein's theory of relativity for example was based on certain key assumptions. Before his ideas were tested through experiment and observation they had the status of hypotheses. Today they are recognized as fundamental truths about the nature of time and space. It is less easy to test the idea that UFOs are visitors from the stars. However, it is possible to make some kind of objective assessment of the reasonableness of the idea. The American astronomer Carl Sagan has worked out what the extra-terrestrial hypothesis really implies about the universe.

Sagan explains that an estimate can be made of the total number of technologically advanced civilizations existing in the Milky Way. The number turns out to be one-tenth of the average lifetime in years of a technological civilization. Assuming this is around 10 million years, there should be about one million advanced civilizations. This sounds like a good argument in favor of the hypothesis. Surely out of so many civilizations, some would have probed as far as the Earth. But Sagan points out that there are many more interesting places than Earth for interplanetary spacecraft to visit. He goes on to calculate how many probes every one of the civilizations would have to launch in order to account for an average of just one visiting Earth each year. It appears that 10,000 launches would be needed by each civilization every year throughout the galaxy. This is a staggering amount of space activity the cost of which would destroy even the wealthiest economies. And Sagan's calculation is based on fairly optimistic estimates of the numbers of civilizations able to launch starships. If their numbers were considerably less the launch rate would have to be much higher.

The obvious objection to Sagan's argument is that it is statistical. The Earth after all is a place of special interest that other civilizations would specifically want to seek out and visit. But this objection actually reveals the most important fallacy in the extraterrestrial hypothesis. If we accept that life is not unique to Earth and that a fairly large number of star-going civilizations exist, there is actually little reason to expect such advanced races to pay much attention to creatures of so few skills as mankind. Far from paying Earth any special attention they would be more likely to ignore it altogether and concentrate on contacting their technological equals from whom they may be able to gain some advantage in terms of trade or intellectual exchange.

The real question may not be whether or not we are alone in the universe but whether other civilizations would want to contact us even if they knew we existed.

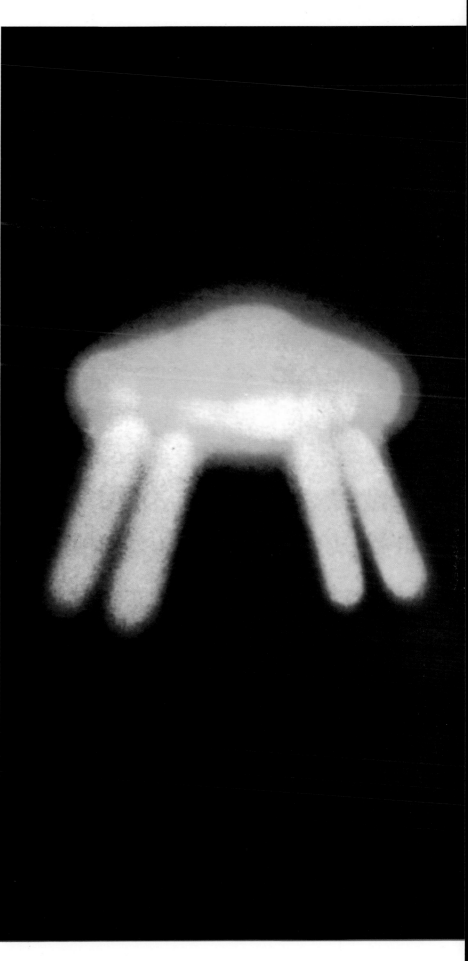

Index

Page numbers in *italics* refer to captions

A

absolute magnitude, 48
absorption spectra, 27
acceleration of star ships, 256
achromatic lenses, *28*, 29
Adams, John Couch
  discovery of Neptune, 108
Adamski, George, 310
Adonis
  orbit, *112*
aerials *see* antennae
agriculture
  link with sunspots, 42
  necessity for calendar, 10
  on Mars, 251–2
Alcor, 63
Aldebaran, 71
  eventual flypast by Pioneer 10, 287
  luminosity, 56
  name from Arabic, 13
  size, *57*
Aldrin, Edwin "Buzz," *159*, 183
  moonwalk, *184*
Alfonso X, King of Spain (1252–1284), 13
Alfvén, Hannes, 79
  concept of contraction of universe, 156
Algol (Beta Persei)
  brightness varies, 62, *62*, 63
  name from Arabic, 13
The Alphonsine Tables, 13
Amalthea, 101, *102*
Amen-Ra (Egyptian sun-god), 38
amino acids, *278*, 279
Amor
  orbit, *112*
Anaxagoras (Greek philosopher, c500–428BC)
  observation of Sun, 38
Anaximander (philosopher, fl. 6th century BC)
  contribution to astrology, 64
  theory of origins of life, 278
Anders, William, 183
Anderson, Carl
  discovery of positrons, 206
Andromeda nebula, 25, *35*, *119*, 124, *219*
  early observations, 120
  observation of supernova, 58
  studied by Hubble, 25
Antares
  luminosity, 56
  size, *57*
antennae
  Bell Telephone Laboratories array, 150, *150*
  dish
    used for Grote Reuber's radio-telescope, 32
    using natural hollow in ground, 33
  Mills Cross type, 33

antideuterium, 206
antimatter, 147, 206–7
  problems of applying to propulsion systems, 262–3
antineutrons, 206
antiprotons, 206
Apollo
  orbit, *112*
Apollo Applications Program *see* Skylab
Apollo Mission Control, 180, *186*
Apollo Project, 92, 94, 180
  Apollo 4 tragedy, 182–3, *182*
  Apollo 5: unmanned flight, 183
  Apollo 7 mission, *183*
  Apollo 8: first lunar mission, 183
  Apollo 9 photograph of California, *186*
  Apollo 10: flight time compared with Verne's story, *160*
  Apollo 11: first moon landing, 183–5
  Apollo 13: problems, 185–6
  Apollo 14 on moon's surface, *186*
  Apollo 17: last moonflight, 186
  costs, 182, 228
  modular design of spacecraft, 180–1
  use of Saturn V as booster, 182
Aquarius
  Persian star map, *68*
Arabs
  contribution to astronomy, 13
  influence on astronomy of Ptolemy's work, 12
  many astronomical terms from language, 13
  naming Algol, 62
Arcturus
  size, *57*
Arecibo (Puerto Rico) dish antenna, *32*, *33*, *318*
  message beamed into space, 289, *289*
  survey of surface of Venus, 84–5
Arend-Roland comet, *115*
Argo Navis
  subdivisions, 69–70
Ariel, *106*, 107
Ariel 6 satellite, *214*
Aristarchus of Samos (Greek astronomer, fl. 270 BC), 11
  calculation of distance from Sun to Earth, *11*
Aristotle (Greek philosopher, 385–322 BC)
  belief concerning movement of stars and planets, 11
  explanation of comets, 114
  theories as basis of medieval cosmology, *138*
  theory of origins of life, 278
  theory of speed of descent of a body, 17
arms race, 166
  see also weapon systems
Armstrong, Neil, 183
Arnold, Kenneth, 300, *300*, 302–3
Arp, Halton C., *144*, 145
artificial satellite *see* satellites, artificial
asteroids, 112–13, *112*

asteroid belt used for research, 253
  orbits, 112, *112*, 253
  possible future in research, 113
  possible origin, 113
  possibly mineral-rich, 113, *113*
  use for building materials, 253
astrolabes
  description, 13
  invented by Arabs, 12–13
  Persian, *13*
astrology, 64–5
  Muslim illustration, *64*
  see also Zodiac
Astron magnetic fusion experiment, *249*
astronauts
  effect of space travel, 198–9, *198*, *199*
  training facilities, 180
  see also manned space flights
astronomical instruments, *13*
  problems of use on earth, 234
  quadrants, *20*
  spectroscopes, 26–7, *27*, 140
  use on the Moon, 241
  see also astrolabes; telescopes
Astronomical Unit (UA), 35
Astronomie Populaire (Flammarion), *114*
astronomy, 10–35
  earliest forms, 10
Atlas booster rockets, 169
  for Mercury project. 172
Atlas missile, 166, 167
atmosphere of Earth
  barrier to astronomical observation, 234
  constituents, 86
  filters out harmful radiations, *86*, *87*, 87
atomic bomb, *277*
  Manhattan missile, 166
atomic nuclei, initial formation, 147
aurora, 42
Australian aboriginal figure, *295*
Avro aircraft, *315*
Aztec sun-god, *38*

B

BL Lacertae (galaxy type), 135
BSOs (Blue Stellar Objects), *218*, 219
Baade, Walter
  upsets Hubble's galactic theories, 127
Babylonians
  record of positions of Jupiter, *10*
Baghdad astronomical center, 12
Barnard, Edward, 49
Barnard's Star
  distance from Earth, 48
  measurement of proper motion, 49, *49*
  starship project, 262
Barnes, Harry, *304*
Bean, Alan, 193
Bell, Jocelyn, 33
Bell Telephone Laboratories, 150, *150*

Belyayev, Pavel, *174*
Bender, Alfred K., 313, *313*
Bergedorf Observatory (Hamburg), 30
Bernal, J. D., 242
Bernal Sphere space colonies, *242*
beryllium, *203*
Bessel, Friedrich, 56
Beta Persei *see* Algol
Betelgeuse
   luminosity, 56
   size, *57*
big bang theory of creation, 146, *148*
   Alfvén's disagreement with, 156
   alternative propounded, 152–3
   seen as part of cycle of creation, 155
Big Bird military satellites, 170
binary stars, 63
Biston betularia, 283
black dwarfs, *54*, 55, 57
black holes, 59, *60*, *155*, 208–15
   as gateways to other universes, 214
   at center of galaxies, 133
   behavior of light, *212*
   detection in our galaxy, 215, *215*
   eventual use as energy stores, 270
   gravitational pull, 59, *209*
   imaginary physical investigation,
     210, 212
   impression of a probe into, *213*
   mathematical investigation, 210
   means of detection, 215
   possibly in association with
     Cygnus X-1, 223
   rotating, 214, 270, 271
   theory not applicable to BL Lacs, 135
Blue Book project, 304, 307
   deposited in public archive, 308–9
blue giants, 54
blue shift
   few galaxies of this type observed,
     141
Blue Stellar Objects (BSOs), *218*, 219
Bode, Johann, 69
Bode's Law, 112
Boeing 747 aircraft
   used as space shuttle carrier, *228*
Bondi, Hermann, 152
boosters *see* rockets
Boötes, 219
Borman, Frank, 183
boron, *203*
Brahe, Tycho (astronomer, 1546–
   1601), 20, *20*
   supernova observation, 58
brightness contour map, *59*
British Interplanetary Society
   starship study, 261
bubble chamber, *206*
Buffon, Georges, 77
Bumper rocket booster, 166, *166*
Bussard, R. W., 260

C

C-field, 153
calendars
   early forms, 10
   Egyptian, 64

Callisto, 101, *103*
   possibility of establishing base on,
     253
Cambridge (England)
   chart of radio sources in sky over,
     218
   new radiotelescope with dipole
     antennae, 220, *221*
   radiotelescope, *216*
Capella
   size, *57*
Carnac megalithic stones, *299*
Carr, Gerald, 193
Cassegrain telescope, 30
Cassini, Jean Dominique, 105
Cassini's division (Saturn's rings), 105
Cassiopeia A
   x-ray map, *223*
Centaurus, 71
Cepheid stars, 63, *63*, *124*
Ceres, 112
   orbit, *112*
Chaffe, Roger B., 182–3
Chamberlin, Thomas Chrowder, 77
Charon, 111
chemical composition of stars
   use of spectroscope, 26
Chichen Itza observatory, *13*
Chinese
   early use of rockets, 161
Chiron, 113
Christy, James
   discovery of Pluto's moon, 111
chromosphere, 40, 44
chronometer
   astrolabe used as, 13
church *see* religions
civilizations, 276
   self-destructiveness, 277
Clark, Alvan, 56
Claudius Ptolemaeus *see* Ptolemy
clocks
   Galileo's discovery of pendulum
     motion, 17, *17*
"close encounters," 310
clouds, *86*
Coal Sack nebula, 124
Coconni, Guiseppe, 287
Collins, Michael, 183
collisions, high-energy, 205
colonization of space, *226*, 242–5
   construction of facilities, *244*
   political implications, 245
   psychological problems, 244
   Space Ark concept, 255–6
   torus-shaped colonies, 244–5, *245*
   see also extraterrestrial life
coma (head of comet), 114
Coma Berenices cluster of galaxies,
   119, 125, *125*
comets, 114–15
   development of tails, 114–15
   formation, 114
   Newton's work on, 23
   nomenclature system, *115*
   seen as omens, 114
   theory of origin, 115
command module of Apollo spacecraft, 181

communication satellites, 170, *171*
communications
   between living creatures, 275
   development for Apollo program, 180
   on surface of Mars, 251
   satellite subsystem, 171
   with extraterrestrial life, 286–9
   with interplanetary probes, 188
Condon, Edward U., 307
Condon Committee on UFOs, 307–8
Congreve, Sir William
   conceived military use of rockets,
     160, 161
Conrad, Charles "Pete," 193
constellations
   chart on mercator projection, *72–3*
   used as signposts, 70
   visual imagery, 68
"construction shacks" in space, 243–4
continental drift, 89
continental plates, 89
continuous creation process, 153
control systems
   gyroscopic, 163, *163*
   of satellites, 171
Copernican theory
   Brahe's version, *20*
   see also Copernicus, Nicolaus
Copernicus (satellite), *135*, *234*
Copernicus, Nicolaus (1473–1543),
   14–15, *14*
   chart of heavens, *15*
   data verified by Kepler, 21
   Galileo convinced by theories,
     18–19
   theory of solar system, 76
   views opposed by Tycho Brahe, 20
Coriolis effect, 244
corona, 40, *40*, *44*
"cosmic background radiation," *150*
"cosmic egg" theory, 146
cosmic rays, 204–5
   impact on space helmet, *205*
   photograph of track, *205*
cosmology
   following Hubble's Law, 141
   investigating philosophical issues of
     creation, 143
   medieval, 138
Cosmos satellites
   maneuvering in orbit, 246
Crab nebula, *58*
   as a source of cosmic rays, *204*
   detection of visible pulsar, 220, 221
   seen as supernova in 11th century,
     58
   Soyuz investigation, *194*
Creation Field, 153
Crick, Francis, *274*
crust of the Earth, 88
Crux Australis *see* Southern Cross
crystals
   fission fragment tracks, *260*
Cyclops project, 287, *287*
Cygnus
   Veil nebula, *117*, 123
Cygnus A
   radio source, 33, *33*, 133, 134, *135*
Cygnus X-1, *210*, 215, 223

D

DNA (deoxyribonucleic acid)
   structure, *274*
   synthesis, 279
Daedalus project, 261–2, *262*
Damascus astronomical center, 12
d'Arrest, Heinrich, 108
Darwin, Charles, 278–9
   Beagle voyage, 280–1
   On the Origin of Species, 283
Darwin, George, 92
Davis, Raymond, Jnr., *203*
   neutrino detection research, 202
day length
   Jupiter, 100
   Mars, 97
   Mercury, 80
   Moon, 92
   Neptune, 108
   Saturn, 104
   Uranus, 106
   Venus, 85
De Revolutionibus Orbium Caelestium
   (Copernicus), 14–15
deferent (circular orbit of the center of
   an epicycle), 11, *12*
degeneracy pressure, 54
Deimos, *98*, 99
   possible use as radio relay station,
   251
Delle Macchie Solari (Galileo), *18*
Delta Cephei, 63
"Demon Star," Greek name for Beta
   Persei
Deneb
   absolute magnitude, 48
density
   effect on escape velocities, 208
deuterium, *146*
   antideuterium produced momentarily,
   206
*A Dialogue Concerning the Two Chief
   World Systems* (Galileo), *17, 19*
dimensions, to define a position, 200–1
Dione, *104*
Dirac, Paul, 206
dissected mountain formation, 89
distance
   measurement, 34–5, 48
     of nearest star to Earth, 46, 47
docking facilities
   of Skylab, 192
   of space colonies, 244
Dog Star *see* Sirius
Dollond, John, *28*, 29
Dollond, Peter, *28*, 29
Doppler, Christian Johann, 140, *140*
Dopper Effect, 140–1, *140*
Doradus
   most luminous star, *124*
double stars, 63
Drake, Frank, *286*, 287, *318*
Driggers, Gerald W., 243
Dumbbell nebula, *35, 119*
Dürer, Albrecht
   star maps, 68
dwarf stars *see* black dwarfs; white
dwarfs
Dyson, Freeman J., 261, 267
Dyson Sphere, 267–8

E

Eagle lunar module, 181, 184, *184,
   185*
early warning systems, 247
Earth, 86–91
   appearance from space, 86, *86*
     areas of water, 90–1
   atmosphere, 86, *86*, 87, *87*, 234
   condensed mass, 148
   crust, 88
   diameter, 86
   distance from Sun, 39
     basis of Astronomical Unit (AU), 35
     early Greek estimates, 38–9
   escape velocity, 208, 241
   mantle, 88
   orbit, 14, 86
     through four dimensional space
      time, *200*
   pointing to age of our galaxy, 145
   problems of population and
     pollution, 91, *91*
   quantity of neutrinos crossing, 202
   reasons for ability to support
     advanced life forms, 86
   representation of geological eras,
     281
   roundness suspected by Thales, 11
   structure, 88, *88*
   study of radiation belts, *171*
   visits by aliens, 292–3, 310–13
earth resources satellites, 170, *186*
earthquakes
   caused by geological faults, 89
   use in research into Earth's
     interior, 88
Easter Island stones, 295, *295, 297*
eclipses
   early Chinese calculations, 10
   lunar, 95
   possible use of Stonehenge, *38*
   solar, 40, *40*
eclipsing binary stars, 62–3, *62*
Eddington, Arthur, *138*, 201
Escher, Maurits, *139*
Egerov, Boris, *173*
Egyptians
   Amen-Ra (sun-god), 38
   astrology, 64
   construction of pyramids, 298
   representation of cosmos, *11*
   records of eclipses and comets, 10
Einstein, Albert, 44, 76, 138–9, *138*,
   147, 224
   impact of relativity theory, 198–201
   theory of static universe disproved,
     141
electrical storms
   related to solar flare activity, 42
electricity
   converted from microwaves, 237
   turbogenerators in space, 238, *238*
electromagnetic radiation
   initial dominance of universe, 148
   wavelengths changing with time,
     151
electron acceleratots, *248*
   see also particle accelerators
electrons, 147
   acted on by cosmic rays, 204
   tracks in bubble chamber, *206*
Elektron satellite, *171*
elements
   absorption and emission of light, 27
elliptical orbits, 76
embryology, *274*
emission spectra, 27
energy
   collection from space, 236–9
   consumed by technological societies,
     264
environmental control
   in artificial satellites, 171
epicycle (small circle having its center
     on the circumference of a greater
     circle), 11, *12*
Equator
   underground complex, 295
equatorial telescopes, *28*
Eros, 113
   orbit, *112*
escape velocity, 208
   for rockets, *160*
   of Earth, 208, 241
   of Moon, 241
Eta Ursa Majoris, *321*
Europa, 101, *103*
event horizons, 210, 212, 270
evolution of living organisms, 280–3
exosphere, 88
Explorer I satellite, 168, *168*
   boosted by Juno I, *169*
   cost, 228
extraterrestrial life, 252
   chances of contact with Earth, 319
   communication with, 286–9
   estimate of number of civilizations,
     276
   fantasy creatures, *273*
   in our galaxy, *118*
   see also colonization of space
Ezekiel's vision, 298, *298*

F

Fabricius, David, 63
Feoktistov, Konstantin, *173*
Fermi, Enrico, 277
Flammarion, Camille
   Astronomie Populaire, *114*
flares, solar, 42
flying saucers
   recorded sightings, 300
   see also unidentified flying objects
     (UFOs)
fourth dimension, 200. *200*
Fraunhofer, Joseph von (1787–1826),
   26, *26*
Fraunhofer lines, 26
Freedom 7 spacecraft, 172
From the Earth to the Moon (Jules
   Verne), *160*
fusion pulse rockets, 261

G

Gagarin, Yuri Alexeyvich, 168, 172, *172*
Galapagos Islands, 281
    finches, *281*
galaxies
    classification, 126, *127*, *128*
        into Population I and II stars, 127–8
    collisions, 133–4
        with antimatter galaxies, 207
    differences between constituent
        stars, 127
    distances from our own, 140
    ellipsoidal, 126, *127*, *128*, *128*
    evidence of upheaval, 132
    evolution, 128, 131, 153
        from "cosmic egg," 146
    final form as quasars, 219
    globular clusters, 128, 131, *131*
    Hubble's Law, 141, *141*, *142*
    in an expanding universe, 139
    irregular, 127
    light reaching earth from earliest
        days of universe, 144
    movement in cluster, 124
    moving as space expands, 146
    mutual gravitational attraction, 124
    observations up to 9000 million
        light-years away, 35
    photograph of a chain, *144*
    radio emissions, 134–5
    receding from each other, 141
        speed of recession, 27
    speculation on unlimited numbers,
        123
    spiral, 126, *127*, *128*
    structure, 126–31
    superclusters, 125
    youngest are probably spirals, 128
    see also Seyfert galaxies
Galileo Galilei (1564–1642), 16–19,
    16, 234
    discovery of Jupiter's moons, 101
    observation of the Milky Way, 118
Galle, Johann, 108
gamma radiation, *203*, 204
    produced by matter/anti-matter
        reactions, 206, *207*
    satellite-borne detectors, 207
Gamow, George, *146*
Ganswindt, Hermann, 260
Ganymede, *75*, 101, *102*
    possibility of establishing base on,
        253
Garriott, Owen, 193
geiger counters
    to detect cosmic rays, *204*
Gemini program, 169
    advance on Mercury program, 174
    Gemini 7, *174*
General Theory of Relativity, 200, 224
    applied to escape velocities, 208
geons (particles of gravitational
       energy), 201
Georgian planet *see* Uranus
Georgium Sidus see Uranus
germanium selenium crystal, *253*

Germany
    development of rockets, 163
    see also Peenemünde rocket site
giant stars *see* red giants
Gibson, Edward, 193
Glaser, Peter, 237, *238*
Glasgow University
    gravitational radiation detection
       equipment, *225*
Glenn, John
    first American in space, 169, 173
globular clusters in galaxies, 128, 131,
    131
Goddard, Robert H.
    advocate of use of sounding
       rockets, 162–3
    banned from launching in
       Massachusetts, 163
    influenced by Jules Verne, 162
    inventor of liquid fuel rocket, 160,
       161
Goddard Space Flight Center, *161*, 180
Godwin, Francis, 161
Gold, Thomas, 152, 157
    on source of signals from pulsars,
       221, *221*
Goodricke, John
    detailed study of Algol, 62
gravity, 21
    effect on industrial processes, *253*
    effect on tides, *23*
    envisaged by Kepler, 160
    formation of ripples in, 224
    in Einstein's General Relativity
       Theory, 201, 208, 224
    influencing planetary orbits, 76
    law of universal gravitation, 23
    Newton's discovery, 22
    of mascons on Moon, 179
    on Mars, effect on human life, 252
    provision in space stations, 243, 244
    role in formation of stars, 52
    surrounding black holes, 59, *209*
gravity waves, 224–5, *224*
Great Bear
    known to Egyptians as Sarcophagus,
       68
    subdivision of Big-Dipper, 69
Great Comet (1843), 114, *114*
Great Nebula, 52, 123
Grechko, Georgi, *194*, 195
Greeks
    astrology, 64
    interest in astronomy, 11
    Phoebus-Apollo (sun-god), 38
    view of Earth's position in universe,
       11, *11*
Greenwich Observatory, 24
Grissom, Virgil, 182
Gubarev, Aleksey, *194*, 195
guidance systems of satellites, 171
gyroscope control systems, 163, *163*

H

H-R (Hertzsprung-Russell) diagram,
    52, *53*
Haise, Fred, 186
Hale Observatory *see* Mount Palomar

Hall, Chester Moor, 28–9
Halley, Edmund, 23
    catalog of nebulae, 25
Halley's Comet, 23, 114, *114*
Harder, James, *316*
Hardy, David, *246*
Hawaiian volcanoes, 90
Hazard, Cyril, 216
heat energy, 154
    dissipation, 154–5
        from suburban house, *154*
heat shields, 173
heliostats, *237*
helium
    proportion of first atomic nuclei
       formed, 147
helium-3, 261
Henderson, Garry, 316
Hennessy, Basil, 298
Hercules
    globular cluster, 131, *131*
Hermes, *112*
Herschel, William (1738–1822), 24–5,
    24, *234*
    discovery of Uranus, 106, 107
    plotting frequency of stars in
       relation to Milky Way, 118–19
    reflecting telescope, *29*, 30
Hertzsprung, Ejnar, 52, *52*
Hertzsprung-Russell (H-R) diagram,
    52, *53*
Hess, Victor, 204
Hewish, Antony, 220
Heyerdahl, Thor, 295
Hidalgo
    orbit, *112*
Hill, Betty and Barney, 311, *311*
Hipparchos (astronomer, fl. 146–127
    BC)
    catalog of stars by brightness, 46
    contribution to astrology, 64
    estimate of distance from Sun to
       Earth, 38–9
Hooke, Robert, 100
Hörbiger, Hans, 77
horoscopes, 65
Horsehead nebula, 124, *125*
Hoyle, Sir Fred, 79, 152
Hubble, Edwin, *122*
    discovery of Andromeda composi-
       tion, 25, *35*, 123
    effect on other astronomers, 140
    theory of galactic evolution, 126,
       127, *128*
Hubble's Law, 141, *141*, *142*
Huggins, William, 49
Huizilopochtli (Aztec sun-god), *38*
human life
    at early stage of development, 276
    Biblical explanation of origin, *280*
    captured by aliens, 311
    emergence on Earth, 280
    evolution into space life, 245
    extending to extraterrestrial colonies,
       162
    only supportable in present
       conditions, 156–7
    suggestion of cross-breeding with
       aliens, 293–4

Humason, Milton, *115*
Humason Comet, *115*
Huygens, Christian, 234
  discovery of Saturn's rings, 18
  discovery of Titan, 105
hydrogen
  basic ingredient of most new stars,
    52
  basis of ramjet propulsion concept,
    260
  in continuous creation theory, 153
  proportion of first atomic nuclei
    formed, 147
hydrological cycle, 91
hydroponics, 252
  use in Israeli desert area, *252*
Hynek, J. Allen, 310, 316, *316*
hypersphere, 139

I

Icarus, 113, 253
  orbit, *112*
Ice Ages
  correlation with Sun's neutrino
    output, 203
ice fish, *282*
Icelandic volcanoes, *90*
ilmenite, *241*
Incas
  Inti (sun-god), 38
  system of roads and carvings, 293,
    293, 295
infrared radiation
  from Seyfert galaxies, 132
instruments *see* astronomical
  instruments
Integrated Missile Early Warning
  Satellites, 247
intelligence, 274–5
  level difference between man and
    animals, *275*
Intelsat communication satellite, 170,
  171
intercontinental ballistic missiles
  (ICBMs), 166
  use by Russians as basis for space
    program, 168
International Astronomical Union
  system of celestial divisions, 69
International Flying Saucer Bureau, 313
International Geophysical Year (1955),
  167
International Ultraviolet Explorer
  Spacecraft, *321*
interplanetary probes, 188–91
Inti (Inca sun-god), 38
Io, 101, *103*
ion-drives, 165
ionosphere, 88
Irwin, James, *186*
Italy
  medieval astronomers, *13*

J

Jansky, Karl, 32, *32*
Jantar Mantar observatory (India), *34*
Janus, 105

jet stream, 88
Jodrell Bank radiotelescope, 32–3
Juno booster rocket, 169, *169*
Jupiter, 100–3, *100*, 191, *191*
  atmosphere, 100
  Babylonian record of positions, *10*
  data from Mariner probes, 191
  day length, 100
  diameter and mass, 100
  discovery of ring, *100*, 101
  dismantled for materials for Dyson
    Sphere, 267–8
  Galileo's discovery of satellites, 18,
    19
  Great Red Spot, 100, *101*, 253
  interplanetary missions, 252
  magnetic field, 100, *101*
  pattern of movement, *76*
  possibility of primitive life forms,
    100–1, 285, *285*
  probable structure, 100
  source of helium-3, 261
  thirteen moons, 101
  ultraviolet emissions, 234
Jupiter missile, 167
  booster for Explorer I, 168

K

Kant, Immanuel, 78
  theory of nebulae, 122
Kennedy Space Center, 180
Kepler, Johannes (1571–1630), 20–1,
  20
  casting horoscopes, 65
  laws of planetary motion, 21
  Second Law, *21*
  science fiction *Somnium*, 160
  supernova observation, 58
  verification of Copernicus' theory, 76
Kerwin, Joseph, 193
Keyhole nebula, *52*
Kirchhoff, Gustav Robert (1824–1887),
  26, *26*
Kitt Peak observatory
  brightness contour map, *59*
Komarov, Vladimir, *173*, 175
Kosmolyot, 233
Kowan, Charles
  discovery of Chiron, 113
Krakow University, 14
  Geometry Room, *14*
Kristian, Jerome, 219
Kuiper, Gerard, 79, 189
  discovery of fifth moon of Uranus,
    107
  discovery of Nereid, 109

L

Lagoon nebula, 123, *124*
Lamarck, Jean Baptiste, 280
landing routines, 173
  of space shuttle, 231
Landsats, 170
La Paz, Lincoln, *308*
Laplace, Pierre-Simon de, 78, 122
laser galleon concept, 257–60, *258*
lasers, *247*

  application to nuclear pulse engines,
    261
  Sun-pumped, *289*
  weapons, 247
Lassell, William
  discovery of two moons of Uranus,
    107
launch pads, 182
lava, 90
  from lunar volcanoes, 93
Leavitt, Henrietta, 63, *124*
Lee, R. B., 275
Lemaître, Georges, 146
lens systems of telescopes, 28
  achromatic, *28*, 29
  compound, 29
  problems of achieving size, 29
Leonov, Aleksey A.
  first walk in space, 174, *174*
  painting of Vostok spacecraft, *172*
Lexell, Anders, 25
Lick observatory
  first picture of pulsar, *220*
  refractor telescope, 29, *29*
life-support systems
  need first realized by Tsiolkovsky, 162
light
  behavior in a plasma, 147
  behavior in vicinity of black holes,
    212
  bent as it passes Sun, *138*
  in relation to space travel, 198–9,
    198, 199
  in relativity theory, 198
  speed of, limiting information
    transmission, 289, *289*
  waves stretched by expansion, 142
light-years, 48
  as unit of measurement, 35
lightning
  mistaken for UFOs, *298*
Linnaeus, Carolus, 280
Lippershey, Hans, 28, 234
  builder of first telescope, 17
living organisms
  ambiguous results from Mars soil
    samples, 284–5
  as yet found only on Earth, 274
  behavior, 275
  evolution, 280–3
  increasing complexity, *279*
  likelihood of evolution on other
    planets, 276
  Linnaeus' classification, 280
  origins, 278–9
  six criteria, 274
  see also extraterrestrial life
long-period variable stars, 63
Looped nebula, *124*
Lousma, Jack, 193
Lovell, Sir Bernard, 32
Lovell, James, 183, 186
Low, Robert, 307–8
Lowell, Percival, 96, *97*
  laid foundations for discovery of
    Pluto, 110, *111*
Lubbock Lights, *304*
Lucian of Samos (c.125–190 AD)
  first science fiction writer, 160

Lulworth Cove (Dorset), *89*
Luna (lunar probes), 176, *176, 178*
   first soft landing, 178
Lunar module
   of Apollo spacecraft, 181, 184, *184,*
   185
lunar probes, 176, *176,* 178, *178*
lunar rover vehicles, *94, 186,* 241, *251*
lunation, 95
Lunik spacecraft
   lunar circumnavigation, 92
Lunokhod, 179, *179*

M

M13 *see* Hercules: globular cluster
M31 nebula *see* Andromeda
McDonald, James, 309
Magellan, Ferdinand
   observation of small galaxies, 124
Magellanic Clouds, 63, *119,* 124, *124*
   classed as irregular galaxies, 126
   size compared with our galaxy, *144*
magma, 90
Malthus, Thomas, 282
man *see* human life
maneuverability
   of Apollo spacecraft, 181
   of Cosmos satellites, 246
   of Gemini spacecraft, 174
   of interplanetary probes, 188
   of space shuttle orbiter, 229, 230–1
   training astronauts, *181*
Manhattan missile, 166
Manned Space Flight Network, 180
manned space flights
   first Soviet, 168, 172
   see also Apollo program;
      astronauts; Gemini program
Mantell, Thomas, 301–2, *302*
mantle of the Earth, 88
Mariner space probes, 80
   to Mars, 98, *98,* 99, 190
   Venus fly-by mission, *188,* 189
Marius, Simon
   observation of nebula in Andromeda,
      120
Mars, 96–9
   atmosphere, 97
   climatic changes, 96, 97
   converting to habitable condition,
      266
   discovery of elliptical orbit, 21, *21*
   first soft-landing by Soviet probe,
      190
   hostile environment, 97
   impression of a landing, *266*
   landscape, 97–8, *98, 99*
   legend of "canals", 96, *97*
   length of day, *97*
   Olympus Mons, 3
   pattern of movement, *76*
   photographs from Mariner probes,
      190, 251
   polar regions, 98–9
   popular fantasies, 284, *284*
   possibility of life forms, 97, 284
   probes, 190–1
   problems of human communication

      on, 251
   program for manned landings, 99
   projected manned mission, 250–1
   soft-landings of Viking probes, 191
   sunset, *250*
   twin moons, 99
   water periodically on surface, 97
Mars (spacecraft), 190
mascons, 179
mass
   effect on escape velocities, 208
   influence on momentum, 164
   influence on stars' life cycles, 55
mass drivers, *240,* 241
Maxwell, James Clerk, 78–9
Mayan observatories, *13*
medicine
   uses of astrology, 65, *65*
   see also space medicine
Mengel, Howard, 312, *312*
Menzel, Donald
   on Ezekiel's vision, 299
Mercury, 80–1
   atmosphere, 81
   day length, 80
   discovery of magnetic field, 80
   high density, 81
   hostile environment, 80
   not immediately suitable for
      exploration, 252
   pattern of movement, *76*
   photographs of Mariner 10, *188,* 189
   similarity of landscape to Moon, 81
   surface, *80, 81*
Mercury project, 172–3, *173*
mesons
   creation from collisions of protons
      and antiprotons, *207*
Mesopotamia
   astrology, 64
   familiarity with constellations, 68
Messier, Charles
   catalog of nebulae, 25, 122
meteorite fragment, apparently
      patterned by algae, *276*
meteors
   Aristotle's explanation, 114
   causing Moon's craters, 92, 93
   collisions with Moon, 94
   crater in Arizona caused by, *112*
   effect of impacts, 87
micrometeors
   protection of probes against, 188
microwaves
   rock-splitting energy, *249*
   seen as echo of big bang, 146, 150
   transmission of power to Earth,
      236–7
   weapons applications, *249*
military applications of space
      technology, 246–7
military satellites, 170
Milky Way, 118, *118, 119, 120, 121*
   diameter, 218
   distribution of neutral hydrogen, *157*
   estimate of number of stars, 120
   estimates of size, 119
   Galileo's observations, 18
   Herschel's scheme of, *24*

hot younger stars, 234
movement with local group of
   galaxies, *119,* 124
radio signals received from, 32
resource utilization, 209
Miller, Stanley, 279
Millikan, Robert
   coined phrase "cosmic rays," 204
Mills, Bernard, 33
Mills Cross antenna, 33
mineral resources
   of asteroids, 113, *113,* 253
   of Mercury, 252
   of the Moon, 240, *240*
Mira (Omicron Ceti), 63
Miranda, *106,* 107
mirror systems in telescopes, 30, *30*
missiles
   V-1 rockets, *162*
   V-2 rockets, *162,* 163, *163*
   see also weapon systems
Mizar, 63
Molniya communication satellites, 170,
   171
momentum of a body, 164
Moon, 92–5
   advantages for astronomical
      observations, 241
   appearance of craters, 93
   appears geologically inert, 94
   data on structure from Surveyor
      probe, 178
   Earth's only natural satellite, 92
   eclipses, 95
   escape velocity, 208, 241
   establishing bases, 240–1
   experiments set up by Apollo 11
      crew, *184*
   explanation of orbital variation, 23
   first close-up photographs of
      surface, 176
   first orbital flight, 183
   first studies through telescope by
      Galileo, *16,* 17
   full disk, from Apollo 11, *95*
   gravitational field variations, 179
   importance of manned expeditions,
      94, 95
   lack of atmosphere, 92
   landscape, 93
   low mass, 92
   mining resources, 240, *240*
   Orbiter survey of radiation levels, 179
   painted as seen through Galilean
      telescope (*c*.1700), 92
   phases, *93,* 94–5
   photographs of "dark side," *93,* 176
   pictures from Ranger probes, *176,*
      177–8
   provision of raw materials for space
      stations, 243
   rock samples brought back to
      Earth, *94, 186*
   soil sample, *240*
   speed of rotation and day length, 92
   structure of core, 94
   theories of formation, 92–3
   theories on structure, 176
   time lapse exposure of apparent

positions, *34*
used in detection of radiostars,
216–17
volcanic action, 93, 94
see also lunar probes; lunar rover
vehicle
moon landings
Apollo 11 mission, 183–5
first imagined by Lucian of Samos,
160
first landings, *159*, 178
foreshadowed in Goddard's 1919
paper, 163
Moore, Charles, *314*
Moricz, Juan, 295–6
"Morning Star" *see* Venus
Morrison, Philip, 287
motion, laws of
Galileo's discovery, 17
Newton's work, *22*
Third Law, 164
Moulton, Forest Ray, 77
Mount Palomar observatory, *9*, 30–1,
30, *124*, 127
photographs of Cygnus A, 133
Mount Wilson observatory, 30, *122*
observations of Andromeda nebula,
123
mountain range formation, *88*, 89–90,
89
on the Moon, 93–4
mutations, 282

N

NASA (National Aeronautics and
Space Administration)
biosatellite studies, *285*
concentration on lunar exploration,
180
financial cutback, 186
lunar probe failures, 176–7
Space Telescope project, 235, *235*
voted funds to develop space
shuttle, 228
work on Orbiting Astronomical
Observatories, 234
NGC 4151 (galaxy), 132, *132*
natural selection process, 282–3
nebulae
Andromeda, 25, *25*, *35*, 58, *119*,
120, 122–3, 124, *219*
catalogued by Halley, 25
catalogued by Messier, 122
Coal Sack, 124
Crab, 58, 194, *204*, *220*, 221
Cygnus, *117*
dark, 123–4
Dumbbell, *35*, *119*
early observations, 120, 122
Great, 52, 123
Herschel's explanation, 25
Horsehead, 124, *125*
Kant's theory, 122
Keyhole, *52*
Lagoon, 123, *124*
Looped, *124*
Lord Rosse's observations, 126
Magellanic Clouds, 63

Orion, *137*
possible birthplace of stars, 52
recycling stellar material, *137*
spiral structure discovered, *28*
Triangulum Spiral, *119*
Veil, *117*, 123
see also planetary nebulae
nebular hypothesis, 78–9
Laplace's, 123
Neptune, 108–9, *108*, *109*
atmosphere, 108
day length, 108
mathematical discovery, 108
orbit crosses that of Pluto, 108, 110
size and probable structure, 108
two known moons, 108
Nereid, 108–9, *108*
neutrinos, 54
extraordinary qualities, 202
generated by the Sun, 202
measurement of strength of flux, *203*
only directed by reaction with other
particles, *202*
neutron stars, 55, 59, *155*
beginnings of black holes, *208*, 209
possible explanation of pulsars, 221,
221
Soyuz investigation, *194*
neutrons
first combination with protons, 147
Newcomen, Thomas, 154
Newton, Sir Isaac (1642–1727),
22–3, *22*
ideas of gravity upset by Einstein,
200
investigation of spectrum, 26
Third Law of Motion, 164
Newtonian telescopes, 30
Nicolson, Iain, 268
Nimbus (meteorological satellite), 170
Norem, Philip, 257, *258*
North Star *see* Polaris
nozzles of rocket engines, 164
nuclear fusion
application to starship propulsion,
260–1
in formation of stars, 52
possible source of cosmic rays, 204
sequence in degeneration of stars,
54–5
Nut (Egyptian sky-goddess), *64*

O

Oberon, 106–7, *106*
Oberth, Hermann, 163
observations
Galileo's insistence on, 16
Herschel's work, 24
observatories
Arecibo, *32*, 33, 84–5, 289, *289*, *318*
Bergedorf, 30
Chichen, Itza, *13*
Greenwich, 24
Jantar Mantar, *34*
Kitt Peak, *59*
Lick, 29, *29*, 220
Mount Palomar, *9*, 30–1, *30*, *124*,
127, 133

Mount Wilson, 30, *122*, 123
orbiting astronomical observatories,
234–5
siting, 31
Yerkes, 29
oceans
originating life on Earth, 91
possibilities of using resources, 91
oil, off-shore fields, 91, *91*
Old Fashioned missile, 166
Omicron Ceti *see* Mira
On the Origin of Species (Charles
Darwin), 283
O'Neill, Gerard K., 242
Oort, Jan Hendrik
theory of origin of comets, 115
work on dynamics of the Milky
Way, 119–20
oppositions
used in Kepler's calculations, *21*
Orbiter lunar probe, 178–9, *178*
Orbiting Astronomical Observatories
(OAOs), 234–5
ore extraction, 243–4
Orion, 70
Great Nebula, 123, *137*
Horsehead Nebula, 124, *125*
in Galileo's *Siderius Nunctus*, *18*
Osiander, Andreas (theologian,
1498–1552), 15
oxygen for space stations
by-product of lunar ore extraction,
243–4
Ozma project, *286*, 287
ozone, 88
ozone layer
strength affected by solar wind, 43

P

parachute landings *see* landing
routines
parallax
used in calculating distances from
stars, 47
parhelia (meteorological phenomenon),
299
parsec (unit of measurement), 48
Parsons, William *see* Rosse, William
Parsons, 3rd Earl of
particle accelerators, 205
use as weapons, 247, *248*
Peebles, P. J. E.
work on electromagnetic radiation,
150–1
Peenemünde rocket site, *162*, 163, 166
pendulum
Galileo's discovery of motion, 17, *17*
Penrose, Roger, 270, *270*
Penzias, Arno, 146
measurement of radio emissions,
150, *150*
Peru
system of roads and carvings, 293,
293, *295*
*Philosophiae Naturalis Principia
Mathematica* (Isaac Newton), 23
Phobos, 99
possible use as radio relay station,

251
Phoebe, 105
Phoebus-Apollo (Greek sun-god), 38
photosphere, 39, *41*
photosynthesis, *265*
   to reduce atmospheric carbon
     dioxide, 265
Piazzi, Giuseppe, 112
Pickering, Edward C., 63
Pioneer space probes, 176
   extremely long journeys planned,
     191
   Jupiter fly-by missions, 100
   photographing Saturn, *104, 105*
   survey of Venus, 85
Pioneer 10 probe
   cruising speed, 254
   identification plaque, *213, 286, 286*
   slingshot technique, *188*
Pisa University, 16, 17
planetary nebulae, 55, *119*
planets
   Bode's Law, 112
   colonization possibilities, 242
   effect of discovery on mysticism, 10
   elliptical orbits, 76
   Galileo's observations, 18
   Kepler's discovery of elliptical orbits,
     21
   possibility of one undiscovered
     beyond Pluto, 111
   Ptolemy's research into movements,
     11
   theory of formation, 77
   unable to support advanced life, 318
   use in astrology, 65
   visits by unmanned probes, 250
plasma, 146–7, *147*
plasma era in evolution of universe,
   146–7
Pleiades, 71, *131*
   constituent stars, 18
   typical cluster formation, 52
Pluto, 110–11
   discovery in 1930, 110, *111*
   eccentric orbit, 110
   effect on other planetary orbits a
     mystery, 110–11
   environment hostile, 111
   extent of orbit, 35, 254
   impression of view from, *110*
   may not be a planet, 111
   orbit crosses that of Neptune, 108,
     110
Pogue, William, 193
Polaris (North Star), 70
political implications of space
   colonization, 245
Ponnamperuma, Cyril, 279
positrons, 147, *203*
   discovery, 206
   tracks in bubble chamber, *206*
power supply
   of artificial satellites, 170
   of interplanetary probes, 188
powersats, 237–8, *238*
   obtaining raw materials from
     Moon, 240
Price, P. B., *260*

prominences, solar, 42
proper motion, 49, *49*
propulsion systems
   envisaged by science fiction
     writers, 161
   for space ships to remote stars, 254
   laser galleon concept, 257–60, *258*
   matter/antimatter reaction
     application, 262–3
   of Orbiter spacecraft, 228
   ramjet concept, *257*, 260
   satellite on-board systems, 171
   Tsiolkovsky's ideas on rockets, 162
   variable thrust, 163
protogalaxies, 131
proton-proton chain, *203*
protons
   first combination with neutrons, 147
   proton-beam weapons, 247, 249
   querying weight, 156
Proxima Centauri, 71, 254
   distance from earth, 48
   laser galleon plan, 257, 260
   parallax, 47
psychology of space station life, 244
ptarmigans, *282*
Ptolemaic system, 12, *12*
   Copernicus' disagreement, 14
   seen as wrong by Tycho Brahe, 20
Ptolemy (Claudius Ptolemaeus,
     fl. 2nd century AD), 11
   Galileo sceptical about theory, 17,
     18–19
   star catalog, 68
   System of Mathematics, 12, 65
pulsars, *60*, 220–1, *220*
   detected in 1967, 33
   prediction of slowdown in rotation,
     221
   remains of supernovae, 59
   Soyuz observation, *194*
   see also supernovae
The Pup *see* Sirius B

Q

quadrants
   used by Tycho Brahe, *20*
quantum era, 148
quarks, 148
quasars (quasi-stellar sources), 33,
     153, 216–19
   as final stage in life of galaxies, 219
   BL Lacs as embryos, 135
   central "hot spot" of 3C-273, *145*
   discovery of 3C-273, 217–18, *217*
   estimates of size, 218
   luminosity, 218
   possible explanations, 219
   throwing doubt on Hubble's red
     shift theory, 145
Quintanella, Hector, *317*

R

radar anomalies, 306–7
radiation, types reaching Earth, *46*
radio emissions
   affected by solar flares, 42

   from artificial satellites, 171
   from Cygnus A, 33, *33*, 133, 134,
     135
   from Io, *103*
   from other galaxies, 134–5
   from our galaxy, 150
   from quasars, 216
   interference from pulsars, 220–1
   most common wavelength, 287, 289
   scintillation measurement, 220
radiotelescopes, 32–3
   dish antennae at Cambridge, *197*
   in extraterrestrial communication
     experiments, 287
ramjet concept for space propulsion,
     257, 260
Ranger lunar probes, 176–7, *176*
red giants, 54
   in galaxies, 128
   place in life cycle of stars, 56
   size depends on initial mass of star,
     56
red-shift (spectroscopy, 27, *48*, 49
   analogy with Doppler Effect, 140,
     140
   correlation with quasars, 217–18
   explained in terms of an expanding
     universe, 142
   Hubble's theory discounted by Arp,
     145
   in light from distant galaxies, 140,
     141
Redi, Francesco, 278
Redstone missile, 167, *167*
   launch vehicle for first US manned
     space flight, 172
refraction, used in telescopes, 28
relativity theory, 76–7, 198
   see also General Theory of
     Relativity; Special Theory of
     Relativity
religions
   Copernicus' work seen as heresy,
     14–15
   Galileo's breach with church, 18
   importance of heavenly bodies, 10
   rejection of theory of evolution,
     280–1
   sun worship, 38
Reuber, Grote, 32
Rhaeticus, Georg (Georg Joachim van
     Lauchen, 1514–1576), 15
Rhea, *104*
rilles on the Moon, 94
Ritter Crater, *176*
rock drawings and paintings, 294, *295*
   Australian, *295*
rockets
   carrying x-ray detectors, 222, *222*
   design principles, 164–5
   disadvantages as boosters, 228
   early forms, 161
   early military use, *160*, 161
   first liquid-fuel, 160, *164*
   fuels containing own oxygen, 164
   multistage, 163, 164, 166, *166*
   solid-fuel, *164*
   space shuttle reusable booster,
     229–30

Tsiolkovsky's ideas, 162
Rosse, William Parsons, 3rd Earl of, *28*,
    30, 126, *126*
Russell, Henry, 52, *52*
Russia *see* Soviet Union
Ryle, Martin, 216, *216*

S

S Andromedae, 58
SAS (Small Astronomy Satellite), *222*
Sagan, Carl, 97
    enthusiasm for extraterrestrial
        communication, 289
    estimate of advanced civilizations,
        318, 319
    suggestion for inhabiting Venus, 265
    theory of ice volcanoes on Titan, 105
Sagittarius
    Lagoon nebula, 123, *124*
    location of x-ray source, *222*
Salyut space stations, 194-5
    astronomical observatory, 234-5
    compared with Skylab, 194
    Salyut 1 tragedy, 194
SAMOS military satellites, 170
Sandage, Allan, *218*, 219
satellites, artificial, 162, 170-1
    as telescope platforms, 31, *31*
    communication, 170, *171*
    earth resources, 170, *186*
    environmental control, 171
    for astronomical observation, 234-5
    number in orbit (1978), 246
    number launched, 170
    power-generating, 237-8, *238*
    secret launchings, 246
    similar to interplanetary probes, 188
    structural design, 171
    subsystems, 170-1
    US/Soviet race to launch, 167
    use of space shuttles to place in
        orbit, 233, *233*
    used for spying, 246
    weapon platforms, 246-7
Saturn, 104-5, *104*, 191
    atmosphere, 104
    day length, 104
    diameter, 104
    Galileo's observation of triple image,
        18
    interplanetary missions, 252
    pattern of movement, *76*
    rings, 105, *105*
    size and length of orbit, 104
    sixth and seventh satellites
        discovered, *24*
    ten moons, 105
Saturn launch vehicles
    ultrapowerful boosters, 169
    Saturn I, 169
    Saturn V three-stage booster, 164-5
        on crawler transporter, *169*, 182
        use as Apollo booster, 169, 182
    third stage used in Skylab, 192
Scheiner, Christoph, *39*
Schiaparelli, Giovanni, 96, *97*, 234
Schmidt, Bernhard, 30
Schmidt telescopes, 30, *31*

Schwarzschild, Karl, 208
Schwarzschild radius, 209, *209*, 213
science fiction
    first story by Lucian of Samos, 160
    Francis Godwin's story, 161
    Kepler's *Somnium*, 160
Sco X-1 x-ray source, 222
Scorpius
    intense source of x-rays, 222
Sculptor
    galaxy, *123*
seismic waves, 88
service module of Apollo spacecraft,
    181
Seven Sisters *see* Pleiades
Seyfert, Carl
    discovery of chaotic galaxies, 132,
        132
Seyfert galaxies, 132, *132*
    comparisons of wavelength
        distribution, *133*
    see also BL Lacertae
Shapley, Harlow, 63
Shepard, Alan B.
    first US manned space flight, 172,
        173
Shere, Pat, 309
Shklovskii, I. S., 99
short-period variable stars, *63*
Siderius Nunctus (Galileo), *18*
singularity (mathematical concept), 214
Sipapu program, 249
Sirius
    absolute magnitude, 48
    brightness, 46, *56*
    detection of binary companion, 56,
        56
    distance from Earth, 48
    size, *57*
    spectrum, *27*
Sirius B ("The Pup"), 56-7, *56*
Skyhook balloons, *304*
Skylab, 192, *192*
    astronomical observations, 234-5
    compared with Salyut, 194
    interior design, 192, *193*
    problems following launch, 192-3
    solar cells, *236*
    weightlessness experiment, *252*
Smith, E. J., *301*
Smith, F. Graham, 133
soil samples from Moon, *177*, 178
solar cells
    applications, 236
    built into powersats, 237-8
    in artificial satellites, 171, *171*
    of Skylab, 192
    on interplanetary probes, 188
    on Orbiter space probes, 179
    use in space stations, 243
solar energy, 264
    to penetrate metal, *247*
solar flares, 42
solar furnaces, *237*, 238
solar power, 236, *237*
solar system, *77*
    Copernican revolution, 14
    exploitation of energy resources, 264
    formation, *78-9*

theories, 77
    illustration drawn to scale, *76*
    Ptolemy's theories, *76*
solar wind, 40
    effect on Earth's environment, 42-3
    effect on tails of comets, 115
solstices
    possible use of Stonehenge, *38*
Somnium (Johannes Kepler), 160
Southern Cross, *69*
    Coal Sack nebula, 124
Soviet Union
    development of reusable orbiter, 233
    ICBM program, 166
    interplanetary probes, 189-91
    manned space station program, 194
    236-inch telescope, 31
    work with lunar probes, 179
Soyuz program, 175, 194
    Soyuz 1 crash, 175
    transfer of crews between two
        craft, *194*
space
    Einstein's theory of its shape, 138,
        224
    Euclidean and non-Euclidean
        geometry, 138-9, *138*
    Kepler's idea of vacuum, 160
Space Ark concept, *254*, 255-6, 261
space flight
    making available to non-astronauts,
        233
    to remote stars, 254-63
    see also manned space flight
    space medicine, 162
    adaptation to Martian environment,
        252
    loss of calcium in weightless
        condition, 244
space mission control room (Houston),
    186
space shuttles, 228-33, *228*, *230*
    achieving cost reductions, 232
    aerodynamic features, 228, 231
    cargo-carrying capability, *229*
    cyclical use, *229*, 230
    emergency routines, 231
    production capability, *233*
    three main elements, 228
space shuttle main engine (SSME),
    228-9, *230*, *231*
space stations, 192-5, 242
    NASA's objective, 186
    O'Neill's designs, 243
space-time, 138, 200, *200*
    flexibility, 146
    geometry resulting in gravity, 201
    under conditions of singularity, 214
space travel, instantaneous, 201
    see also space flight
space tugs, 231-2, *254*, 267
    to tow asteroids out of orbit, 253
space walks, 174
space war, *246*
Spacelab, *229*, 232
Special Theory of Relativity, 198, *198*,
    199, 256
spectral sequence, 49, *50-1*
    see also Hertzsprung-Russell

diagram
spectrometers
  in orbiting observatories, 234
spectroscopes, 26-7, *27*
  used by Hubble to demonstrate red
    shift, 140
spectroscopic binaries, 63
spectroscopy, 26
  in analysis of chemical composition
    of stars, *47*, 48
spectrum
  Newton's investigations, 26, *26*
  of BL Lacs, 135
spiders, *275*
splashdown *see* landing routines
spontaneous generation, 278, *279*
Sputnik satellites, 168
spy satellites, 246
stabilizing devices in V-2 rockets, *163*
Stanford, Ray, 317
star catalogs
  of Hipparchos, 46
  Third Cambridge Catalog of Radio
    Stars, 216
  Tycho Brahe's work, 20
"star-gauging" technique, 25
star maps
  Albrecht Dürer, 68
  of northern sky, *70*
  of southern sky, *71*
Starlight International Project, 317
stars
  ancient identification with mythical
    beings, *66-7*
  brightness contour map, *59*
  brightness scale, 46, 62
  classification, 49
  comparison of luminosities, 48
  determination of temperature of
    chemical constituents, 26
  distance related to brightness, 46-7
  final evolutionary stages, *155*
  formation, 52
  gravitational collapse, *209*
  Herschel's study of distribution, 25
  lifespan, 45, 52-5, *54*
  list of 20 brightest, *73*
  mass dictating final form, *155*
  names from Arabic, 13
  observation, 46-51
  spectral lines, 48
  spectroscopic analysis, 26
  total collapse creating black hole,
    208, *208*
  velocities, 49
starships, 254-63, *256*
steady state model of the universe,
  152-3, *152*
steam engines, 154
Stonehenge, *38*
Stratoscope II project, *30*
stratosphere, 88
Sun
  absolute magnitude (brightness), 48
  color-coded TV picture, *36*
  constitution, 39
  Copernicus' discovery of true place,
    14
  cross-section, *44*

diameter and comparative mass, 39
distance from Earth, 39
  early Greek estimates, 38-9
early observations, 38
eclipses, 40, *40*
effect on earth of sunspots and
  solar wind, 42
energy created by nuclear fusion, 44,
  44
engraving (1635), *39*
evidence of changes in core
  reactions, 202
Fraunhofer's observations of
  spectrum, 26
gravitational field deflects light from
  stars, *138*
harnessing wasted energy, 267-9
hydrogen distribution surface, *45*
layers of gas, 39-40
major events, 42
mass, 44
orbit, 119-20
position in main sequence of H-R
  diagram, 54
possible sequence of degeneration,
  54
radiations, *87*
safe methods of observation, 39, *39*
satellite-based observations, 40, *42*
setting over Mars, *250*
source of life on Earth, 39
source of radiant energy, 236
spectroscopic analysis, 27
spectrum, *27*
surface temperature, 49
sustained energy output, 44
temperature, 39
will probably become white dwarf,
  155
sun-gods, *38*
sunspots
  effect on earth's environment, 42
  fluctuations in activity, *40*
  Galileo's observations, 18, *18*
  numbers vary in 11-year cycles,
    41-2
  probable causes, 40
  Scheiner's discovery, *39*
  size, 40-1
  structure, 40, *41*
supernovae, 58-9
  core of exploded star becomes
    pulsar, 59
  evolution, 60-1
  final explosion of degenerated star,
    55
  Galileo's observation, 17
  on brightness contour map, *59*
  possible source of cosmic rays, 207
  process used to dismantle stars, 270
  rare event in our galaxy, 58
  x-ray map of remnant of Cassiopeia
    A, *223*
  see also pulsars
superspace theory, 201
Surveyor lunar probe, *177*, 178
  soft landings, 178
Swigert, Jack, 186
System of Mathematics (Ptolemy), 65

T

T-1 rocket
  development from V-2, 166
tachyons (elementary particles), 205
Tassili rock drawings, 294, *295*
Taurus, 70
  Chinese observation of supernova in
    11th century, 58
  see also Pleiades
technology
  growth by energy resource
    exploitation, 264
Teetotaler missile, 166
telescopes, 28-31
  "cup-and-saucer" type, *235*
  Galileo's use of, *16*, 17, *19*
  invented by Lippershey, 28
  on orbiting satellites, 31, *31*, 234
  radio, 32-3, *197*, 287
  reflecting, 30
    Herschel's, 24, *24*
    Newton's, *23*
  refracting, 28
  see also lens systems
Telstar II communications satellite, *170*
temperature
  gradients, 154
  of Earth and its atmosphere, 87
  of stars
    use of spectroscope, 26
    related to color analysis, 48-9
  of the Sun, 39
temperature inversion, possible cause
  of UFO sightings, *316*
Tereshkova, Valentina, 174
terraforming, 265-6
Tethys, *104*
tetrachloroethylene, *203*
Thales (Greek philosopher, fl. 6th
  century AD), 11
thermodynamics, Second Law of, 154
thermogram, *154*
Thor missile, 167
thrust from rocket engines, 164
  of Saturn V staged engines, 165,
    169
tidal movements, *23*, 212
  Newton's explanation, 23
time
  concept for fast star ships, 256
  effect of relativity, *198*, 199-200
  measurement, 34-5
  theory of reversal, 157
  traveling backward, 205
  see also space-time
Tiros meteorological satellites, 170
Titan, *104*, 105
  atmosphere, 105
  possibility of establishing base on,
    252-3
Titan booster rockets
  in Gemini program, 169
Titania, 106-7, *106*
Tokyo, 91
Tombaugh, Clyde
  discovery of Pluto, 110, *111*
torus-shaped colonies, 244-5, *245*

trenches in structure of Earth, 89
Trent, Paul, *308*
Triangulum Spiral, *119*
Triton, 108–9, *108*
Trojans (group of asteroids), 113
tropopause, 88
troposphere, 87
Tsiolkovsky, Konstantin, *160*, 161–2
Tyndall, John, 278

U

UK 6 satellite, 215
Ulam, Stanislaw, 261
ultraviolet radiation, 87
   absorbed by ozone layer, 88
   International UV Explorer spacecraft,
     321
Umbriel, 107, *106*
unidentified flying objects (UFOs),
     292–9
   Adamski's photographs, *311*
   data reliability, 314–16
   hoaxes, 302, *309*
   mistaken sightings, *314*
   possibly confused with ball
     lightning, *298*
   post-World War II sightings, 300–1,
     301
   public's wish to believe in, 318
   radar anomalies, 306–7
   scientific investigation, 314–17
   sighting in France (1974), *319*
   suspicion of government secrecy,
     302–4
United States of America arms
     program, 166
universe
   concept of contraction, 156, *156*
   cooling to around absolute zero, 155
   cycles of creation theory, 155,
     156–7, *156*
   dating the expansion, 145–6
   early theories of its arrangement, 118
   Escher's portrayal, *139*
   expansion, 141–5, *143*
   formation, 35
   four-dimensional, 200, *200*
   inevitable heat death, 154
   initial density, 143, 148
   investigation into rate of expansion,
     157
   mathematical investigation, 138–9
   plasma era, 146–7
   steady-state model, 152–3, *152*
   temperature drop after big bang, 147
   two-dimensional model, *143*
   vastness of scale, 123, 138
   see also big bang theory
Uranographia (Johann Bode), 69
Uranus, 106–7, *100*, *147*, 191
   atmosphere, 106
   day length, 106
   discovered by Herschel, 25
   discovery of rings, 106, *107*
   five moons, 106–7, *106*
   motion affected by Neptune, 108
   orbit affected by Pluto, 110

   peculiarities of orbit, 106, *107*
   possibility of sixth moon, 107
   probable structure, 106
   rotation of satellites, *24*
Urey, Harold, 279
Ursa Major
   brightness of galaxy in, *59*

V

V-1 rockets, *162*
V-2 rockets, 163
   first Russian firing, 166
   post-war development in USA, 166
V-173 experimental aircraft, 314
Vanguard project, 168
variable stars, 62–3, *63*
Vega
   size, 57
Veil nebula, *117*, 123
Venera probe, *188*, 189–90
   landing on Venus, 84, *85*
Venus, 82–5, *264*
   anticlockwise rotation, 85
   atmosphere encloses Sun's heat, 82,
     85
   cloud blanket, 82–3, *83*
   conversion to habitable condition,
     265, *265*
   day length, 85
   Galileo's discovery of phases, 18
   high-altitude wind storms, 85
   impression of a landing on, *266*
   impression of surface, *85*
   inhospitable atmosphere, 82
   investigation by space probes, 82
   landing of Venera 9 probe, 84, *85*
   length of orbit, 85
   not immediately suitable for
     exploration, 252
   pattern of movement, *76*
   phases, *82*
   popular fantasies, 284
   probable environmental conditions,
     84
   probes, 188–90
   radiotelescope survey of surface,
     84–5
   reflection from sulfuric acid in
     atmosphere, 83
   single photograph from Venera 9,
     190
   sketch of supposed inhabitant, *310*
   Soviet attempts to land instrument
     packages, 190
   surface surveys, 84–5
Verne, Jules
   influence on Goddard, 162
   propulsion ideas, *160*, 161
Viking rockets
   as first stage for Vanguard project,
     168
Viking space probe
   investigation of Mars, *96*, 97, 99
   Mars touchdown, 284
   soft-landings on Mars, *190*, 191
Villas Boas, Antonio, 311
Virgo

cluster of galaxies, 125
   in 18th-century celestial atlas, *69*
volcanic activity, 89
   in Iceland, *90*
   on Io, 101, *103*
   on the Moon, 93, 94
   possibility of ice volcanoes on
     Titan, 105
volcanoes, 90
   on Mars, 98
von Braun, Wernher, 163
   involved with first US operational
     rocket missile, 166–7
   leader of Jupiter team, 168
   sees need for ultrapowerful booster,
     169, 182
von Däniken, Erich, 293–8
Voskhod spacecraft, 174
Vostok rocket
   boosters, *165*
   dimensions, 174
   first grouped flights, 173
   first manned flights, 168, *168*, 172,
     173
   painting by Leonov, *172*
Voyager probes
   extremely long journeys, 191
   Jupiter fly-by mission, 100, 101
   projected fly-by of Neptune, 109
   projected fly-by of Uranus, 106, *107*
Vulpecula
   Dumbbell nebula, *119*

W

Wac Corporal rocket booster, 166, *166*
Wallace, Alfred Russel, 283
   drawings of fish, *283*
Washington airport
   UFO scare, 304–6, *304*
Watson, James, *274*
Whittle, Frank, *275*
weapon countermeasures, 249
weapon systems, 247
   Manhattan missile, 166
   on-board satellites, 170
weather
   effect on troposphere, 87
weather balloon, *314*
Weber, Joseph, 215, *215*
   detection of gravity waves, 224–5,
     225
Wegener, Alfred, 89
weightlessness
   expressed in Godwin's science-
     fiction story, 161
   Skylab experiment, 193, *252*
Weitz, Paul, 193
Weizsacker, Carl, 79
Wells, H. G.
   antigravity idea, 161
   War of the Worlds, 284, *284*
Wheeler, John, 201
Whirlpool galaxy, *122*, *126*
White, Edward H.,
   killed in Apollo 4, 182
   space walk, *175*
white dwarfs, *54*, 55, *60*, *155*, *208*

extreme density, *56*, 57
Wilson, Robert, 146, 150, *150*
Wollaston, William Hyde, 26, *26*
Woodcock, Gordon R., 238, *238*

x-ray stars, 222–3
satellite detection, 222, *222*
theories of stellar upheaval, 223

X

x-ray detection, 234

Y

Yerkes observatory, 29

Z

Zelenchukskaya telescope, 31
Zodiac, 10–11, *64*
formed in Egypt and Greece, 64
correlation with human body, *65*
twelve houses, 64–5
Zond IV spacecraft
mistaken for UFO, 316

Picture Credits

334

173(R)   Novosti
174–   Fotohronika Tass
175(T)
174–   NASA
175(B)
176(L)   Novosti
176(C)(R)   NASA
177(T)   © 1969 by Scientific American, Inc. All rights
              reserved
177(BL)   NASA
177(BR)   Hughes Aircraft Company
178(T)   © 1968 by Scientific American, Inc. All rights
              reserved
178(B),   Novosti
179
180–181   NASA
182   Popperfoto
183–186   NASA
187(TR)   United States Information Service, London
187(B)   NASA
188(T)   Photri
189(T)   Novosti
189(B)   NASA
190(B)   © Aldus Books
190(T),   © California Institute of Technology and Carnegie
191(T)   Institution of Washington
191(B),   NASA
192
193   Photri
194–195   Novosti
196   Robert Estall
198–199   © Aldus Books
200(L)   Paul Brierley
200(R),   Alan Holingbery © Aldus Books
201
202   Argon National Laboratory
203(T)   Alan Holingbery © Aldus Books
203(B)   Photri
204(T)   Popperfoto
204(B)   Space Frontiers Ltd.
205(L)   From The Study of Elementary Particles By the
              Photographic Method by Powell, Fowler and
              Perkins (Physics Department, University of Bristol),
              Pergamon Press
205(R)   Courtesy of The General Electric Company. Comstock,
              G. M. et al, Cosmic Ray Tracks in Plastics: The
              Apollo Helmet Dosimetry Experiment, in Science,
              Vol. 172, pp. 154–157, April 9, 1971.
206   Photo CERN
207(T)   Alan Holingbery © Aldus Books
207(B)   Hale Observatories
208–209   Alan Holingbery © Aldus Books
210   Royal Greenwich Observatory
211   Kip S. Thorne, The Search for Black Holes, December,
              1973. © Scientific American, Inc. All rights
              reserved
212(B)   Alan Holingbery © Aldus Books
213   Andrew Farmer, from The Road to the Stars, David &
              Charles, 1978
214   Marconi Space and Defence Systems Ltd.
215   Eric E. Becklin and Gerry Neugebauer, The California
              Institute of Technology and the Hale Observatories
216   Camera Press
217(L)   Hale Observatories
217(R)   Reproduced by permission of the University of
              Manchester, Jodrell Bank
218(T)   The Mullard Radio Astronomy Observatories,
              University of Cambridge, England
218(B),   Hale Observatories
219
220   Lick Observatory
221(T)   The Mullard Radio Astronomy Observatories,
              University of Cambridge, England
221(B)   Alan Holingbery © Aldus Books
222(L)   © H. W. Schnopper, 1979
222(R)   Photri
223(T)   Dr. Richard Borken

223(B)   Space Frontiers Ltd.
224   © Aldus Books
225(T)   University of Maryland
225(B)   University of Glasgow
226   © 1977 Future Concepts Ltd. Published and
              distributed by Portal Publications Ltd., Corte
              Madeira, Calif.
228   Rockwell International, Space Division
229   Erno Raumfahrttechnik GmbH
230–   Rockwell International, Rocketdyne Division
231(T)
230–   Rockwell International, Space Division
231(B)
232(L)   Erno Raumfahrttechnik GmbH
232(TR),   Rockwell International, Space Division
233
234   Photri
235(T)   Lockheed Missiles & Space Co.
235(B),   Photri
236(L)
236(R)   Robert Harding Picture Library
237   Keystone
238   Photri
239(T)   Boeing/Space Frontiers Ltd.
239(B)   Photri
240(B)   Space Frontiers Ltd.
241   Photri
242   Martin Marietta Corporation, Bethesda, Md.
243   Photri
244   Artwork by David Hardy from The New Challenge of
              the Stars, Mitchell Beazley, 1977
245(T)   G. & E. Bradley Ltd.
245(B),   Keystone
246
247(T)   Popperfoto
247(B)   Lawrence Livermore Laboratory
248(T)   Popperfoto
248(B)   Space Frontiers Ltd
249(T)   Photri
249(B)   John Zimmerman/Camera Press
250   Photri
251(T)   Space Frontiers Ltd.
252(T)   Robert Estall
252(B)   Photri
253   Space Frontiers Ltd.
255   Artwork by Andrew Farmer from The Road to the
              Stars, Westbridge Books (David and Charles), 1978
256   Don Dixon/Space Frontiers Ltd.
257   Artwork by Andrew Farmer from The Road to the
              Stars, Westbridge Books (David and Charles), 1978
258–259   Chris Foss © Aldus Books
260   Professor P. B. Price
261   © Aldus Books
263   Artwork by Andrew Farmer from The Road to the
              Stars, Westbridge Books (David and Charles), 1978
264   Keystone
265(T)   Hilda Canter-Land/Natural History Photographic
              Agency
265(B)   Keystone
266(T)   Grumman/Space Frontiers Ltd.
266(B)   Michael Freeman/Bruce Coleman Ltd.
267   Photri
268–271   David Jackson/Young Artists
272   Design: Chris Foss, © Verkerke GmbH, 1978
274(T)   Cavendish Laboratory, University of Cambridge,
              England
274(B)   Heather Angel
275(T)   John Markham/Bruce Coleman Ltd.
275(B)   BBC Hulton Picture Library
276–   © Aldus Books
277(T)
276(B)   Microphotograph by G. Claus and B. Nagy
277(B)   United Kingdom Atomic Energy Authority
278   Sigurgeir Jonasson/Frank W. Lane
279(L)   Aldus Archives
279(R)   Rudolph Britto © Aldus Books
280   Ann Ronan Picture Library

281(T)   Edward Poulton © Aldus Books
281(B)   © Aldus Books
282(T)   D. G. Bone/British Antarctic Survey
282(C)   Strobino/Jacana
282(B)   Frédéric/Jacana
283   British Museum (Natural History)/Photos Mike
          Busselle © Aldus Books
284   Anthony Frewin, *One Hundred Years of Science
          Fiction Illustration, 1840–1940*, Jupiter Books,
          London, 1974
285(T)   Snark International
285(B)   Keystone
286(T)   NASA
286(B)   Camera Press
287(T)   Photri
287(B)   Camera Press
288   Carl Sagan and Frank Drake, *The Search for
          Extraterrestrial Intelligence*, May, 1975. © Scientific
          American, Inc. All rights reserved
289   Popperfoto
290   *Radio Times*
292–293   Chris Foss © Aldus Books
294(T)   Picturepoint, London
294(B)   UPI Photos, New York
295(T)   David South/Camera Press
295(B)   British Museum/Photo R. B. Fleming © Aldus Books
296–297   Chris Foss © Aldus Books
298(T)   Photo J.-L. Charmet
298(C)   Camera Press
298–   Roy Jennings/Frank W. Lane
299(CB)
299(R)   Georg Gerster/Rapho/Photo Researchers Inc.
300–   Popperfoto
301(TC)
300(B),   Associated Press
301(TR)(B)
302–303   Chris Foss © Aldus Books

304(T)   Keystone
304(B)   Albert Fenn *Life* © Time Inc, 1976
305(TR)   Rex Features
305(B)   Popperfoto
306–307   Chris Foss © Aldus Books
308(T)   Popperfoto
308(B)   Peter Tomlin © Aldus Books (based on the original
          by Leora La Paz)
309(T)   Photo Gordon Faulkner/Neville Spearman Ltd.,
          London
309(BL)   Photo Alan Smith/*The Oklahoma Journal*
309(BR)   Photo courtesy Robert Chapman
310–   Neville Spearman Ltd., London
311(T)
310(B)   Peter Tomlin © Aldus Books
311(B),   Philip Daly
312(T)
312(B)   Ralph Crane *Life* © Time Inc., 1976
313(TL), Courtesy Albert K. Bender
313(TC)
313(TR)   W. Gordon Allen, *Space-Craft From Beyond Three
          Dimensions*, © 1959 by W. Gordon Allen.
          Reprinted by permission of Exposition Press, Inc.,
          Hicksville, New York
313(B)   Saucerian Press, Inc., Clarksburg, West Virginia/
          Courtesy August C. Roberts
314(T)   Philip Daly
314(B)   Associated Press
315(T)   © King Features Syndicate, 1975
315(B)   Camera Press
316(T)   UPI Photos, New York
316(B)   Associated Press
317(T)   Stan Wayman *Life* © Time Inc., 1976
317(B)   Black Star, London
318(T)   Dick Frank/*New York Times* Pictures
319   Courtesy R. Veillith
320   Keystone

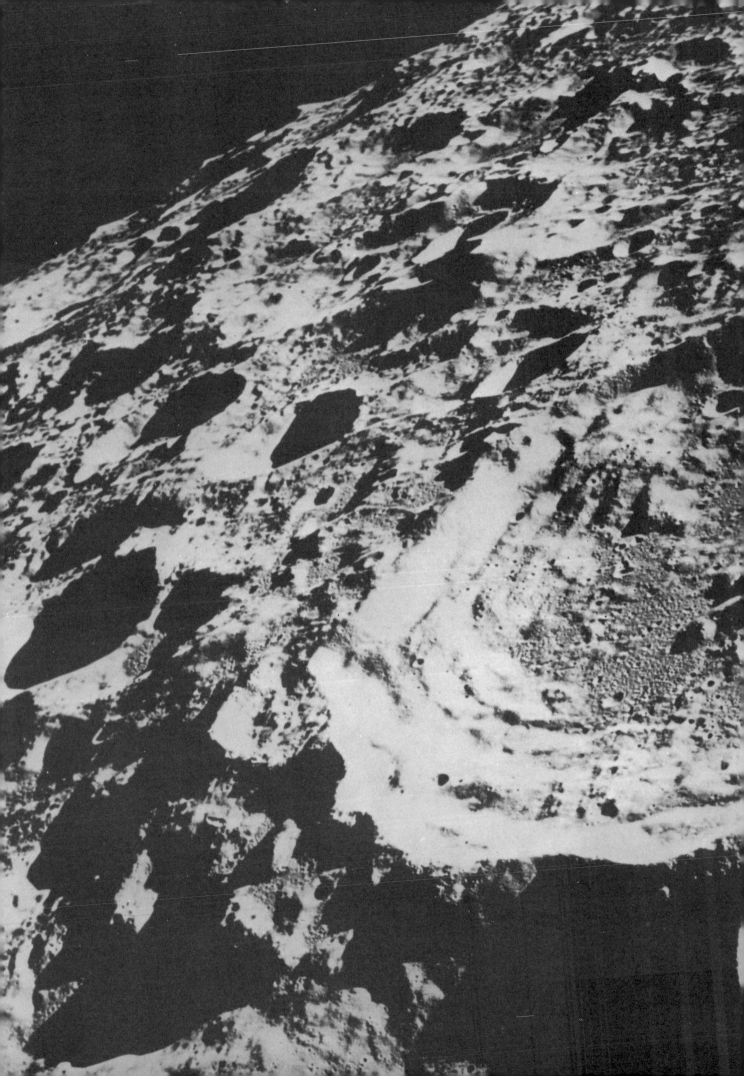